Geography of Water Resources

Geography of Water Resources

Ram Kumar Gurjar
B.C. Jat

RAWAT PUBLICATIONS
Jaipur • New Delhi • Bangalore • Hyderabad • Guwahati

ISBN 81-316-0183-8 (HB)
81-316-0184-6 (PB)

Reprinted, 2012

Published by
Prem Rawat for **Rawat Publications**
Satyam Apts., Sector 3, Jawahar Nagar, Jaipur 302 004 (India)
Phone: 0141 265 1748 / 7006 Fax: 0141 265 1748
E-mail: info@rawatbooks.com
Website: www.rawatbooks.com

New Delhi Office
4858/24, Ansari Road, Daryaganj, New Delhi 110 002
Phone: 011 2326 3290

Also at *Bangalore*, *Hyderabad* and *Guwahati*

Typeset by Rawat Computers, Jaipur
Printed at Nice Printing Press, New Delhi

Contents

Preface

Water is a basic resource on earth for all living organisms including mankind and for development and survival of plant community. Environmental processes of biosphere are also regulated by water. Evidence of importance of water is found in the form of human settlements near water. Availability of water motivates development, whereas absence of water leads to destruction. During last century, man has exploited this resource very fast through various activities as a result of which many water scarcity areas of the world have come up as hot spots of water crisis resulting in danger to renewable and non-renewable sources of fresh water. In this context, the importance of Water Resource Geography seems evident. It is a new development. This subject developed as a branch of Physical Geography during later half of the twentieth century and has established as an independent subject at present. Its main subject matter is sustainable and restorative use of water after getting rid of water crisis as a result of increasing demand for water.

The present book contains all such new subjects in complete form in this context. It comprises of syllabus as per directives of the University Grants Commission as well as situation of water crisis faced throughout the world. It is based on integrated plan of different syllabi prescribed for degree and postgraduate courses of Indian universities. Its language is simple, plain and forceful along with world level maps, pictures and graphs at appropriate places to make it more informative and useful. In the beginning of the book, its development has been explained linking with the present context.

Thereafter distribution of water resources world over have been presented with reference to India. General description of universal movement of water has been explained with reference to main water reservoirs.

In the middle portion of the book, after describing various uses of water resources and consequent problems, different aspects of its management have been clarified which include traditional as well as modern techniques. In the end, effects of environmental disasters on quantitative and qualitative aspects of water resources have been highlighted in detail.

Both the writers feel highly obliged to all the Indian as well as foreign scholars who directly or indirectly cooperated for it. Full assistance was received from Prof. (Smt.) Lakshmi Shukla, Jaipur; Prof. Abha Lakshmi Singh, Department of Geography, Aligarh Muslim University, Aligarh; Prof. R.B. Singh, Delhi School of Economics, Delhi University, New Delhi; Dr. R.D. Doi, Reader, Geography Department, University of Rajasthan, Jaipur; Mamta Siddharth, S.L. Chopra and Manju Yadav. We are thankful to Kailash Rawat of Rawat Publications with whose efforts this book could be published at appropriate time. We also feel obliged to G.S. Narwani IAS (Rtd.) for taking personal pains for the English rendering of this book which was originally published in Hindi.

Effort has been made to present this book named 'Geography of Water Resources', in quite correct and contextual form which is a summary of researches and writings of both the authors in the last two decades. It is hoped that this edition will fulfill the expectations of the readers. It is also expected from the readers to give their suggestions from time to time so as to make future editions more effective and, for which authors would feel grateful.

Ram Kumar Gurjar & B.C. Jat

1

Definition and Scope of Water Resources Geography

In view of increasing specialization in geography during the second half of the 20th century, importance has been accorded to independent study of resources, as a result of which different branches of geography have come up. Among them Water Resources Geography is an important one. As a result of technical development, man has increased the use of water in different forms, which has led to quantitative and qualitative deterioration of water resources. In such a context, while giving importance to distribution, use and conservation of available water in nature, Association of American Geographers (AAG) has included Water Resources Geography as an independent branch in the series.[1]

Study of water resources has been done since ancient past. In the ancient Vedic era, Greek and Roman geographers had described the use and conservation of water resources. This study remained of normal importance till mediaeval times because there was balance between its availability and use. During the renaissance period, many discoveries and inventions took place and resources were identified in various countries. The Industrial Revolution that took place in the 18th century resulted in excessive exploitation of water resources in the whole world. At that time, study of water resources was done as a part of Physical Geography. Due to the increasing demand and decreasing availability of water during the 20th century, study of

Water Resources Geography was started as a separate subject in different countries of the world to save the people from the resulting crisis and to give sustainable basis to the use of water. Many environmentalists started calling the earth famous as 'Blue Planet' as 'Endangered Planet' in view of the decreasing availability of water. They believe that due to the worldwide water crisis, many species on earth are on the verge of extinction.[2]

Definition of Water Resources Geography

Study of Water Resources Geography as a separate branch is being done for the last five decades. It was considered as a study of hydrosphere in the beginning, which included groundwater besides oceanic water. Subsequently, after climatic and meteorological studies, the quantity of water existing in the atmosphere was also considered as a part of Water Resources Geography since it was an important link of the hydrological cycle. Thus, along with distribution and movement of water on the earth, its uses in different forms by human beings were also included.

In India, many scholars and organizations have worked on distribution, use and management of water resources but they have defined some aspects only. In this context, the works of Dakshinamoorty (1972), Nag and Kathpalia (1972), K.L. Rao (1968), Suraj Bhan, Lakshmi Shukla and D.P.Nag are important. The task of defining it as one subject has been done by all these writers.

"*Water Resources Geography is the study of nature, spatial distribution, utilization and conservation of water on the earth. It consists of all the phenomena of hydrological cycle that passes through all the spheres: hydrosphere, atmosphere, lithosphere and biosphere on the earth.*"

It is clear that the definition includes quantitative and qualitative nature of water, distribution in different forms (snow, water vapour and water) at different places, and its conservation along with its use by man. All these facts are organized in a complete chain by the hydrological cycle, which along with maintaining availability of water for human beings at different places, also works as a link in the water cycle. At present, the definition of Water Resources Geography has considerably expanded, whereas water available in large quantity on this planet is going beyond the reach of man.

H.J. DeBlij and Peter Muller have (1993) considered Water Resources Geography as a developed branch from hydrology and a new branch of Physical Geography. According to them, it is a

Figure 1.1
Water Resources Geography as a New Branch of Physical Geography

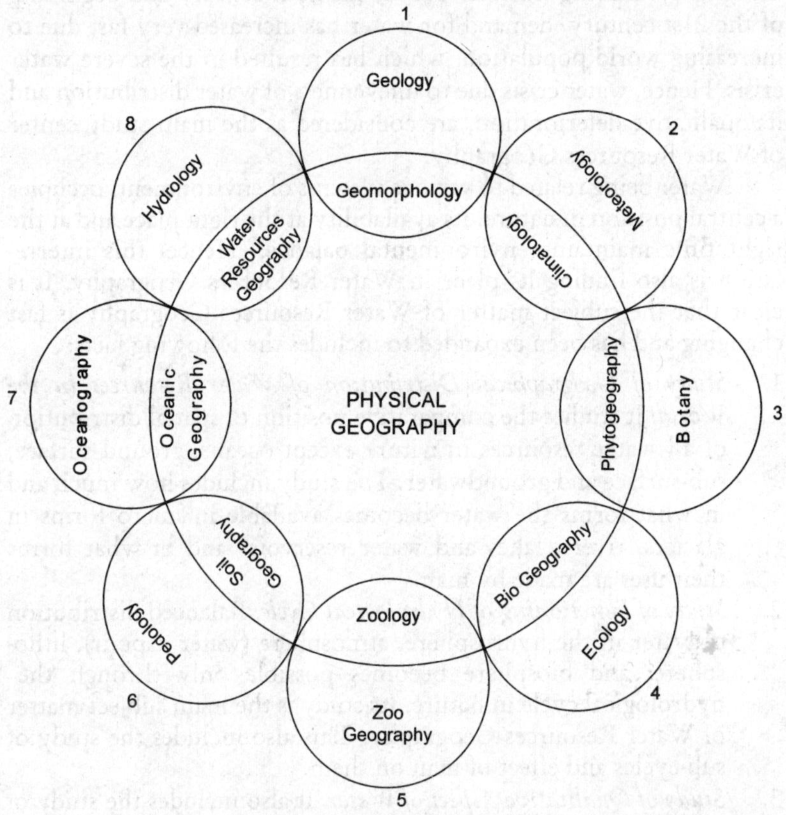

sub-area of Physical Geography, which along with physical constituents, also includes interaction of human resources. They have considered economic and topographical study of the uses of water resources by man on the ground in different forms (surface, sub-surface and groundwater) as important.[3]

Scope of Water Resources Geography

The subject matter of Water Resources Geography has been changing over time. In the beginning, it was the study of distribution of water in different forms on the earth and the hydrological cycle only. Presently, the increasing importance of water, unevenness in its distribution, increasing demand in its various forms and decreasing availability as well as future strategy for its conservation, have become important elements in the study of Water Resources Geography. During the later half of the 20th century and beginning of the 21st century, demand for water has increased very fast due to increasing world population, which has resulted in the severe water crisis. Hence, water crisis due to unevenness of water distribution and its qualitative deterioration, are considered as the main study center of Water Resources Geography.

Water being related to various systems of environment, occupies a central position in nature. Its availability at the right place and at the right time maintains environmental balance.[4] Hence, this interrelation is also finding its place in Water Resources Geography. It is clear that the subject matter of Water Resources Geography is fast changing and has been expanded to includes the following facts:

1. *Study of Geographical Distribution of Water Resources in the World:* It studies the comparative position of spatial distribution of all water resources in nature except oceans, ground surface, sub-surface and groundwater. The study includes how much and in what forms the water becomes available in micro forms in glaciers, rivers, lakes and water reservoirs and in what forms their uses are made by man.
2. *Study of Functioning of Hydrological Cycle:* Balanced distribution of water in the hydrosphere, atmosphere (water vapour), lithosphere, and biosphere becomes possible only through the hydrological cycle in nature. Its study is the main subject matter of Water Resources Geography. This also includes the study of sub-cycles and effect of man on them.
3. *Study of Qualitative Aspect of Water:* It also includes the study of qualitative deterioration of water due to water pollution and mixing of undesirable elements in water because of reducing availability of fresh water.
4. *Study of Water-Borne Problems:* Many problems are created due to uneven distribution of water by man. The important ones among them are salinity, alkalinity, flouride, arsenic, and waterlogging. These new problems are also studied in Water Resource Geography.
5. *Study of Water Management in Flood-Prone and Drought-Prone Areas:* This includes the study of providing sustainable base for water excess, flood-prone, and scarce, water affected drought-prone areas.

6. *Study of Uses of Water by Man:* Apart from for own domestic use, man uses water available in nature for economic use in various forms mainly for agricultural and industrial uses. Because of continuously increasing demand for water in these spheres, this subject has gained importance. For this reason, the cyclical use of water has also become important.
7. *Geographical Study of Watershed:* Since the last decade, particularly since 1994, watershed is considered as a geographical unit for water management because it involves activities of physical and ecological regeneration. It is a community participation programme which includes various activities of water conservation.
8. *Study of Effects of Natural Calamities on Distribution and Availability of Water:* Materialistic culture of man has brought many changes in nature in the 20th century. The effect of this culture can be seen on the distribution and quantitative aspect of water. Among them climatic changes, global warming, melting of snow and acid rain etc. are important.
9. *Study of Water Crisis and Water Conservation:* Water crisis has been created since the last century due to population increasing at a fast rate. Along with the study of main reasons for water crisis, study of finding solutions is also an important subject matter. Simultaneously, a strategy has to be evolved which can conserve water in different forms. At present, stress is being laid on sustainable management of water, whose origin is considered to be a report '*Limits to Growth*' by a research group headed by Denis Meadoz in 1968. Published in 1972, it stressed on improving the quality of life by preventing deterioration of resources including water, and thereby creating a healthy environment. Thereafter, in 1987, World Commission on Environment and Development (WCED) published its report named '*Our Common Future*', also called "Braudtland Commission" after the name of its chairman, which also publicized the sustainability viewpoint. Sustainable water management is essential for the fulfillment of the needs of the present generation, without reducing the capacity of fulfilling the needs of future generations.

Water Crisis: A Scenario

A vast amount of water is found on the earth, which is why it is also called 'Water Planet'. However, due to its uneven distribution, all the living organisms on earth are not able to get water resources in equal proportion. Out of a total quantity of 1.5 billion cubic km in nature, only 12,500 to 14,000 billion litre water is available annually for use by human beings. Thus, whereas about one-third population of the world at present is not able to get fresh water, a population of 2.8 billion of 48 countries up to 2025, and a population of 4 billion of 54 countries up to 2050, may face water crisis. These figures are 40 per cent of the projected population of the world.

The present water crisis scenario has clarified that 1.1 billion population of the world does not get sufficient fresh water, or one out of every six persons faces water crisis. Apart from supply of fresh water, from the qualitative point of view also it is clear that 2.5 billion population of the world does not get sanitary facilities and 20,000 children die every year of water-borne diseases. In India, 170 million people do not get fresh water in sufficient quantity. Water crisis started in the beginning due to unevenness of distribution and increasing demand of the population, but in present times the qualitative aspect has also been added to the quantitative aspect. If the total quantity of water available in India is made pollution-free, it is sufficient for the existing population, but along with rivers and lakes, even groundwater is becoming polluted. In the beginning, pollution was considered as limited up to ground and oceans only but now atmospheric water (vapour) is also taking the form of acid rain. Thus, all living organisms are suffering from the problem of water crisis.

The seemingly vast amount of water on earth is thus, finding itself unable to quench the thirst of the inhabitants. The layer of white sheet (snow) spread on the earth, which is a huge reservoir of fresh water, is slowly melting away. Along with it, changes in quantity and nature of rain are taking place. All these changes have been made by the human community demanding water. On one side, there was uneven distribution of water in nature since the very beginning, and on the other side, man has made this invaluable gift of nature beyond the reach of living organisms by his selfish behaviour. Hence, due to water crisis, some day, man too would also face extinction like some animals.

Distribution of Water

On the earth, 361 million sq km area of the whole earth is completely covered with water. While the remaining 149 million sq km ground area also has water in form of rivers, glaciers, lakes, marshy land and groundwater. The hydrological cycle also plays an important role in distribution of water on earth. After evaporation, water from oceans, seas, lakes and rivers and transpiration from plants reaches the atmosphere and again converts into water and reaches the ground. Temperature plays an important role in changing the form of water. Hydrological cycle maintains the water balance on the earth.

In nature, only 2 per cent of the total water on earth is in the form of fresh water, out of which two-third is in the form of snow caps and glaciers on poles. Of the remaining quantity of water, a little less than one per cent is directly available for living organisms. If the total ice of the earth melts, all the rivers of the earth can flow at the current speed for the coming 1,000 years, but this cannot happen all of a sudden. It is a long-term process of climatic change in nature whose balanced implementation depends on many geographical factors. Water remains stable at different places for different periods. Stability of water in plants and organisms remains stable up to one week. In the atmosphere it remains stable for for 8 to 10 days, in rivers for two weeks, in soil from two weeks to one year, in lakes and humid soils for years and in oceans for thousands of years.

Table 1.1

Main Natural Water Reservoirs

Reservoir	*Percentage of Total Water*
Oceans	97.41
Snow Caps and Glaciers	1.9
Underground Water	0.5
Soil Moisture	0.01
Lakes and Rivers	0.009
Atmosphere	0.0001

In the system of hydrological cycle, distribution of water goes on changing with time. Such changes occur at different rates in different environmental systems. During the Snow Age, the maximum part of water was frozen in the form of snow, due to which the water level of seas had gone down. After a certain period, on completion of Snow

Age, the water in seas rose high due to the melting of snow cover. In the present times also, due to rise in temperature, the snow cover is constantly reducing resulting in expansion of seas.

Due to rise in world temperature, the snow of Antarctica and Greenland is melting. Snow of the highest and most important mountain peak Kilimanjaro is melting fast in Africa. About 90 per cent of world's fresh water is found in a vast area of 1,40,00,000 sq km of Antarctica. This plays a dominant role in regulating the marine ecological system and climate of the world. Thus, the present water distribution in nature is also not of permanent nature but has been changing. Three decades ago, there was sufficient water available but a water crisis is being faced today. On the other side, excess water is also a problem at some other places. Every year 40,000 cubic kl. water flows from the ground towards the seas, out of which man can use only 9,000 cubic kl. water. Similarly, about 5 lakh cubic kl. water evaporates from the seas and goes into the atmosphere. It is clear that though water is distributed over a vast area but still it is only partially available for mankind.

Effect of Population on Decreasing Water on the Earth

Quantity of water in nature neither decreases nor increases but only changes its form with time. With increase in population, there has been continuous increase in demand for water. In 1940, only 1000 cubic kl. water was used during a year. In 1960, consumption of water became double in comparison to 1940 and up to the year 1990, its consumption was double of what it had been in 1960. In 2001, the population of the world was about 613.17 crores but the quantity of water available is the same. Due to the qualitative deterioration, per capita availability of fresh water has decreased considerably.

Nature has not provided water to many parts of the world. There was not even partial rain for four years from 1989 to 1994 in Atakama desert of Africa. According to geographical norms, areas having less than 300 mm rain during a year are called arid areas. About 60 crore population of the world lives in such arid regions. While 40 per cent population of Africa suffers from scarcity of water, about 20 per cent population of India resides in such arid regions.

Table 1.2

Availability of Water and its Consumption in India as Compared to other Countries

Country	Available Water		Use of Water		Use in Different fields (In Percentage)		
	Total (Cubic Kl.)	Per Capita (Cubic metre/ Year)	Total (Cubic Kl.)	Per Capita (Cubic metre/ Year)	Domestic	Industries	Agriculture
India*	1869	1869	380.00	610	5	15	80
Australia	343	21300	17.80	1310	16	6	78
England	120	2140	28.35	510	20	79	1
Germany	96	1100	50.53	600	14	72	14
France	170	3090	37.20	460	16	71	13
China	2800	2580	460.00	920	6	7	87
Japan	547	4480	107.80	1060	17	33	50
Argentina	694	22030	27.60	2190	9	18	73
America (USA)	2478	10223	472.00	1500	10	49	41
Canada	2901	111740	36.15	400	18	70	12
South Africa	50	1470	9.20		17	0	83

* *Estimate of Central Water Resources Ministry.*

Due to increase in population along with urbanization and industrialization, changes have also occurred in the agricultural sector. Due to increasing urbanization leading to increasing demand for water, qualitative deterioration is also happening very fast. This has rendered water unfit for use by human beings. In the beginning, partial impurities of water were cleaned automatically but today the position is that if 450 cubic km balanced water flowing every year in springs and rivers is made reusable, it would destroy 6,000 cubic km fresh water.

Due to increasing population per capita availability of water has also gone down. Per capita water availability per person per day is 8.5 litre in Asia and 4 litre in Africa. In Great Britain, it is 334 litre and in United States of America it is 587 litre which is much more as compared to Asian and African countries. The main reason for such

Figure 1.2

Water Availability in the World

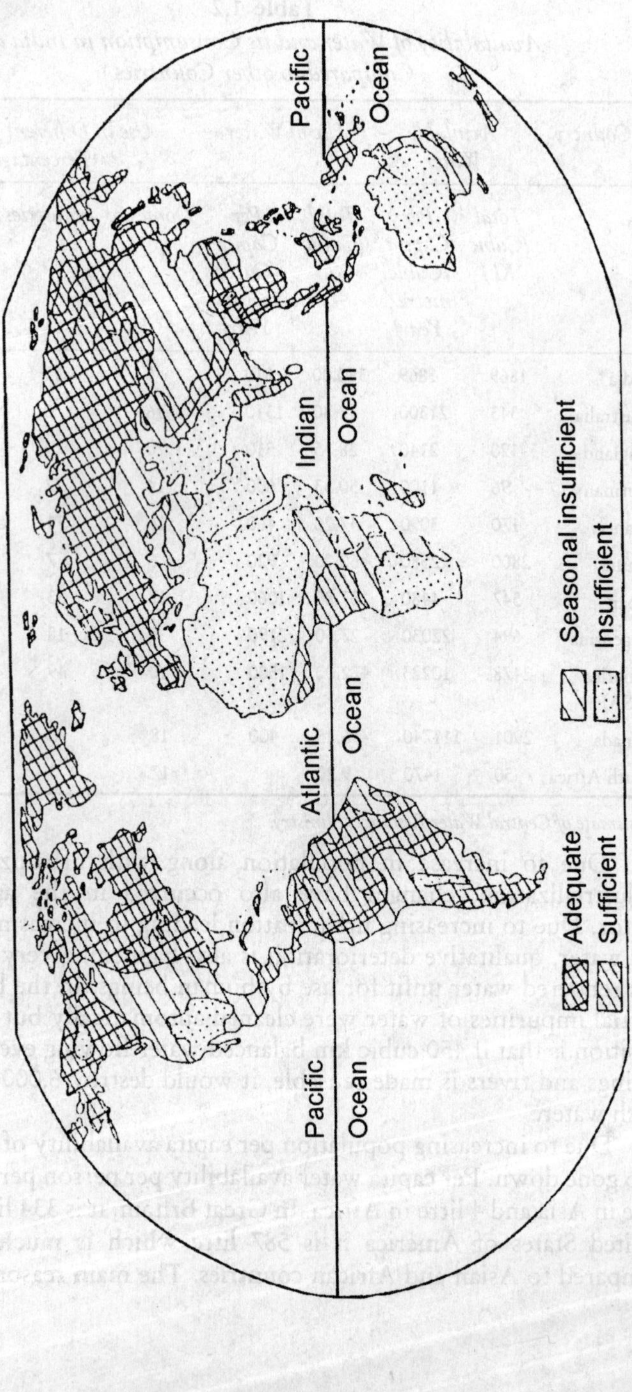

uneven distribution is increase in population. Population of the United States of America is one-fourth of that of India, whereas water resources are more than double as compared to India.

Hot Spots of Water Crisis

Former Vice President of the World Bank Ismail Seregaldein had predicted many years ago that in the coming century, wars would be waged not for oil, but for water. His prediction is coming true to a great extent because there are five such places in the world where disputes have been raised many times on account of water. In India also, during the last decade, human losses have been caused on account of water disputes at many places. It is said that in case a solution to these problems is not arrived at soon, war might start on account of water. Hence, they have been termed as 'hot spots' of the water crisis. They are as follows:

Aral Sea Area

Aral Sea was the world's fourth biggest water reservoir once upon a time. During the decade of 1960s, its water content reduced to only one-third. This happened because water of the rivers Amudariya and Sardariya that supplied water to the sea, started being used for cultivation of cotton in desert area. Amudariya meets the demand of water for countries of Afghanistan, Uzbekistan and Turkistan, and if Afghanistan utilizes more water of Amudariya river for its own development, tension would naturally be created in the rest of the countries.

Indus Water Area

Indus river flowing from Kashmir initially in India and later in Pakistan, has been the cause of dispute between the two countries on account of distribution of its water. It is assumed that Kashmir dispute between the two countries has increased this problem.

Jordan River Area

Many a time war also disturbs the balance of supply of water. This is happening in the Middle East, where distribution of river water has been affected on account of war. Israel captured Golan Hills (Syria), the catchment area of Jordan and Beka (Lebanon), and 80 per cent water of this river is being used by Israel whereas, only 20 per cent water is given to Palestine. Water supply to Jordan is also made from this very river.

Tigris and Euphrates Rivers Area

Turkey has constructed dams over Tigris and Euphrates Rivers flowing there, which have resulted in lowering down water level of the rivers. Water supply of Iraq and Syria also depends on the same rivers. Iraq and Syria are alleging breach of agreement regarding distribution of this water source against Turkey.

Nile River Area

Disputes have always existed regarding water of river Nile among Egypt, Sudan and Ethiopia. It is said that if increase in population of Africa continues, there would be a possibility of war on account of water.

Water Crisis in India

Assessment of water crisis is made on the basis of per capita per year availability of water. If per capita per year availability of water in a country is less than 1,000 cubic metres per year, such areas are considered to be suffering from water crisis. At present, about 26 countries where more than 24 crore population of the world resides, come in this category. Maximum countries suffering from water crisis are in Africa while nine countries are in the Near East. Two-third part of India also suffers from water crisis. Even the Cherapunji area, which is famous for maximum rainfall in the world, is suffering from water crisis. Due to fast deforestation, rain water of hills cannot be stored and most of the surface water flows away to saline seas. Due to mining of soap stone, deforestation has taken place on a large scale.

Maximum exploitation of water in India is being made for irrigation in agriculture. At the world level, 69 per cent of pure water is used for agricultural purposes, 23 per cent for industries and 8 per cent for domestic consumption. During the last four decades, due to revolutionary changes in agriculture, demand for water has increased and this demand is being met from fresh water. Unscientific methods too were adopted for selfish motives, which required more water. In Maharashtra, tubewells were installed in such a large number in fields of sugarcane and sugar mills that about 2,000 wells of Talgaon became dry on that account. About 23,000 villages are affected by this problem due to exploitation of groundwater in agriculture, resulting in scarcity of drinking water in the area. Similarly, the situation in Rajasthan is also grave. Out of the total area (3,42,239 sq km), 63 per cent (2,15,142 sq km) is groundwater probability area. About 29 per cent of this is located in western Rajasthan, which is of saline nature, while 8 per cent is hilly area. Due to excessive exploitation of groundwater in this area, the water level has gone down to the extent of 5 to 10 metres. Consequently, where sufficient water was available for agriculture two decades ago in the 1980s, water crisis is being faced today.

Due to decreasing water resources in the country, availability of per capita water has also reduced. At the time of independence of the country, availability of per capita water was 5,326 cubic metre per year which reduced to 2,267 cubic metre in 1991. The demand increased relatively with time. In 1990, per capita demand for water was 550 cubic km, which increased by 75 cubic km in the year 2000 and it is estimated to be 1,050 cubic km by the year 2025. The main reason is increasing population and quantitative as well as qualitative deterioration of water. India gets 4,006 cubic km water from snow and rainfall but due to uneven distribution of water, on one side Cherapunji gets 11,000 mm rainfall annually while, Rajasthan gets only 150 mm rainfall annually. Thus, most of the water received from rainfall flows away unused. Surface water sources of rivers, ponds, lakes, wells, baoris etc. have only 1,869 cubic km water, out of which only 690 cubic km water becomes available for different uses. This also includes 360 cubic km underground water which is used for irrigation. During the last decade, more than 85 per cent water was used for irrigation but in the coming two decades it would reduce to 60-70 per cent. For the remaining purposes only 15 per cent water is being presently used, which is likely to increase to 27 per cent by the year 2025. Thus, by the year 2010, water crisis would be visible in the whole country.

Among the river basins of the country, maximum water is available in Brahmaputra Basin. Here, per capita water available is 18,081 cubic km whereas, it is only 360 cubic metre in Sabarmati, 366 cubic metre in Pennar and Kanyakumari. Availability of water in Brahmaputra basin is sufficient but it is not used in such proportion and most of the water flows waste. It is interesting to note that the tradition of conservation of such water has been in practice since the ancient times. Yamuna canal and Agra canal were constructed in the 16th century, which carried water of Himalayas to Punjab, Rajasthan and Uttar Pradesh. In the 19th century, Kurnool, Kuddappa and Periyar Begad canals were constructed. Thus, wherever there is excess water in the country, it is taken to water scarcity areas. For this purpose, it is proposed to form a National Water Grid in India.

Water Crisis in the World

Water crisis is continuously increasing in the world for many reasons. One reason is that due to deforestation, rain has disconnected itself from soil leading to water scarcity. Scientists of the world however consider undesirable changes in climate as the reason for the water crisis. There is a total of 140 crore cubic metres water on earth, out of which, 97.2 per cent is in saline form and 2.15 per cent is stored in snow peaks and glaciers. According to a recently published statistics by United Nations Organization (UNO), about 12 crore population of developing countries of the world does not get fresh and potable water and every year 2.50 crore persons die because of polluted water in these countries. Only 75 per cent people in urban areas and 40 per cent people of rural areas in the world get fresh water. It can be estimated on such a basis that during the first decade of the 21st century, grave water crisis would be faced by around 25 countries of the world. According to UNO, such water crisis would be not of quantity only but of quality and distribution also. In most of the areas, this crisis is the result of excessive exploitation and unsystematic use of water. From the universal angle, water crisis is assuming dangerous dimensions. Out of 25 countries , 19 are African countries where people do not get fresh water, though in whole of the continent, availability of water per year is 4,00,000 crore cubic metres. Unfortunately due to lack of basic facilities, and lack of technical and financial knowhow, only 4 per cent water is utilized. In the whole of the Asian continent, 1,30,000 crore cubic metres water is available. In India, due to lack of industrial balance and suitable water management, water is being wasted. In Russia, 90 per cent water flows to the seas.

In the whole world, 60 per cent water of total available quantity is used by urban areas and for industrial purposes, but due to excessive exploitation, grave water crisis is being faced. According to one estimate, only 21 per cent water of the 245 rivers of Europe can be considered fresh and free from bacterial pollution. About 13 per cent water of the world is available in American and Carribean countries, but most of the rivers are polluted. There is acute imbalance in per

Figure 1.3

Water Excess and Water Scarcity Areas of the World

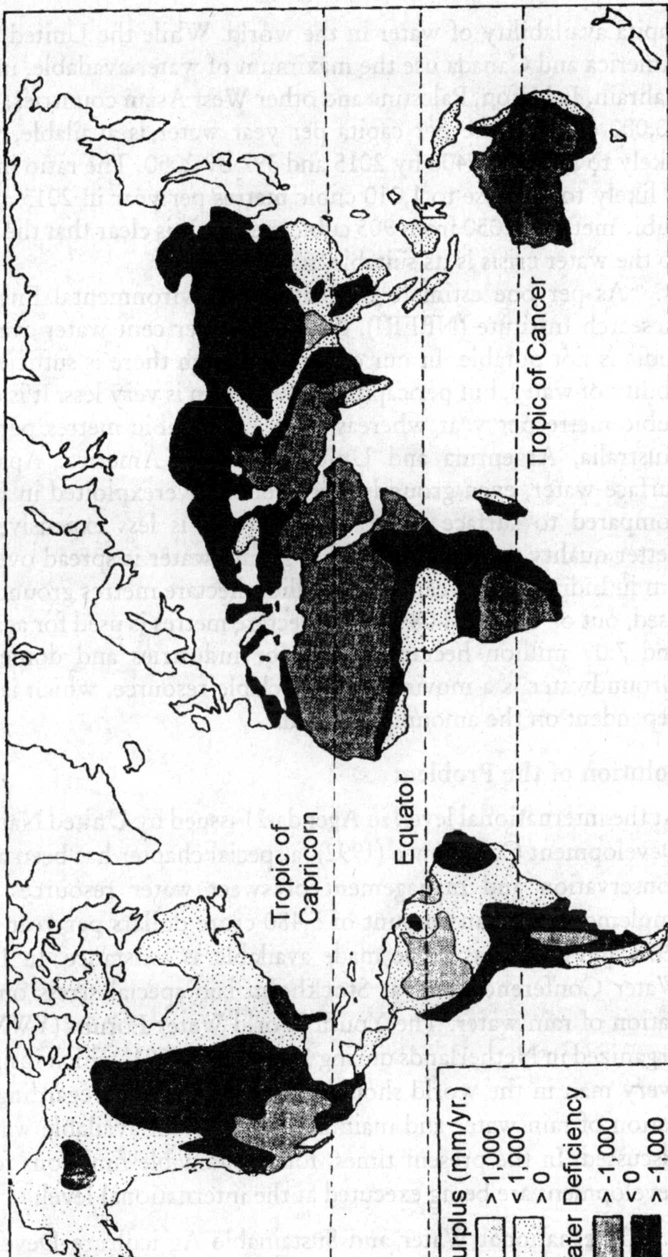

capita availability of water in the world. While the United States of America and Canada use the maximum of water available, in Jordan, Bahrain, Lebanon, Palestine and other West Asian countries, less than 10,000 cubic metre per capita per year water is available, which is likely to reduce to 406 by 2015 and 397 by 2050. The ratio in Europe is likely to increase to 1,010 cubic metres per year in 2015 and 1,020 cubic metre in 2050 from 905 cubic metres. It is clear that the solution to the water crisis is its suitable management.

As per one estimate by National Environmental Engineering Research Institute (NEERI), Nagpur, 70 per cent water available in India is not potable. In our country, though there is sufficient availability of water, but per capita consumption is very less. It is only 610 cubic metre per year whereas, it is 1,000 cubic metres per year in Australia, Argentina and United States of America. Apart from surface water, even groundwater is being overexploited in India. As compared to surface water, groundwater is less expensive and of better quality. It is estimated that groundwater is spread over 452 sq km in India. Every year, 43.13 million hectare metres groundwater is used, out of which 38.08 million hectare metres is used for agriculture and 7.09 million hectare metres for industries and domestic use. Groundwater is a moving and recyclable resource, which is directly dependent on the amount of rainfall.

Solution of the Problem

At the international level, in Agenda-21 issued by United Nations and Development Conference (1992), a special chapter has been added on conservation and management of sweet water resources. For its implementation, an amount of 5,480 crore dollars per year between 1993 and 2000 was to be made available as assistance. In 1999, the Water Conference held at Stockholm laid special stress on conservation of rain water. The Youth World Water Forum (YWWF) was organized in Netherlands during 25-28 June, 2001, which insisted that every man in the world should get water. Matters regarding conservation of rain water and maintaining purity of available water were discussed. In the present times, four sustainable functions for water development are being executed at the international level:

1. International Water and Sustainable Agriculture Development Programme (FAO)
2. Water Supply and Sanitation Programme (WHO)

3. International Technological Research Programme in Irrigation and Water Development (World Bank)
4. Panel of Experts on Control of Parasites (WHO)

The United Nations Sub-Committee on water resources is coordinating these programmes. In the Indian context, the following measures can be useful to combat the water crisis:

1. Stabilization and control of increase in population.
2. Sustainable use of land and water management.
3. Development of alternative sources of energy.
4. Control on water pollution.
5. Giving importance to community participation in water management.
6. Making special security arrangement, at exrtaordinarily polluted areas and disputed rivers.
7. Ensuring suitable control in industrial areas and encouraging re-use of water.
8. Preventing disposal of harmful solid waste of industries, agriculture and domestic areas into rivers.
9. Construction of pakka ghats around water reservoirs (ponds etc.) to prevent soil erosion.
10. Ensuring balanced exploitation rather than excessive exploitation of groundwater.
11. Adopting traditional water conservation practices.

In order to get rid of future water crisis, the whole world will have to adopt a unified view for water conservation and sustainable development. Presently, different organizations are working in different domains, of which the main are water supply, sanitation, irrigation management, soil conservation, and forest development. All these organizations will have to join and work on single platform with water as their main base, otherwise civilizations surrounded by water will vanish due to water scarcity. In 1999, in the World Water Conference at Stockholm, emphasis was laid on conservation of rain water. In World Environment Summit held at Johannesburg in 2002 also, water crisis was made an important agenda, and by 2015, providing pure drinking water and sufficient sanitary facilities to the world was emphasized. But on the one hand, United Nations Organization talks about facing water crisis by providing fresh water to 2.80 billion people by 2025, while on the other hand, the World Environment Summit talks about providing drinking water for all by 2015. Let us see what comes true.

Notes

1. Bergman, E.F. and Renwick, W.H., (1999), *Introduction to Geography*, Prentice Hall, p. 4.
2. Barlow, M. and Clarke, T., (2002), *Blue Gold, The Fight to Stop the Corporate Theft of the World's Water*, Leftword Books India, p. 26.
3. De Blij, H.J. and Muller, P.O., (1993), *Physical Geography of the Global Environment*, John Wiley, pp. 7-8.
4. Park, C., (1997), *The Environment, Principles and Applications*, Blackwell. p. 345.

2

Inventory and Distribution of World's Water Resources

In nature, water is distributed in different forms at different places from where it moves to other places, since water does not remain permanently anywhere. It evaporates and again reaches the earth in the form of rainfall. Water on earth remains stable for different durations. During the movement of water, its form keeps changing. In oceans, water is found in the form of liquid. From solid (snow) and liquid (water), it evaporates and reaches atmosphere in the form of vapour. Heat of the sun plays a predominant role in changing the form of water depending upon climate and location. Heat of the sun is the maximum on the equator and minimum on the poles. Due to this reason, the snowline remains above 5,000 metre on the equator and near ground level on the poles.

Hydrosphere includes the total amount of water in all the three spheres lithosphere, atmosphere and hydrosphere. Normally it is presumed that only oceans are included in the hydrosphere, but along with oceanic water, surface water on the ground, underground water and water vapour in the atmosphere etc. also constitute hydrosphere.

Though vapour exists in the atmosphere on a small scale, it is very important because it forms the hydrological cycle of hydrosphere, atmosphere and lithosphere. After evaporation from oceans, lakes, rivers, snow cover and transpiration from plants, it reaches the atmosphere from where it distributes in the form of rainfall to different water sources. During the last century, the movement of water has been affected due to unplanned disturbance with nature.

This has resulted in changes in nature and the time of distribution of rainfall on the surface of the earth. The latest examples are regular droughts in India and floods in Europe during the year 2002, which affected the area from Britain to Romania.

About 13,84,120,000 cubic km water is found in different forms in the hydrosphere (Table 2.1). Its biggest part is found in oceans, While 77.23 per cent is in snow caps, icebergs and glaciers. Of the whole hydrosphere, 2.6 per cent (36,020,000 km^3) is fresh water, 2.21 per cent is groundwater located up to a depth of 4 kms. The remaining water remains in soil, lakes, living organisms and atmosphere in small quantities.

Table 2.1

Quantity of Water in Hydrosphere

Water	*Quantity (km^3)*	*Quantity (in percentage)*
Total Water (Pure and Saline)		
Oceans	1,348,000,000	97.39
Snow caps, Icebergs, Glaciers	227,820,000	2.01
Groundwater and Soil Moisture	8,062,000	0.58
Lakes and Rivers	225,000	0.02
Atmosphere	13,000	0.001
Total	*1,384,120,000*	*100.00*
Pure Water (2.6 % of total water)		
Snow caps, Icebergs, Glaciers	27,818,246	77.23
Groundwater up to 0-0.8 km Depth	3,551,572	9.86
Groundwater up to 0.8-4 km Depth	4,448,470	12.35
Soil Moisture	61,1234	0.17
Fresh Water Lakes	126,070	0.35
Rivers	1,080	0.003
Aquatic Minerals of Earth	360	0.001
Atmosphere	14,408	0.003
Life	1,081	100.00
Total	*36,020,000*	

Source: Baum Garther and Reichel from R.J.Hugget, *Environmental Change* (1997).

Distribution of Water

Water found in hydrosphere is distributed in different forms on earth. Its major portion (97.39) is distributed in saline form in seas, whereas fresh water is available in a small quantity. Water vapour is an important part of the atmosphere which plays an important role in the energy circle. Thus in normal water distribution, surface water, groundwater and oceanic water are included, whereas water vapour in the atmosphere completes many processes. Universal distribution of water remains in balance through the water cycle in which water vapour has its own geographical importance.

Distribution of water has not remained in equal proportion since ancient past. During the snow age, major portion of atmospheric moisture and oceanic water became frozen as ice which was converted into liquid form after climatic changes. Thus, distributed in changed form, water resources are presently found in the following forms:

1. Oceans
2. Groundwater
3. Surface Water

Oceans

Almost 70.87 part of the earth is watery out of which 97.39 per cent exists in the form of oceans, which contain 1,34,80,000,00 cubic km water. Thus, earth due to the presence of vast quantity of water, it is also called 'Water Planet'. In the northern hemisphere, 40 per cent area is water and 60 per cent is ground. In the southern hemisphere, there is water on 81 per cent area and 19 per cent is ground. That is why the southern hemisphere is also called 'Water Hemisphere'.

Pacific Ocean is the biggest among all the oceans, which contains 53.9 per cent of water of the world. Atlantic Ocean contains 24.9 per cent and Indian Ocean has 21.1 per cent share of world's water (Table 2.2). Water of all the three oceans is saline. Except these oceans, major portion of water found in North Pole Ocean and South Pole Ocean exists in form of snow. Within the snow area, Arctic Ocean covers 3.97 per cent part of total water area.

Table 2.2

Factual Geography of Main Oceans

Ocean	*Area (102 km²)*	*Percentage of World Oceans*	*Percentage of Land Surface*	*Area (106 km²)*	*Average Depth (Metre)*	*Maximum Depth (Metre)*
Atlantic Ocean	94.3 (82.2)	24.9	17.7	340.3	3570	8381 (Puerto Rico)
Arctic Ocean	12.2	0.9	0.6	13.7	1170	5450 (Eurasian Basin)
Indian Ocean	74.1 (73.4)	21.1	14.9	286.7	3840	8047 (Diamantine)
Pacific Ocean	1713 (165.3)	53.1	37.6	717.8	3940	11033 (Mariana)
World	381.9	100.0	70.8	1358.5	3730	11033 (Mariana)

Note: In south, three southern oceans have been included up to southernmost borders. All coastal seas have been included in all oceans, their actual area is given in brackets.

Source: After Smithson, P., Addison, K. and Atkinson, K., 2002:214

Pacific Ocean is the most expanded water portion on earth which encircles almost one-third part of the earth. Its area is 16,53,84,000 sq km and average depth is 4,280 metres. The main coastal seas of this ocean are Bering, Okhotsk, Japan Sea, China Sea, Sulu Sea, Yellow Sea, Java Sea, Banda Sea, Celebes Sea, California and Aleutian sea.

The second largest part of the hydrosphere exists in the Atlantic Sea. It extends up to 8.2 crore sq km. Its main coasts are Grand bank, George bank, St. Pierre bank, Seville island bank and Dagger bank. Its main bordering seas are Mediterranean Sea, Baltic Sea, Northern Sea, Black Sea, Aegean Sea, Adriatic Sea, Baffin Bay, Hudson Bay, Gulf of Mexico and Caribbean Sea.

The third great water area of the world is the Indian Ocean. This is bordered by Asian lands in the North, Africa in the west and Antarctica in the south. Pacific Ocean is in its east. Sixty per cent portion of Indian Ocean is between 4,000 to 6,000 metres deep. Its main bordering seas are Bay of Bengal, Gulf of Aden, Andaman Sea, Gulfs of Oman, Kutch and Cambay etc.

Table 2.3

Area and Quantity of Water in Important Seas

Sea	*Area (km²)*	*Volume (km³)*
Malaya Sea	8143100	9873000
Carribean Sea	4319500	9573000
Mediterranean Sea	2966000	4238000
Baltic Sea	422300	23000
Gulf of Hudson	12333000	158000
Red Sea	4379002	15000
Persian Gulf	238800	6000
Bering Sea	2268200	3259000
Okhotsk Sea	1427600	1279000
Japan Sea	100300	1361000
Eastern China Sea	1249200	235000
Andaman Sea	797600	694002
California Sea	168200	132000
Northern Sea	575300	54000
English Channel and Irish Sea	178600	10000
Laurentian Sea	237800	30000

Large amount of water exists in the bordering seas of the oceans also. In the Pacific oceanic area, the main seas are Bering Sea, Okhotsk Sea, Japan Sea, China Sea, Zulu Sea, Yellow Sea, Java Sea, Banda Sea, Celebes Sea, Gulf of California Aleutian. In the Atlantic ocean, Baltic Sea, Northern Sea, Mediterranean Sea, Black Sea, Aegean Sea, Adriatic Sea, Baffen Bay, Hudson Bay, Gulf of Mexico and Caribbean Sea are found. Main bordering seas of Indian ocean are Bay of Bengal, Mozambique Channel, Arabian Sea, Red Sea, Persian Gulf, Gulf of Adan, Andaman Sea etc.

Groundwater

Water existing in voids and fissures of rocks is called groundwater. It is regulated by quantity and speed of rainfall, timings of rain, quantity of evaporation, temperature, slope of land, dryness of air, voids and non-permeability of rocks, vegetation cover, and water absorbing capacity of the soil. Surface flow goes underground through

Figure 2.1

Oceans and Seas of the World (names are shown in the map and list of names follows)

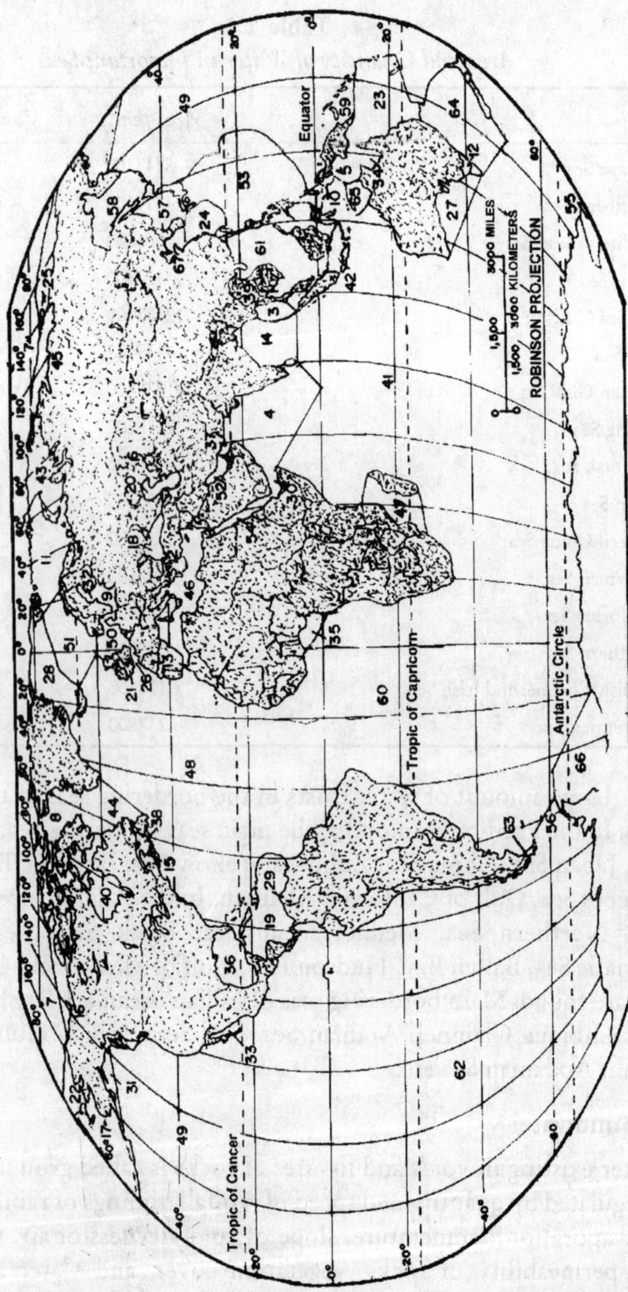

1. Adriatic Sea
2. Aegean Sea
3. Andaman Sea
4. Arabian Sea
5. Arafura Sea
6. Aral Sea
7. Arctic Ocean
8. Baffin Bay
9. Baltic Sea
10. Banda Sea
11. Barents Sea
12. Bass strait
13. Bay of Biscay
14. Bay of Bengal
15. Bay of Fundi
16. Beaufort Sea
17. Bering Sea
18. Black Sea
19. Caribbean Sea
20. Caspian Sea
21. Celtic Sea
22. Chukchi Sea
23. Coral Sea
24. Eastern China Sea
25. East Siberian Sea
26. English Channel
27. Great Australian Bight
28. Greenland Sea
29. Gulf of Venezula
30. Gulf of Aden
31. Gulf of Alaska
32. Gulf of Bothania
33. Gulf of California
34. Gulf of Carpentaria
35. Gulf of Guiana
36. Gulf of Mexico
37. Gulf of Oman
38. Gulf of St.Lawrence
39. Gulf of Thailand
40. Hudson Bay
41. Indian Ocean
42. Java Sea
43. Kara Sea
44. Labrador Sea Ocean
45. Laptev Sea
46. Mediterranean Sea
47. Mozambique Channel
48. North Atlantic Sea/Ocean
49. North Pacific Ocean
50. North Sea
51. Norwegian Sea
52. Persian Gulf
53. Philippinesea Sea
54. Red Sea
55. Ross Sea
56. Scotia Sea
57. Sea of Japan
58. Sea of Okhotsk
59. Solomon Sea
60.South Atlantic
61. South China
62. South Pacific
63. Strait of Magalan
64. Tasman Sea
65. Timor Sea
66. Weddell Sea
67.Yellow Sea

percolation. Groundwater is 0.58 per cent of total water resources and out of total fresh water (2.6%) it is 22.1 per cent, which is located up to 4 kilometres (in-depth). Since it is found below the ground, it is also called sub-surface water.

Storage of Groundwater

Water received from different sources on surface of the earth goes underground after entering through voids, cavities, fissures and bedding planes and takes the form of groundwater. After percolation downwards when water collects in rock voids, it is called saturation of rocks. When rock is fully saturated, it is called *saturated zone*. The part of saturated rocks below the surface is called saturated zone and its upper part is called groundwater or water table.

According to location, groundwater below the surface is found in three zones, which are as follows:

Non-Saturated Zone

The upper most part of groundwater is called non-saturated zone where air is filled in voids of rocks. Hence, it is also called '*Zone of Aeration*'. In this zone, the water exists only during rainy period. In the remaining periods, rocks cannot be saturated with water.

Figure 2.2

Various Areas of Groundwater Storage

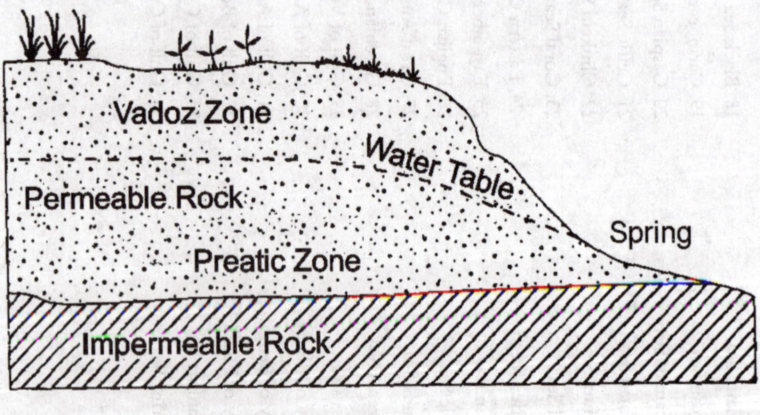

Saturated Zone

After percolation from the non-saturated zone, water reaches the middle portion and fills voids of rocks completely, resulting in rise of the water level. This zone gets maximum water during rainy period, whereas in the dry season the water table again goes down. Thus, this zone having higher water table during rainy season and lower water table during dry season, is also called '*Intermittent Zone of Saturation*'. The zone below it remains permanently saturated, which is called '*Zone of Permanent Saturation*'. The saturation zone is also called '*Preatic Zone*'.

Rock Flowage Zone

Below the surface of earth, there is a definite area where voids of rocks are closed due to the weight of rocks. Because of this the water cannot infiltrate further. Because the water cannot percolate further, it collects there. This collected water is known as '*Aquifer*'. Such water is assumed to be at the depth of 16 kilometres from the ground. The water existing at this great depth can flow, and this zone is called 'Flow Area'.

Sources of Groundwater Availability

Groundwater exists below the ground in different zones due to layers of non-permeable rocks. The upper part of groundwater is known as *surface water level* from where water enters through voids. The portion where water flows is known as *aquifer*. Here, sources of groundwater implies such sources from which groundwater is received on the ground, whereas in Chapter 4, sources of groundwater are those source points like meteoric water, connate water, and magma water sources, from which groundwater is stored. Here, water cannot move so fast that it becomes sufficient for springs or other forms. It is clear that due to different nature of internal structure ground level, groundwater exists at different places in different quantities, which comes to the ground from the various sources. There are two broad ways by which groundwater may be discharged. One is by artificial methods and the other is by natural methods.

First, the natural methods. These include:

- Evapo-transpiration, particularly in the low lying areas where the water table is close to the ground surface.

- Effluent seepage into surface water bodies, especially where the aquifer flows into a water body such as a lake or a river.
- Leakage through aquacultures into the adjacent aquifers, and
- Spring flow

Second is the artificial abstraction.

As for artificial abstraction, it is a story all of you must be familiar with. It has adversely affected storage in many areas. Groundwater when pumped from wells and boreholes causes a decline in the height of the water surface in the wells and results in the formation of a cone of depression around it. Successive abstraction from a large number of wells over a long period of time results in the gradual lowering of the water table or the piezometric surface.

Springs

That part of underground water which comes out on the ground from some natural void, is called a spring. Springs move according to the structure of layers of rocks. Rain water infiltrates through permeable rocks under the ground but when non-permeable rocks come in the way, that water cannot move and it starts flowing upwards by the side of non-permeable rocks. This water appears on the surface of the earth in the form of springs. In hilly or sloppy areas, this collected water sprouts out from raised permeable rocks in the form of a water source on the surface of the earth. On the basis of permanency of water, these springs are divided into two categories:

1. *Permanent Springs:* When non-permeable rocks below the surface of the earth are in inverted position and the water table is high, groundwater flows out permanently due to availability of slope.
2. *Temporary Springs:* These springs exist due to unequal water level below the surface of the earth and they depend on rainfall.

Based on nature of water, springs are of following types:

1. *Common Springs:* These water sources are of normal depth which give fresh, cold and potable drinking water.
2. *Thermal Springs:* They are deeper in size which emit out hot water. Such sources are found in California Park.
3. *Warm Springs:* Temperature of these springs is even higher than boiling point because they are very deep. About 4,000 such springs are located at Kulu and Manikarn in Himachal Pradesh (India) as well as Yellowstone National Park of United States of America.

4. *Mineral Springs:* These sources are mainly found in volcanic areas from where water mixed with minerals like sulphur, salt, etc. comes out, is of medicinal value. Sahstradhara near Dehradun in Uttaranchal, Atari in Orissa and Chhindwara in Madhya Pradesh are such sources of water.

Springs are an interesting feature. The groundwater discharge through springs occurs in areas where the upper surface of the zone of saturation intersects the ground surface. Springs are different from seepage. Springs appear in the form of concentrated discharge of groundwater. A seepage area on the other hand, is the slower movement of groundwater of the surface.

K. Bryan has classified springs into several categories. First is the 'Depression Spring' in which the groundwater flows to the surface from a permeable aquifer. Then there is the 'Contact Spring' that forms where the contact plane between a permeable and impermeable rock intersects the ground surface in such a way that the groundwater is deflected to the surface. The Contact Spring is found at the foot of limestone or chalk escarpment, foot of a scree slope and also along a fault line. 'Fracture Springs' occur when a system of interconnected minor faults lead the groundwater to the surface. Another type is the 'Tubular Spring' in which groundwater moves through lava tubes or interstices in solution enlarged in limestone. And, last but not the least, is the spring which arises from hydrostatic pressure.

Well

To get water from the collected water area located below the surface of the earth, wells are dug in permeable rocks below which, due to non-permeable rocks it is full of water. Wells based on rains are less deep and they become dry in summer. Deeper wells are helpful for water supply for a longer period.

Artesian Well

Artesian wells are natural water sources from which water automatically appears on the surface of the earth. For their construction, two types of favourable structures are required. First, a synclinal or dome shaped inverse structure, and second, a Monocline structure. The region from where the water of artesian wells moves upwards, is called aquifer. There should be non-permeable rocks, below as well as above aquifer, whose nature is similar to an aquiclude. Its lower level prevents seepage of water, whereas the upper level prevents water

Figure 2.3

The Great Artesian Basin of Australia

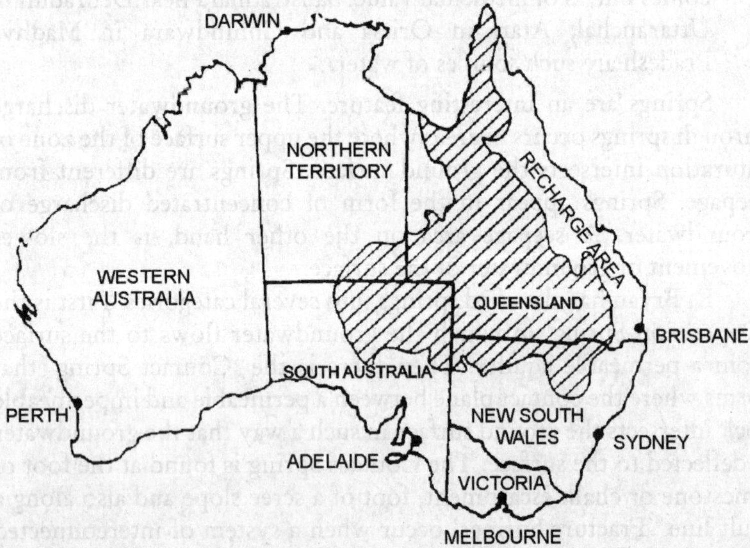

from evaporation. The part of the aquifer opening on the surface is called drainage area. From where the water entering through the permeable layer, reaches the central portion. By, and by as soon as the permeable layer is full, the 'Hydrostatic Pressure Head' is formed. Now, if a well is dug from the surface to the depth of the aquifer, then water of the aquifer would appear automatically on the surface as water source due to hydrostatic pressure. This forms an artesian well.

Geysers

In the language of Iceland, *'Gosir'* means *'Sponter'*. According to Holmes, "Geysers are sources of hot water from which after some intervals hot water and vapour come out with speed." Temperature of such water remains up to 75°C to 90°C. There are 100 geysers in the world including 'Old Faithful' in the world famous Yellowstone National Park situated in Wyoming the state of United States of America. Many geysers are found in Iceland and New Zealand.

Surface Water

Water on the surface of the earth is found both in fixed as well as moving forms. Water is found in different shapes on the basis of form.

On high mountainous areas it is found in the form of snow caps and icebergs, and in lower parts it is found in liquid form. Surface water is distributed on different continental parts. More than two-third part of fresh water is found in the form of snow Antarctica, and that is why the geographical position of the continent is different among the seven continents of the world. On the other hand, there is no permanent snow area in Australia. Except Antarctica and Greenland, snow is found in other places which are above the snow line.

The height of the snowline is different at different places. It is 5,000 to 6,000 metres at the equator, 4,300 to 5,300 metres at the Himalayas and 666 metres at South Chile and South Greenland. Thus, water on earth is distributed in different forms, the description of which is as follows:

Snow Fields

Above the snowline that part is called snow area where snow remains permanently frozen. Such parts are called 'Permanently Snow Fields or Permanent Frost'. Except the continent of Australia, permanent snow fields are found on all the remaining continents of the earth. At present, 24 per cent area of the whole earth is permanently snow covered area which extends to about 15 million sq kms. Ninety-six per cent of glaciated ice cover is found in thick masses in Antarctica and Greenland. All these areas are spread over the Arctic Ocean and the Antarctic continent, where polar type climate (E type climatic group presented by Koppen) is found. Summer temperature here remains below 10°C, which is the hottest month here. The coldest and hardest snow covers are in parts beyond the Tundra, which are called permanent snow caps. Being located in Polar parts, they are also called 'Polar Ice Caps.' The Arctic portion is called Internal of Greenland. Arctic and Antarctic regions are included in it. Regional snow cover here is visible in the form of mountains which are called 'Nunatak'.

The Arctic

This region includes Arctic Ocean and the nearby areas in which the extended northern part of Canada, Russia, Greenland, Scandinavia, Iceland and Alaska are the main areas. Permanent snow cover or pack ice is found here. Icebergs break from glaciers and the reach the sea, which float at 100 metres above sea level. In April 1912, The Titanic

Box 2.1

Snow and Ice

Snow and ice both are related to cold regions where permanently or for sometime the temperature reaches below freezing point (0°C) or remains at it. Ice is solid form of water and it forms due to freezing of water. It remains frozen at any place where temperature remains below freezing point. In glaciers or seas, or by condensation of water drops in form of ice crystals, it can stay in the atmosphere. Its density remains 0.9166, which floats in water due to its density being lower. Formation of snow is there when the dew point reaches below the freezing point. In this condition, atmospheric vapour in hexagonal shape condenses in the form of ice crystals and comes on the ground level. In case temperature goes below freezing point, snow accumulates on ground level and by effect of air etc. spreads as close cover. Snow melts again as soon as temperature rises above the freezing point.

ship of United States of America had collided with such a pack ice and drowned.

Antarctica

This is the fifth great continent of the world which is situated in the midst of the Antarctic circle. It is the coldest continent of the world. The lowest temperature was 88.3°C as recorded at Vostok station on 24th August 1960. About 95 per cent of its part is snow covered which is about a 2,000 metre thick layer. Maximum thickness of snow is 4270 metres. Ninety per cent of the total fresh water resources of the world are situated in the Antarctica continent in the form of snow. Due to global warming, this snow is slowly melting, which will bring changes in the sea level also. Antarctica plays an important role in regulating world climate and oceanic ecology system where first forest life habitation is found on the earth.

Glaciers

Above the snow line, there is located a vast mass of ice which is called glacier, which moves due to gravitational force. The movement remains in the snow area on high latitude and high mountainous parts. Glaciers are formed due to accumulation and compaction of ice above snow line. Glaciers of all types on the whole earth have been estimated to be between 70,000 to 2,00,000.

Types of Glacier

Glaciers are of different types according to process of formation, size, form, situation and extension. As per size and location glaciers are of four types :

1. Ice Cap Glaciers
2. Alpine or Valley Glaciers
3. Piedmont Glaciers
4. Continental Glaciers

1. *Ice Cap Glaciers:* These are combinations of Ice Cap Glaciers, Alpine Glaciers and Continental Glaciers, which are encircled by mountains and high plateaus. Alpine glaciers flow on steep slopes. They are also called mini continental glaciers. Snow lumps extending for less than 50,000 sq km are called snowcaps. 'Svalbard (Spits bergan) Island' of Norway is such a type of glacier.
2. *Alpine or Valley Glaciers:* In mountaineous regions, when 'Ice Cap Glaciers' move to lower slopes because of gravitational force, they are called 'Valley Glaciers'. Because its study was done first on Alps Mountain, it is also called 'Alpine Glacier'. Such glaciers are found in abundance in the Himalayas (Asia) and the Rocky Mountains (North America). These glaciers become narrower as and when they flow downwards. Valley Glaciers are found above the snow line. Valley Glaciers start with accumulation of snow at the peak of the valley in Cirque that coalesce into glaciers by and by.
3. *Piedmont Glaciers:* They are extended Glacier sheets, which are formed at the foot of mountains by merging and coalescence of Alpine Glaciers. Such types of Glaciers are found specially in Alaska. Malaspina is the main Piedmont Glacier of Alaska which is spread over in 3,900 sq km.
4. *Continental Glaciers:* They are vast covers of dense snow, which encircle the entire ground surface of their area. They develop as an extended sheet due to continuous accumulation of snow in a vast area. They are also called 'Ice Sheet'.

In Antarctica, there is an extension of 13,000,00 sq km of Continental Glaciers. An area of 18,00,00 sq km in Greenland is also covered by Ice Sheet. Near Gulf of James, the thickness of ice is 3,000 metres.

The above four types of glaciers that have been described above are normally classified. The remaining minor glaciers are described below:

1. *Ice Sheet:* Extension of Ice Sheet is in more than 50,000 sq km area. It is scattered in a levelled dome shape and formed by accumulation of ice. At present, there are two wide Ice Sheets, namely Greenland Ice Sheet and Antarctica Ice Sheet, where 11 and 85 per cent ice is situated respectively. Apart from these many minor ice sheets exist in Arctic Canada, Iceland and Norway.
2. *Ice Cap:* It is a small ice sheet which extends to less than 50,000 sq km area. It is a permanent ice mass that is found in high mountains and high altitude. Burnace Ice Cap and Baffin Ice Cap are the main Ice Caps.
3. *Ice Dome:* This is the central portion of ice sheet and ice cap.
4. *Outlet Glacier:* Outlet Glacier is a stream which flows from ice cap and ice sheet. Subsequently, it merges with a Valley Glacier.
5. *Ice Shelf:* This is a floating ice sheet adjacent to coastal parts. It flows independently without any friction with base level.
6. *Ice Field:* This is a mass of pack ice or a wide area of sea ice which is specially found in the polar areas.
7. *Cirque Glacier:* Cirque Glacier is a part of minor ice.

Surface Water Bodies as Snow Fields

Except liquid water found in rivers and lakes on the ground, a large water deposit is also found on the ground in the form of snow. They are found in the Arctic area of the North Pole, and Greenland and Antarctica in the South Pole. Apart from them ice caps on high mountains and glaciers are also important water deposits. Ground areas of Canada, Russia, Greenland, Scandinavia, Iceland and Alaska in the Arctic water region are included in it. Among them, Greenland is the biggest snow-covered area. Antarctica continent in southern hemisphere is the fifth biggest continent where 2,000 metre thick snow-cover exists in 95 per cent of its region on an average. About 90 per cent fresh water of the earth is found here but due to increase in world temperature, this snow is melting. Apart from Arctic and Antarctic, some quantity of water is also found in the form of ice caps and glaciers on Himalayas, Alps, Rocky, Andes, Kilimanjaro and a few other mountains. These ice deposits become source place of many rivers and help in supply of water.

It is clear from the geographical description of the interrelationship between hydrosphere and the process of hydrological cycle that out of the water found in nature only a limited part becomes available for use in the biosphere because its greater part is saline and 80 per cent of total water is frozen as snow. In present times, when the population is growing fast, there is a increasing demand for water on one side while on the other side , there is deterioration in the quantity and quality of water.

Glaciers that contain a large part of the fresh water deposit of the world, are melting due to rise in temperature of the world and these glaciers are directly or indirectly connected with seas. The snow of Antarctica is melting. At a conference held in March 2002, scientists of London reported that Larsen-B glacier linked with Antarctica since past, which was 1,500 years old and vast snow deposit spread over in 1,250 sq miles area had broken, in addition people have made rivers, lakes and oceans as places for disposal of waste. Continuously increasing agricultural area and consequent excessive exploitation of groundwater for irrigation have resulted in acute water crisis. It is clear that looking at the disturbing water balance and deterioration in quality of water, a strategy for its management has to be worked out so that water is able to maintain its balance in nature.

Rivers

Flowing water deposit in any definite way on the surface of the earth is called a river. Rain water flows on the earth in different forms which is called run-off. When this same water starts flowing from a higher plane to a lower plane in a definite sequence due to gravitational force, it takes the shape of a river. Its development takes place by and by. First of all, the rain water flows in the form of sheet flow. While flowing in the form of sheet flow, it follows the slope very soon and flows in the form of a *rill*. When rills combine and flow together, they become a *gully*. When gully becomes deeper, it becomes a *rivulet* and subsequently with combination of many rivulets, a river is formed. The source of rivers are spring, snout of a glacier, water collected from rain on the earth, lakes etc. Slope of the ground, rock structure and climate decide the nature and form of rivers.

Rivers occupy an important place in the ground level water sources. They originate from high mountain areas, snow covered peaks, lakes etc. and flow in speed over ground level. Except independent flow, they also receive interflow at many places like

Figure 2.4

Important River Basins of the World

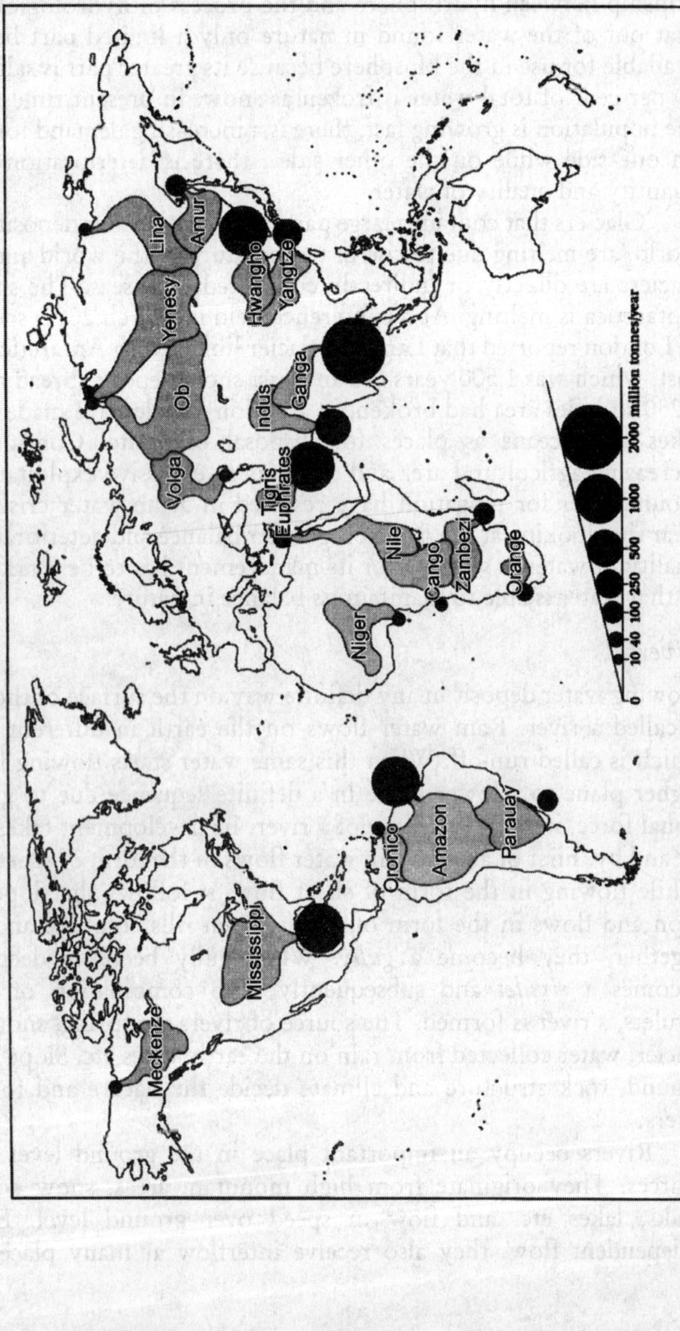

Table 2.4
Important Rivers of the World

S. No.	*Name*	*Source*	*End/Out Flow*	*Length (km)*
1.	River Nile*	Uganda-Kenya-Tanzania	Mediterranean Sea	6695
2.	Amazon River	Lego Bilfere (Peru)	South Atlantic Ocean	6570
3.	Yangtze (Chang Chiang)	Plateau of Tibet	Eastern Chinese Sea	6300
4.	Mississippi	S.Montana (U.S.America)	Gulf of Mexico	6020
5.	Yenisey	Mongolia	Kara Sea	5540
6.	Obe	Mongolia	Gulf of Ob	5640
7.	Hwangho	Jarring Nor (China)	Yellow Sea	5600
8.	Parana	Plateau of Brazil	Rio-De-La Plata	5464
9.	Zaire (Congo)	Bordering Area of Zambia and Zaire	S.Atlantic	4500
10.	Amur	Khingan Ranges	Okhotsk Sea	4416
11.	Lena	Plateau of Russia	Laptev Sea	4400
12.	Mekong	Plateau of Tibet	S.China Sea	4425
13.	Mackenzie	Tattle Lake (Rocky Mountain)	Beaufort Sea	4240
14.	Niger	Kenya	Atlantic Ocean	4185
15.	Murray-Darling	Queensland (Australia)	Gulf of Encounter	3750
16.	Volga	Russia	Caspian Sea	3685
17.	Zambezi	Zambia	S.Indian Ocean	3540
18.	Yukon	British Columbia	Bering Sea	3185
19.	St.Lawrence	Minnesota America	N.Atlantic Ocean	3130
20.	Rio Grande	Son Ja On Ranges	Gulf of Mexico	3033
21.	Indus River	Tibet	Arabian Sea	2850
22.	Brahmaputra	Tibet	Bay of Bengal	2900
23.	Danube	Black Forest (Germany)	Black Sea	2850
24.	Salveen	Tibet	Gulf of Murtban	2810
25.	Tigris-Uphvates	Turkey	Persian Gulf	2740
26.	Orinoco	Venezuela	Atlantic Ocean	2740
27.	Sikiang	Plateau of Yun-nan	S.China Sea	2650
28.	Amoo Dariya	Alai Tag (Pamirs)	Aral Sea	2575
29.	Ural River	Ural Mountain	Caspian Sea	2540
30.	River Ganges	Gangotri (Himalaya Mountains)	Bay of Bengal	2510

* River Nile originates in the form of Blue Nile and White Nile, whose source is respectively the Tana Lake of Ethiopia (Blue Nile) and Victoria Lake (White Nile). Both these rivers meet near Khartoum (Sudan).

water received from percolation, groundwater springs, natural soil lines and infiltration. Thus, many streams combine to take the shape of a river and merge into a sea at the end. Many rivers also fall into lakes, for example, Volga and Ural rivers fall into the Caspian Sea, whereas some rivers are also inland. Ghaghar river emanating from the Himalayas vanishes in the Thar desert. Area of drainage basin, shape of drainage basin, total rainfall, intensity of rainfall and surface of drainage basin control the flow of rivers.

Lakes

Lakes are important water deposits in which 0.017 per cent part of water on earth is stored. Lakes are of different kinds according to nature and quality of water. Among them non-saline and fresh water lakes, saline or saltish lakes and snow lakes are important. According to location, there are mountain lakes, plateau lakes and lakes of plain area. Lakes like Ladoga, Onega, Aral sea and Baikal etc. are of plain area in Western Russia, whereas Kokonor (Ching Hai) of Central Asia, Tania of Africa, Teetikaka of South America, and Mansarovar lake of Tibet are mountain lakes.

Seventy-five per cent water out of the fresh water of the lakes of the world is found in great lakes of North America and Baikal Lake. Caspian Sea is also a closed lake where water is said to have been stationary for the last 200 years. Its main reason is the fast rate of evaporation.

Lakes are formerly glaciers, action of volcanoes and destruction etc. Rudolf lake in the Rift valley of Africa, Albert, Edward and Tanganyika are the longest lakes, and Baikal is the deepest fresh water lake. Caspian Sea is the longest lake of saline water. Sambhar and Chilka are vast saline lakes of India. Lunar lake of Maharashtra is a crater lake.

Great Lakes

The series of five lakes of North America is known as 'Great Lakes' they are located in the United States of America and Canada. These lakes are Superior, Michigan, Huron, Erie and Ontario. All the five lakes are spread over as inter-connected. Superior is the greatest and Ontario is the smallest of these lakes.

Great lakes are spread over an area of 2,44,650 sq km. Fresh water is found in these lakes. Lake Superior is the biggest fresh water lake of the world whose area is 82,350 sq km. It is spread in Minnesota, Wisconsin and Michigan States of the United States of America over 53,000 sq km. The remaining area of this lake is in Canada.

Table 2.5
Important Lakes of the World

S. No.	*Name*	*Location/Country*	*Area (Thousand sq km)*	*Maximum Depth (Metres)*
1.	Caspian Sea	Russia, Kazakhstan, Iran, Azerbaijan	370.9	980
2.	Superior	Canada and America	82.3	406
3.	Victoria	Uganda, Tanzania, Kenya	69.5	80
4.	Aral Sea	Kazakhstan, Uzbekistan	65.5	68
5.	Huron	Canada and America	59.6	228
6.	Michigan	America	58.0	281
7.	Tanganyika	Zaire, Tanzania, Burundi, Zambia	32.9	1435
8.	Gt. Bear	Canada	31.8	82
9.	Baikal	Russia	30.5	1940
10.	Malawi (Nyasa)	Tanzania, Malawi, Mozambique	29.6	678
11.	Gt. Slave	Canada	28.5	163
12.	Erie	Canada and America	25.7	64
13.	Winnipeg	Canada	24.5	36
14.	Ontario	Canada and America	19.5	237
15.	Ladoga	Russia	17.7	225
16.	Balk hash	Kazakhstan	17.4	26
17.	Chad	Niger, Nigeria, Chad, Cameroon	16.3	8
18.	Maracalbo	Venezuela	13.5	—
19.	Onega	Russia	9.6	110
20.	Eyre	Australia	9.5	20
21.	Titicaca	Peru, Bolivia	9.5	278
22.	Athabaska	Canada	8.3	124

Geographers opine that the formation of these Great Lakes about 15,000 years old was due to melting of glaciers. The size of these lakes has changed due to many sudden activities on the earth.

Presently, these lakes are used mostly for internal water transport. Soo canal links Lake Superior and Huron , whereas Wallend canal is in the middle of Ontario lake. Many ports and cities have developed along the banks of these lakes. Due to availability of water and transport, steel manufacturing centers have grown around them. Buffalo, Cleveland and Detroit on the banks of Erie lake, Chicago, Gary etc. on the bank of Michigan lake and Milwaukee port on Ontario have developed as important cities. Toronto port etc. are on the bank of Lake Superior. Duluth and Minnesota are main steel manufacturing centers.

Table 2.6

Great Lakes of North America

Name of the Lake	*Area (In sq km)*	*Depth (In metres)*	*Maximum depth from Sea Level*
Superior	82,100	183	405
Michigan	58,000	176	281
Huron	60,000	177	229
Erie	25,720	174	64
Ontario	19,200	45	244

In between Eurie and Ontario flows river Niagara in which the world famous 'Niagara Fall' is situated.

Atmosphere

About 0.001 per cent water out of the total water found on the earth and 0.04 per cent of the total available fresh water (2.26% of the total water) exists in the atmosphere. It has been estimated that if total water in the atmosphere condenses and falls on earth in the form of rain, a layer of 2 mm will collect on earth. This water is equal to the water available in the rivers of the world. It has become clear from different estimates that duration of water in the atmosphere is about 10 days. On this basis, atmospheric humidity exchanges it with ground level about 40 times. Due to this deficiency, atmosphere remains in mobile position, due to which it remains as an extended linkage with other hydrological systems. In the process of hydrological cycle of the world, atmosphere functions as an important link.

Sub-Surface Water

Minerals and rock pieces scattered over the upper part of the ground surface of the earth in an unorganized group is called soil. Soil

Figure 2.5
Different Sources of Sub-Surface Water

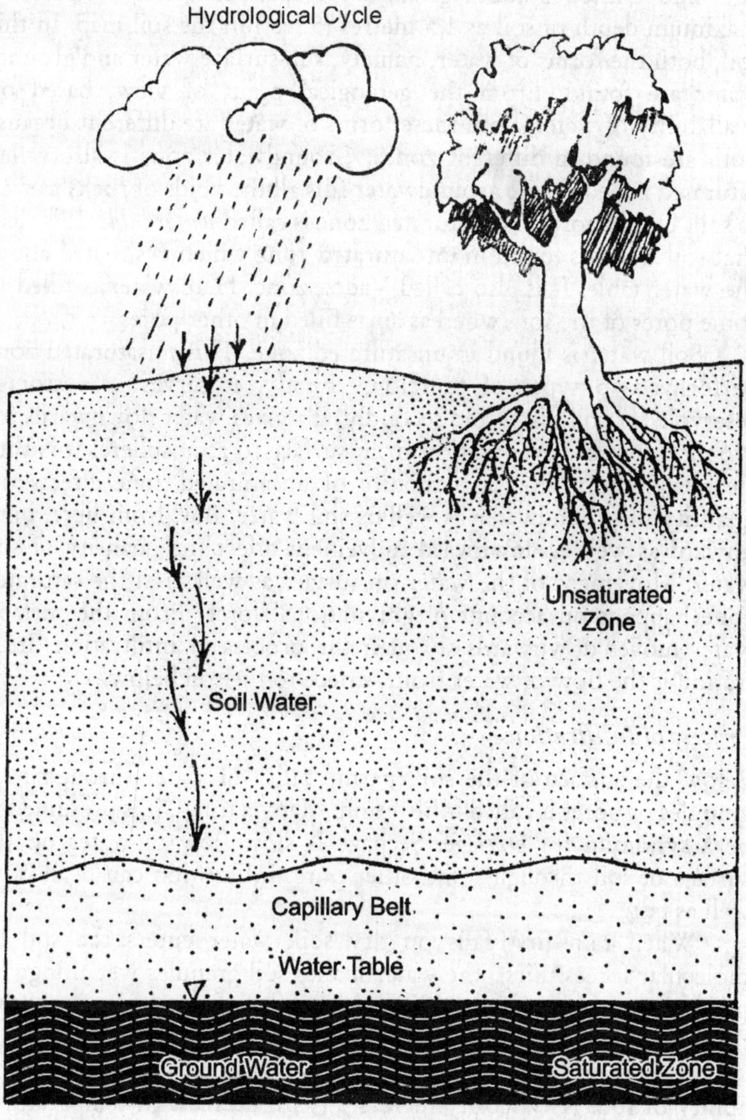

includes fine clay, salt granules and sand particles. Many times pebbles and boulders are also mixed in it. Soil scientists have considered a few feet thick surface of the ground as soil from the point of view of agriculture.

The United States Agriculture Department has decided the maximum depth of soil as 1.5 metres (5 feet) in the soil map. In this soil, both the forms of water, namely, sub-surface water and groundwater are found. From the geological point of view, based on availability of water, both these forms of water are different because both are found in different zones. Groundwater zone is also called saturated zone because groundwater fills all the voids of rocks as well as soil. Upper portion of saturated zone is called *water table*. It is clear that soil water is found in un-saturated zone which is situated above the water table. It is also called Vadoz zone. Here water is filled in some pores of the soil, whereas air is filled in other pores.

Soil water is found in unsaturated zone. This unsaturated zone performs two types of functions. Firstly, it serves as a storage reservoir for soil water and secondly, it moves water downwards for groundwater and evaporation from surface to atmosphere. It is thus clear that the geological structure of unsaturated zone affects the quantity of groundwater as well as soil water. But, from here, both upward as well as downward movement affect the quantity of soil water. Movement of the water on surface as well as in the saturated zone happens on account of gravitational force. From the surface, water moves downwards through the process of infiltration. Infiltration is the flow of water from surface to the soil below.

Process of Infiltration

Below the surface of the soil, permeability of water through rock granules or pores is called infiltration. Many a times, it is presumed to be a synonym for seepage. Seepage is slow flow of water on the surface of soil through some other pore. Infiltration can be slow as well as fast.

When rain first falls on dry soil, water enters the soil by molecular forces. First, the water makes soil granules wet alongwith mineral surface through hydrogen bound. Subsequently, it reaches the minor pores of the soil.

Rate of infiltration depends on viscosity of water and permeability of soil. As viscosity increases (with decrease in temperature), water starts flowing in slow speed. It is clear that with increase in temperature, rate of infiltration is higher and it is less with decrease in temperature. Permeability is that quality of porous materials, which regulates flow of liquid in any material. Water can flow with speed in sand as compared to clay because sand is more permeable than clay.

Effect of Rainfall Intensity on Infiltration

Dry soil immediately soaks drops of water as soon as it rains. During normal rain, rate of infiltration is high and sufficient rain water enters the soil. But in case of intense rainfall, effect of drops of water specially on uncovered soil, is not good. When intense rain drops fall on naked soil, soil granules rearrange themselves and infiltration of water in soil does not taken place. Consequently, water collects on the surface and starts flowing on the surface. Flow of water also expedites soil erosion. Apart from intensity of rain, quality of water also affects soil. After weathering action on the surface, many minerals mix in the soil and along with water they prevent the affect of the downward flow.

Soil Water

Soil water is that part of sub-surface water which is found in the unsaturated zone between ground level and water table. It enters the soil through the process of infiltration. After entering soil this water either is stored there or comes upwards through evaporation or transpiration. Apart from evaporation or transpiration, it can also flow towards the saturated zone. During intense rain or storm, sub-surface water flows through inter-flow also. Horizontal flow of water immediately below soil surface is called inter-flow. But inter-flow is not really soil water. Soil water is an important constituent of hydrological cycle which is helpful for vegetation on ground level and increases the quantity of groundwater below ground level. Unsaturated zone is found in the form of storage for soil water.

The water existing in the pores or voids or open spaces is called soil water. The greater portion of soil water enters by infiltration of rain water. Use of such water is made by plants through root osmosis. From the view point of water-holding capacity, two aspects of soil are important. One aspect is porosity and other aspect is permeability. Porous and permeable soil immediately absorbs rain water and becomes dry early whereas, compact soil has less porosity because its soil granules are very near to each other. Water is a universal solvent which, after dissolving necessary nutritious substances, supplies them to different parts of the plants according to requirement. Famous soil scientist Briggs has classified soil water in three categories:

1. Gravitational Water

2. Hygroscopic Water
3. Capillary Water

1. *Gravitational Water:* Water is found in the soil in different forms. When percolation of water takes place into the soil due to gravitational force, it is called Gravitational Water or Independent Water. Water is filled to a certain level, after which its percolation is not possible. This level is called Zone of Saturation. Quantity of gravitational water in the soil is related to many situations of which quantum of rainfall and place of storage of water are important. Gravitational water works inside the soil in different forms. Firstly, it affects the soil colour, structure and composition. As soon as gravitational water enters and moves into the soil, it takes the fine granules of soil together with it. This it is called 'eluviation' of particles of upper soil. After the water collects clay granules of sub-soil along with it while vibrating. This process of soil is called 'illuviation'. Thus gravitational water is also an agent for transfer and mixing of soil granules from one layer to another layer. Its effect is that the upper soil gets filled with bold particles because the finer granules are transferred to the lower soil after dissolving with water. Due to this, the upper soil becomes more compact and of higher density than the lower soil.

Gravitational water, through affecting chemical composition of the soil, also affects composition, colour and structure of the soil. Thus gravitational water while moving downwards, mixes the nutrients of upper soil, in the lower soil at the depth after dissolving. Taking nutrients of upper soil towards the lower soil is called 'leaching'. Thus, problem of waterlogging is also created by over saturation of the soil with water.

2. *Hygroscopic Water:* This water is received by dry soil through absorption of atmospheric water vapour. If dry soil is heated at 105°C for about 8-12 hours, some part of water would go out from the soil. This minimum quantity of water is called Hygroscopic Water.

This water is in the form of a thin film which remains around soil granules or living organisms by adhesion force. This water destroys evaporation and there is no movement in it. It is also called waste water because living organisms of soil cannot utilize it. The structure of soil, quantity of clay material, humidity and temperature determine its quantity. Clay contains more hygroscopic water as compared to sand. It increases with increase in humidity and reduces with increase in temperature.

Figure 2.6

Water in the soil, water infiltrates in the soil and again Capillary, Surface and Gravitational Water comes (W.M. Marsh and J.M. Grossa)

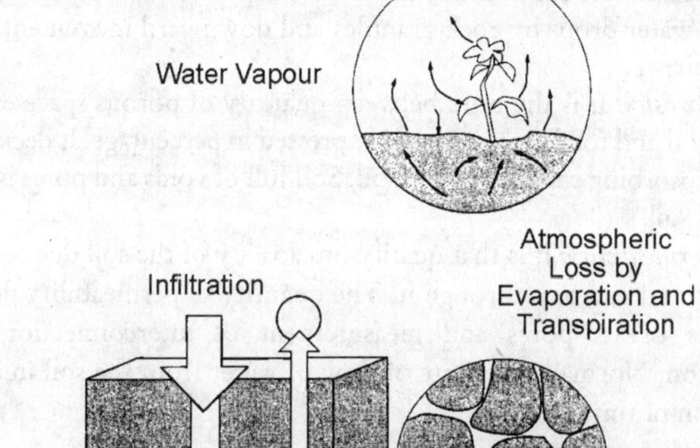

3. *Capillary Water:* Capillary water is attached to soil granules in the form of a thin film on all sides. Such water forms a ring around soil granules.

The size of this ring is affected by humidity because when the quantity of humidity increases, its size also goes on increasing. When this layer of water amalgamates with the nearby soil granules, most of the voids are filled with water, whereas in the remaining voids, air exists. Capillary water is retained by surface tension, and like hygroscopic water, it does not stick to soil granules but it maintains stiffness

Box 2.2

Infiltration, Porosity and Permeability

Infiltration: Permeability of liquid or air in rock granules or inter-vibration is called infiltration. In this process, there is absorption of rain water drops by rock granules and downward movement of the liquid air.

Porosity: It is the ratio between quantity of porous space existing in the soil and total area, which is expressed in percentage. It decides the water absorbing capacity of the soil. Soil full of voids and pores is called porous soil.

Permeability: It is that quality or capacity of the soil due to which liquid or air can pass through it. The quantity of permeability depends on size of its pores and measurement of interconnections, and extension. Normally it is rate of flow of water from the soil in certain duration of time.

to the extent that gravitational force of earth cannot separate it. Atoms of capillary water are independent, mobile and in liquid form. Due to this reason, it evaporates easily. Capillary water is also known as 'easy water', because in spite of being stuck to soil granules, trees and plants can utilize it. Its quantity is decided by the following factors:

1. *Structure of the Soil:* Structure of the soil affects capillary water's retaining capacity because if it is sandy soil, voids and pores are not monocular in it and capillary water is found in little quantity, whereas in clay due to monocular pores, water retaining capacity is greater.
2. *Organic Matter:* Any soil having more organic matter will have more water retaining capacity, because humus itself has more capillary water retaining capacity.
3. *Surface Tension:* Capillary water increases with increase in surface tension and reduces with increase in temperature.
4. *Size of Soil Particles:* If soil granules are more fine, clay will have more capillary water retaining capacity, whereas bold soil granules have less capillary water retaining capacity.

Thus, capillary water is affected by the above factors and maximum quantity of capillary water retaining capacity develops. This can be called maximum capillary water retaining capacity.

Factors Affecting Soil Water

1. *Soil Texture:* Flow of water is affected by size of soil granules. Water can infiltrate easily in sandy soil whereas, it cannot do so in clay soil. Clay soil absorbs more hygroscopic water. If soil particles are finer, capillary water would move faster. Gravitational water can flow in soil having bold particles.
2. *Humidity:* With increase of humidity in atmosphere, absorption of hygroscopic water capacity increases. Its quantity is more in monsoon season and less in dry days.
3. *Rainfall and Temperature:* Intensity of rainfall also affects soils. If rainfall is in slow speed, capillary water can be used by plants but with more intensity of rain, water moves very fast to the lower layers of soil by gravitational force and cannot be used by plants.
4. *Soil Structure:* Soil structure plays an important role in vibration of water in the soil. If soil is more compact, infiltration of water would be slower, whereas in loose soil, water would enter with faster speed. Thus, there is lesser infiltration in soil with fine granules.

Thus, besides the above factors, ice, dew and irrigation also affect soil water. Nature of soil water is changeable. Whenever it is used by plants for transpiration, its place is occupied by air. Thus, soil water and air both are interchangeable.

3

Water Resources of India

India possesses 5 per cent of the world's total water resources. According to K.L. Rao, there are about 10,360 rivers of at least 1.6 km length each with average annual flow of 1,869 cubic kms. From the geographical angle, only 69,000 crore cubic kms (32%) surface water is used due to uneven distribution and other hindrances. Apart from them, 43,200 crore cubic metre water is available at the national level in the form of traditional reserves and by diverting flow of water for re-supply. Maximum flow of surface water is in river Indus, Ganges and Brahmaputra, which comprise 60 per cent of total flow. The existing picture of annual probability of 1,869 cubic kms water indicates that in the country at some places there is water scarcity and at other places there is water in abundance. First of all, according to river basins, Ganges, Godavari and Krishna have many areas where there is heavy rain. Expansion of Ganges basin is up to Chambal, Sindh, Betwa and Ken in the south-west. However, in spite of sufficient rains in Krishna basin area, there are dry areas in it. Most of rivers flowing towards the east originate from Western Ghat and they are sustained by sources of sufficient rain. Even then, in the east of Western Ghat from Dhule to Bijapur Bellari in Maharashtra, there is rain shadow area. Similarly, in the eastern part rivers flow with sufficient water supply.

Surface Water Resources

The annual estimate of surface water in India is 1,86,900 crore cubic metre. The main sources of surface water are rivers, lakes, ponds, tanks etc. Most of the surface water flows through rivers. These are many rivers of large size and length in India which keep the land green and prosperous. In the western part of India having Thar desert however, there is no perennial river, though the artificially constructed Indira Gandhi Canal irrigates large areas in this desert. In southern India also there are many rivers. The total length of rivers of India is 2 lakh miles.

Figure 3.1
Rivers of India

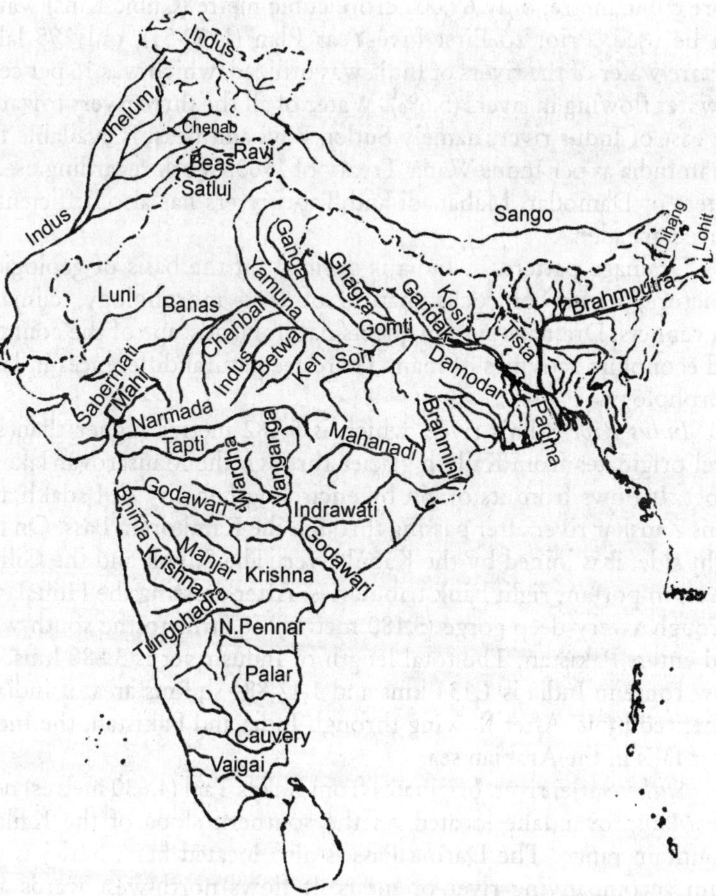

In India, there are two types of rivers: (1) perennial rivers of the north, and (2) Seasonal rivers of southern India. In northern India, 80 per cent of the non-saline water collected on the Himalayas flows in rivers Ganges, Yamuna, Indus and Brahmaputra and many others after melting, whereas in southern India, Mahanadi, Krishna, Kaveri, Godavari etc. are seasonal rivers. All these rivers merge into the Bay of Bengal except Narmada and Tapti rivers which pour their water into the Arabian Sea. Thus, due to the northern and southern rivers, the surface of the country remains green. Annually, 1,86,900 crore cubic metre (cubic km) water flows through the rivers of India. However, due to topographical and other reasons, out of the 1,86,900 crore cubic metre, only 69,000 crore cubic metre (Cubic Kms) water can be used. Prior to First Five-Year Plan (1950-51), only 95 lakh hectare water of the rivers of India was utilized, which was 16 per cent of water flowing in rivers (5.6%). Water of all the three rivers towards the east of Indus river, namely Sutlej, Ravi and Beas, is available for use in India as per Indus Water Treaty of 1960. Work regarding use of waters of Damodar, Mahanadi and Tapti rivers has also sufficiently moved ahead.

Drainage pattern in India is decided on the basis of geological structure, nature of rocks, nature of land morphology, climatic unevenness. Drainage system affects natural landscape of the country and economic activities of man. There are natural differences in land morphology.

Indus River: This river, which is 5,182 metres higher than sea level originates from Kailash glacier through the Mansarovar lake in Tibet. It flows from its origin by encircling Leh city of Ladakh and joins Zanskar river after passing through the Karakoram Pass. On the right side, it is joined by the Kabul river. The Shyok and the Giligit are its important right bank tributaries. After crossing the Himalayas through a very deep gorge (5,180 metres), it turns to the south west and enters Pakistan. The total length of Indus river is 3,880 kms. Its flow route in India is 1,134 kms and 1,17,884 sq kms area of India is benefited by it. After flowing through India and Pakistan, the Indus river falls in the Arabian sea.

Sutlej: Sutlej river originates from Rakas Taal (4,630 metres) near the Mansarovar lake located on the southern slope of the Kailash mountain range. The Darma Pass is also located here. Sutlej is the main accompanying river of Indus. It flows north-west wards and enters Himachal Pradesh at the Shipki Pass. The height of Rakas Taal

(Lake) is 4,630 metres above sea level. Its length in India is 1,050 kms and catchment area is 2,40,000 sq kms. This river is important due to Bhakra and Nangal dams constructed over it. The Sanskrit name of this river is 'Shatdru' or 'Shatudri'.

Chenab: Chenab is the longest river accompanying Indus river. Its origin is the Bara Lacha Pass near Taandi located at Lahul in Himachal Pradesh and is fromed by the joining of Chandra and Bhaga tributaries. Height of Bara Lacha pass is 4,480 metres above sea level. The length of Chenab river in India is 1,180 kms and its catchment area is 26,755 sq kms. Its name in Sanskrit is 'Askani' or 'Chandrabhaga'. Near Kishtwar, it cuts a deep george in the Pir Panjal Range from here it turns southwards and flows in this direction for a short distance.

Ravi: This river is also a companion river of Indus. Its origin is near the Rohtang Pass Bangaleh basin located between Pir Panjal and Dhauladhar Ranges. Ravi is a small river of Punjab and it is also known as river of Lahore. Bangaleh basin is 4,570 metres above the sea level. The length of Ravi river in India is 725 kms and its drainage area is 5,957 sq kms. Its Sanskrit name is 'Parusni' or 'Iraavati'.

Beas: Beas is other companion river of Indus. Its origin place is Pir Panjal which is 4,062 metres above the sea level. The length of Ravi river in India is 470 kms and its catchment area is 25,900 sq kms. Its Sanskrit name is 'Vipasha' or 'Argikiya'.

Jhelum: This is the smallest companion river of Indus in India. Its origin is Verinaag water source located in Jammu and Kashmir. Jhelum is an important river of Kashmir. Its total length is 425 kms, out of which 40 kms is in India. Its catchment area in India is 28,490 sq kms. Its Sanskrit name of Jhelum is 'Vitasta'.

Zanskar and Shyok are other companion rivers of Indus. Zanskar range is the origin of Zanskar river, which meets Indus near Leh. Shyok river originates from Karakoram and it meets Indus river near Kiris. Apart from these rivers, Dras is another a companion river of Indus and its origin lies between 'Jojeela pass' and 'Devsai'.

Saraswati river originates from Shivalik Ranges of Sirmaur on the borders of Ambala district and enters Adhbadri. It disappears between Bhavnipur and Balchhapaar but again appears near Karnaal. Ghaghar river (originating from Shivalik Ranges) in Rasula (Near Patiala). Saraswati is also called 'Haakda' or 'Sutaar'. In Vedic literature, Saraswati river has been described as an even more important river than Indus and Ganges.

The Ganges: The Ganges is the most important river system of India and covers more than half the area of India. The main river of this system is Ganges. From the left side, the rivers Ramganga, Gomti, Ghaghra, Tapti, Gandak and Koshi join the Ganges, whereas from the right side, Yamuna, Chambal, Sindh, Betwa, Tons and Son rivers join it.

The Ganges originates from Gangotri in Uttar Kashi district. The height of Gangotri is 7,016 metres above the sea level. In the beginning, there are two branches of the Ganges namely – Bhagirathi and Alaknanda. Bhagirathi is considered as the main branch. The confluence of Bhagirathi and Alaknanda takes place at Dev Prayag, while the confluence of Mandakini and Alaknanda takes place at Rudra Prayag. The confluence of Alaknanda and Pindar takes place at Karna Prayag and the confluence of Alaknanda and Dhauli takes place at Vishnu Prayag. After Haridwar, the Ganges enters the plain area. After the confluence of Alaknanda and Bhagirathi at Dev Prayag its name changes into Ganges. When the river Ganges crosses Himalayas, it forms a 4,870 metres deep gorge. The length of river Ganges is 2,510 kms and its catchment area is 9,51,600 sq kms.

Yamuna: This is the only river which descends from Himalayas and joins the Ganges. The origin of this river is Jamnotri Glacier located on the Bandarpoonchh Peak in the Kumayun area. The height of Jamnotri Glacier is 6,315 metres above the sea level. After crossing 152 kms in the mountainous track this river enters the plain area from Kalesar. At Allahabad (Uttar Pradesh) it joins river Ganges. The length of Yamuna river between Yamnotri and Allahabad is 1384 kms and its catchment area is 3,59,000 sq kms. Chambal, Sindh, Betwa and Ken are companion rivers of Yamuna.

Son: This is an important companion river of the Ganges. Its origin is the Amarkantak mountains. Amarkantak is located at the height of 600 metres above the sea level. Son river forms many waterfalls on its route. After travelling 780 kms from Amarkantak, it merges into the Ganges in Ramnagar near Patna. The catchment area of this river is 71,900 sq kms.

Ramganga: This is the main companion river joining Ganges from the left side. Ramganga originates from Nainital district in Uttaranchal. The height of its place of origin is 3,110 metres above the sea level. After crossing 144 kms in the hilly region it forms a deep valley in Shivalik ranges and thereafter enters plain areas near

Kalagarh in district Bijnaur. At Allahabad (Uttar Pradesh) it merges into the river Ganges.

Ghaghra: This is an important companion river of Ganges meeting from the left side. It originates from Silam Himani in Kumayun Himalaya. In the beginning, this river was known by the name of Sharda or Kali. After flowing through Pilibhit, Kheeri, Bahraich, Sitapur, Gonda, Faizabad, Basti, Gorakhpur, Azamgarh and Ballia, it merges in river Ganges in District Chhapra of Bihar. The length of Ghaghra river is 1,080 kms and its catchment area is 1,27,500 sq kms. More than half of its area falls in Nepal.

Table 3.1

Available Water Resources in Important River Basins of India

(in cu kms)

Basin	*Average Annual Running Water*	*Water Flow for Use*	*Storage*
Indus	73	46	14.52
Ganga	501	250	37.40
Brahmaputra	537	24	1.09
Godavari	119	76	17.27
Krishna	68	58	32.23
Cauvery	21	19	7.25
Pennar	6.8	6.8	2.37
Mahanadi	67	50	8.93
Brahmani	36	18.1	4.29
Sabarmati	3.8	1.91	1.30
Mahi	11.8	3.1	4.16
Narmada	41	34.5	3.02
Tapti	18	14.5	8.68
Suvarna Rekha	10.8	—	—
Total*	1869	690	—

* It also includes other river systems.

Gandak: Gandak river originates at the height of 7,600 metres on the Nepal-China border and joins river Ganga on its left side as an important companion river. Gandak river enters near Champaran district of Bihar. After flowing through Bihar and Uttar Pradesh it merges into Ganga near Sonpur in district Patna. Kali Gandak, Budhi Gandak and Trishuli Gandak are its companion rivers. The length of

Gandak river is 425 kms and its catchment are is 45,800 sq kms, out of which 9,540 sq kms area lies inside Nepal.

Gomti: This river originates from Gomat Tal at the height of 200 metres in district Pilibhit. After flowing through Pilibhit, Kheeri, Sitapur, Lucknow, Barabanki, Sultanpur, Jaunpur, Gazipur, it merges into river Ganges at Gazipur. Kalyani, Reth, Kundanala, Bednala, Chunha, Mangar, Manjuhi, Sai, Peeli, Basuhi, Barna, etc. are its companion rivers.

Kosi: It is one of the main rivers meeting Ganges on its left side. Originates in the north of Gosainath at the height of 6770 metres, in its begining, it is known as 'Aruna' river. It flows in seven tributaries namely, Milamchi, Bhotiya, Koshi, Tambakoshi, Likhu, Doodh Koshi, Aruna and Tamboor. Its total length is 730 kms and drainage area is 86,900 sq kms. It is also called *'Sorrow of Bihar'*.

Chambal: Chambal river is the sub-companion of the Ganges, since it is the companion river of Yamuna. It originates from Janapav Hills in Madhya Pradesh at the height of 600 metres. After passing through Kota, Jhalawar (Rajasthan), Bhind, Morena (Madhya Pradesh), Agra and Itawa (Utar Pradesh) for 965 kms, it merges into river Yamuna near Itawa.

Damodar: This river originates from Chhota Nagpur Plateau of Palamu district in Jharkhand. Garhi, Konar, Jamuniya and Barakar are its companion rivers. After merging in the Barakar it takes a giant shape. It then travels 543 kms to merges into river Hoogly. The catchment area of river Damodar is 22,000 sq kms. It is called *'Sorrow of Bengal'*.

Betwa: This river originates from the south of Goharganj in the south of Raisen district of Madhya Pradesh. After flowing through Guna, Shivpuri, Jhansi and Hameerpur districts it merges into river Yamuna near Hameerpur.

Tons: Tons originates from Tamasha Kund water body located in Kaimur hills. This river merges into Ganga near Sirsa in the south east of Allahabad. Tons river forms the 114 metre high famous waterfall 'Bihar' in its flow route.

After merging of rivers from the south west side into the Ganga it forms a superimposed water flow image. Due to continuous soil erosion, old rocks having slope against the direction of water flow have come out but these rivers continue to flow on their decided path.

Brahmaputra River System: From the point of view of quantity of water, Brahmaputra river is the biggest among all the rivers flowing in India, but from the the angle of flow area it it is not the biggest one. It originates from the eastern slope of Kailash mountain in the plateau of Tibet at a height of 5,150 metres. Its flow area is in Tibet, Bangladesh and India. It flows in the eastern direction parallel to the Himalayas up to Namchabarva Peak which is known as 'Psang Po'. After Namchabarva Peak it flows towards south and the south-west. In India, this place is known as 'Siang' and 'Dihang'. After crossing Sadiya, it is called Brahmaputra. Suvanshiri, Dhanshiri, Maanas, Sankosh, Raidak, Tista, Diwang, Lohit, Deeshu, Kopili etc. are its companion rivers. The total length of Brahmaputra river is 2,580 kms but its length in India is only 1,346 kms. The total catchment area of this river is more than 5,80,000 sq kms out of which 3,40,000 sq kms lies in India. It is called Jamuna in Bangladesh.

South Indian River Systems

There is a basic difference between northern and southern river systems – The rivers of northern India are perennial whereas, in south India rivers do not flow all the 12 months. During the rainy season water remains in the rivers of the south but during summer as well as winters, there is scarcity of water. Rivers of southern India mostly originate from insular plateaus. Some flow towards the east and others flow towards the west. Rivers flowing towards the east fall in Bay of Bengal. The main rivers among them are as under:

Mahanadi: This river originates from Sihawa in the south of Amarkantak. In the beginning, it flows in the north-eastern direction. After confluence with Sevnath it diverts towards the east and thereafter it flows towards the south-eastern direction. Near Sambhalpur, it takes a giant shape and before merging in Bay of Bengal near Cuttack many companion rivers also join it. The length of river Mahanadi is 885 kms and catchment area is 1,32,000 sq kms, out of which 53 per cent lies in Madhya Pradesh and 46 per cent in Orissa state.

Godavari: Godawari river is the biggest among insular rivers and is the second biggest river of India. This river originates from a place named Triyambak located at a height of 1,067 metres in the east of Sahyadri range in district Nasik of Maharashtra. Pranahita, Indravati, Shabri, Manjra, Penganga, Wardha, Venganga, Taal, Mula, and Pavra are its companion rivers. After Dhavleshwaram, Godavari river bifurcates into two branches – the eastern branch is called Gautami Godavari and western branch is called Vashishta Godavari. In the middle, there flows another river called Vaishnav Godavari. Gautami Godavari falls into the Bay of Bengal at place called 'Yenam', Vashishta Godavari at place called 'Narsapur' and Vaishnav Godavari at the place called 'Nagra'. The total length of river Godavari is 1465 kms whereas, its catchment area is 1,13,839 sq kms.

Krishna: This river is the second largest insular river flowing in the east India. It originates from Mahabaleshwar at the height of 1,327 metres in the Western Ghat of Maharashtra. There are 10 companion rivers of Krishna. Among them Koyna, Malprabha, Ghatprabha, Musi, Panchganga and Doodhganga are important. Bheema meets Krishna before it leaves Maharashtra and enters Andhra Pradesh. Near Kurnool in Andhra Pradesh river Tungabhadra joins it and in the south of Hyderabad river Musi joins it. Near Raichur, Krishna river comes down from the hills of Western Ghat and after traveling for 5 kms, it steps down 120 metres. About 65 kms after Vijayawada, Krishna divides into two parts. The southern part further sub-divides it in different sources and finally near Masulipattam it merges into the sea. The length of this river is 1,400 kms and catchment area is 2,59,000 sq kms.

Tungabhadra: Besides being a companion river of Krishna, it is also an important river system. As a separate river, Tungabhadra has six companion rivers (3 big and 3 small). Tungabhadra is formed by the meeting of *'Tunga'* and *'Bhadra'*. Both originate from 'Gangamool' Peak of Western Ghat located at the height of 1,200 metres in district Chikmangloor of Karnataka. Tunga flows through Shringeri in the northwest; after passing through Teerthalli it joins Bhadra in deep valleys of broken hill ranges at Kudli. Kumudawati, Ouk and Wardha are important companion rivers of Tungabhadra. As a separate river Tungabhadra flows for 645 kms and its catchment area is 71,417 sq kms.

Bhima: This river is also a part of river Krishna as well as an important independent river. It flows in south Maharashtra and Andhra Pradesh. This river forms a delta below Vijayawada. Its catchment area is 76,614 sq kms.

Subarnarekha: Its origin is the plateau of Chhota Nagpur located in south-west of Ranchi in Bihar. Normally, this river flows towards the northern direction. It flows through Singhbhum (Jharkhand), Mayurbhanj (Orissa) and Midnapore (West Bengal).

The length of Subarnarekha river is 395 kms and catchment area is 36,300 sq kms.

Brahmani: This river is born at Koyal and meets Shekh river near Gangpur. It flows in the districts of Bonai, Talchar and Balasor. Before merging into the Bay of Bengal, the river Vaitarni also joins it. The length of Brahmani river is 705 kms and its catchment area is 36,500 sq kms.

Vaitarni: It is an important companion river of Brahmani. Its origin is from Kewanjhar plateau of Orissa. Length of Vaitarni river is 333 kms and its catchment area in Eastern Ghat is 19,500 sq kms.

Penner: The origin of this river is Nandi Durg hill located in district Kolar of Karnataka. This river has two branches. The 560 kms long northern Penner flows through Caddapa, Anantpur and Nellore districts and falls into Bay of Bengal near the south of the city of Nellore. The 620 kms long southern Penner flows through Bangalore district. After flowing through Salem and South Arcot of Tamil Nadu it merges into the Bay of Bengal near Cuddalore. Chitravati and Papavati are its main companion rivers.

Kaveri (Cauvery): The origin of Kaveri river is Brahmagiri Hill of Sahyadri range near Markara district in Coorg of Karnataka. Height of Brahmagiri is 1,341 metres above the sea level. Its important companion rivers are Lakshman Teerth, Kambani, Hemwati, Swarnavati, Lok Pavani, Simsha, Bhavani, Noyal, Amravati, Kanka and Gaajoti. It is a very lean river in Mysore. Before entering Tamil Nadu it is called as 'Meka Datoo'Aadu Thandam Kaveri. It divides into two branches, northern and southern near Shrirangam. The old port of Kaveri Pattanam was at the mouth of the river. Kaveri river forms a big delta before falling into the sea. About 55 per cent portion of this catchment lies in Tamil Nadu, 41 per cent in Karnataka and 3 per cent in Kerala. The length of river Kaveri is 805 kms and its catchment is 80,290 sq kms. During the flow of Kaveri, two waterfalls are formed namely – Shiv Samundram and Hokenangal. Kaveri is called 'Ganges of south India'.

Sharavati: This river originates from a high plateau of Western Ghat. This river falls from a height of 253 metres amidst four hills named Raja, Rorar, Rocket and Rani. This river is world famous because of 'Jog Waterfall' or 'Garsopa Waterfall'

Tamraparni: This river originates from Godigai Hill at a height of 1,838 metres in the Western Ghat. One of the main features of this river is that its origin as well as end is in the same district of Tirunelvelli. The total length of Tamraparni river is 120 kms including 24 kms on the slope of hills. This river also forms a 90 metres high waterfall in Kalyan pilgrim centre. Korkai – the capital of old Pandu regime – had submerged 8 kms down in Tamraparni river. It merges in the Gulf of Mannar.

Periyar: Periyar is the main river flowing in Kerala in South India. Its origin is in the Shivgiri hill in the Western Ghat. It joins with river Sameer Periyar Vagai of Madurai. The total length of this river is 225 kms. It merges into the Arabian Sea near Cochin.

In Kerala, another river named *Pampa* is also famous. It originates from the hills of Kollam district. Its total length is 144 kms and falls into the Arabian Sea. This river is also famous for boat races on the occasion of 'Onam festival'.

Vagai: The origin of this river is in the Rain valley of Cardmom Hills of the Western Ghat. After flowing for 260 kms it merges into Palk Strait.

Most of the rivers flowing towards the west do not form delta but they form estuary. The description of these rivers is as under:

Narmada: Narmada is the biggest river among the peninsular rivers flowing towards the east. The origin of this river is at a height of 1,057 metres from Amarkantak mountain of Maikal category in Madhya Pradesh. Geographically, it is a South Indian river but it is considered as the dividing line between North India and South India. At Kapildhara, it falls from a height of 25 metres. It forms the '*Dhunadhar Waterfall*' on the marble rocks near Jabalpur. It covers a distance of 320 kms between Vindhya and Satpura valleys. Banjar river is main companion of Narmada and it meets at Mandhata. Sher and Shakkar rivers join the Narmada river at Narsinghpur. Tava and Ganwal join it at Hoshangabad near Chhoti and Khandwa. In the north, only one river Hiran joins at Jabalpur. Right from the mouth to 88 kms distance of Narmada river, one can find the tidal effect of sea . After flowing through the rift valley it forms an estuary in Arabian Sea. It then travels for 1312 kms before the falling in the Gulf of Cambay near Bharauch in Gujarat. The catchment area of this river is 93,180 sq kms. This river has been called as 'Reva' in Mahabharata. Its biggest companion river is Orsan, which is 300 kms long.

Tapti: This river is also called Tapi river. It originates from Multai (at the height of 792 metres above the sea level) in district Betul from Satpura hills. Its companion rivers are Lavda, Patki, Gajjal, Bodak, Ambhora, Khursi, Khandu, Kapra, Sipra, Uttauli, Mona, Khekri, Purna, Bhokar, Suki, Maar, Harki, Manki, Guli, Arunavati, Gomai, Nadhur, Gurna, Bori, Panjhra, Amravati etc. This river flows through Berar and before entering Khandesh, it is joined by river Purna. After forming an estuary near Surat this river falls into Gulf of Khambhat. Length of Tapti river is 724 kms and its catchment area is 64,750 sq kms.

Luni: The origin of the river is from Anasagar (Naag Parvat) of district Ajmer located in south-west of Aravali. Sarsuti (originating from Pushkar lake of Ajmer) is its main companion river. Luni river is 300 kms long and its catchment area is 34,866.40 sq kms.

Sabarmati: This river's origin is from Jaisamand lake of Udaipur district in Rajasthan. Savar and Hathmati coming from Idar and Mahikanth are its companion rivers. This river merges into the Gulf of Khambat.

Mahi: Mahi originates from district Betul in Madhya Pradesh. After flowing through Dhar, Ratlam and Gujarat, it merges into the Gulf of Khambat. This river is 560 kms long.

Difference between Himalayan Rivers and Peninsular Rivers

1. Basin or catchment area of Himalayan rivers is extensive and spread over thousands of kilometres. The catchment area of Indus river 115 lakh sq kms and catchment area of Ganges 9 lakh sq kms, whereas the catchment area of peninsular rivers is comparatively lesser.
2. Himalayan rivers do not cut the high Himalayas to make rivers. These rivers are older than the Himalayas itself and have been constantly flowing in all stages of formation of Himalayas. The Himalayas have been rising gradually and these rivers have kept their flow perennial by the eroding base layer. Due to this, the mountains stood like walls on both the sides. Peninsular rivers on the other hand, flow in wide, shallow valleys of balanced slope, which shows that these rivers are older than the Himalayan rivers and have come into existence after the formation of the peninsular plateau. Hence, now they are in adult stage.
3. Himalayan rivers pass through huge depressions which have been formed through soil erosion along with the rising of the Himalayas, for lakhs and crores of years. These rivers are still

eroding soil at a very fast rate but the eroding capacity of peninsular rivers has almost weakened.

4. Water is there in Himalayan rivers throughout the year. They are perennial because the source of water for these rivers is the rain as well as mountains. But the source of water for peninsular rivers is only rain, hence they do not flow for the whole year. Even big peninsular rivers which cause floods during rains, dry up during winters.
5. Himalayan rivers meander in plain areas and change their routes also sometimes, but peninsular rivers cannot meander because of hard rock plateaus at the base and shortage of alluvium.
6. Himalayan rivers are more useful for navigation. Rivers in plateau on the other hand, flow fast during the rains due to steep slope but dry out during winters. Moreover, there are many waterfalls enroute, hence they are not suitable for transportation. Himalayan rivers are well-suited for irrigation.
7. Most of the peninsular rivers flow towards the east (Narmada and Tapti), whereas most of the Himalayan rivers flow towards the west.

Groundwater Resources

About 300 lakh hectare metre water is available in India at present. After complete development of surface water resources, it is expected to become 350 lakh hectare metre. Till the end of the Fourth Five-Year Plan, hardly 160 lakh hectare metre water could be utilized. At present at least 240 lakh hectare metre underground water can be used. Unfortunately, every year quantity of underground water is reducing. Although about 270 lakh hectare metre water is added to underground water every year through infiltration of rain water and seepage from canals, but so much excess water is being used in agriculture and industry that the water level is going lower and lower day by day. According to Geological Survey of India, groundwater in India can be divided into three parts:

1. Rocky plateau area
2. Alluvial area
3. Integrated sedimentary area

About seventy per cent area of the country is rocky plateau area. Here water is available in very limited quantity and only from open wells. In the alluvial area of Indus and Ganges, there is a vast reservoir of water which has its own place in the world. In the south, there is sufficient water in the alluvial area of coastal regions. In the western part of Rajasthan, sufficient water is estimated at 100-150 metres depth.

R.L. Singh, in his book *India: A Regional Geography* has divided groundwater areas of India into eight regions as under:

1. *Precambrian Crystalline Rocky Region:* Almost half of the area of the country is included in this region, which extends to Tamil Nadu, Karnataka, Andhra Pradesh, Dandakaranya, Bundelkhand, Aravali etc.
2. *Precambrean Sedimentary Rocky Region:* This area extends to Cuddappa and Vindhyan Basin.

Figure 3.2

Regions above Underground Water (as per R.L. Singh)

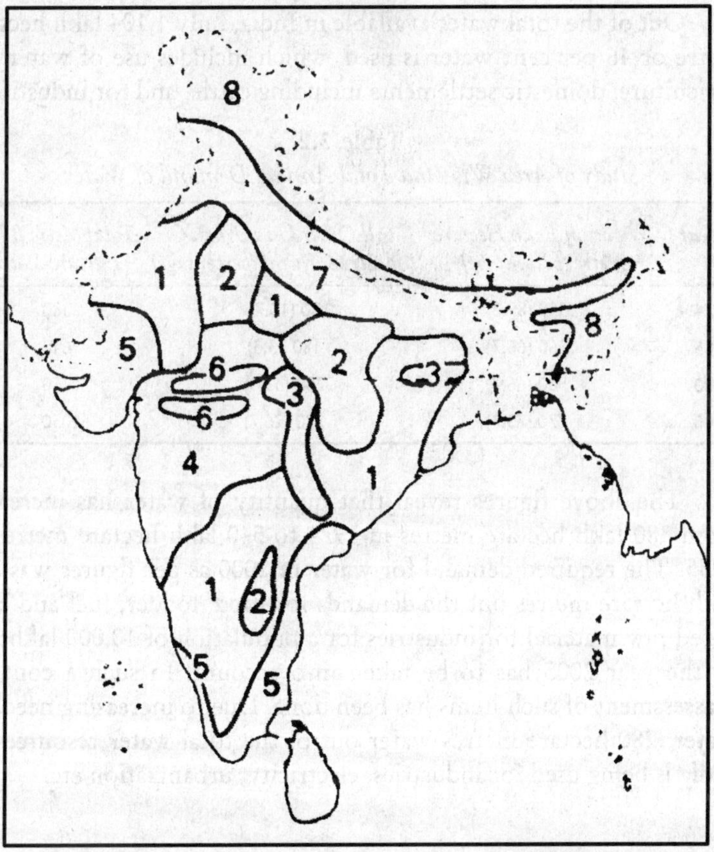

3. *Gondwana Sedimentary Rocky Region:* Sufficient water is found in this region that is in the Barakar and Godavari basin area, which have sand rocks.
4. *Deccan Trap Region:* In this region there is a 1200 metre thick impermeable basalt strip where there is shortage of groundwater.
5. *Cenozoic Sedimentary Rocky Region:* This is the tertiary carpet having sufficient groundwater, which is located in the coastal parts of Tamil Nadu, Andhra Pradesh, Kerala and Gujarat.
6. *Cenozoic Fault Basin Region*: Situated in the fault basins of Narmada, Tapti and Purna rivers, sufficient groundwater from Kamp Jamav is available (80 to 160 metre deep) here.
7 *Ganges Brahmaputra Alluvial Region*: In this region, there is enormous water at the upper level.
8. *Himalayan Region:* This is not important from the point of view of groundwater, the main reason being its compact composition.

Out of the total water available in India, only 1,104 lakh hectare metre or 46 per cent water is used, which includes use of water for agriculture, domestic settlements including cattle, and for industries.

Table 3.2

Study of Area Wise and Total Annual Demand of Water

Year	*Irrigation Lakh Hectare Metre (percentage)*	*All Other Uses Urban/ Industries/Rural (percentage)*	*Total Metre (Lakh Hectare Metre)*
1974	350 (89.3)	30 (10.7)	380
1985	360 (66.7)	180 33.3)	640
2000	500 (66.7)	250 (33.3)	750
2005	770 (73.3)	280 (26.7)	1050

The above figures reveal that quantity of water has increased from 380 lakh hectare metres in 1974 to 540 lakh hectare metres in 1985. The required demand for water in 2000 as per figures was 750 lakh hectare metres but the demands for food, fodder, fuel and land based raw material for industries for a population of 10,000 lakhs up to the year 2005 has to be taken into account. In such a context, reassessment of such items has been done. Due to increasing needs of water, 280 hectare metres water out of the total water resources of India is being used for industries, electricity, urbanization etc.

Table 3.3

Estimated Demand of Water in Non-Agricultural Sector in India during the year 2002

	Use	*Water (Lakh Hectare Metre)*
1.	For Domestic Consumption	
	(a) Urban	49.9
	(b) Rural	22.2
	Total	72.1
2	For Use in Industries	4.7
3.	For Electricity and Mining	60.0

Thus, only 913.2 lakh hectare metre water will be available for agriculture. As per National Water Policy, water available for irrigation would be able to irrigate at least 25 per cent of sown area. Total sown area is 1,500 lakh hectares out of which only 11,250 lakh hectares should be irrigated.

As compared to surface water, groundwater is less expensive and more qualitative. It has been estimated that underground water extends to approximately 452 sq kms area in India. At present, our country is utilizing about 45.22 million hectare underground water, of which 38.38 million hectare is used for irrigation and 6.94 million hectare for industries and domestic consumption.

Underground water is a moving and renewable resource and its estimate is mainly based on annual availability of water. Thus, availability of groundwater depends on geological and climatic conditions. Annual demand of usable groundwater for the country has been estimated to be 4.5 crore hectare metres. Out of this estimated demand 3.85 crore hectare groundwater is available for irrigation, whereas 68.3 lakh hectare metres is for drinking water, industries and other essential uses.

In eastern Uttar Pradesh, Bihar, West Bengal and Orissa there is thick layer of marshy soil spread over the ground , hence they are most suitable for under groundwater development because of lesser depth. The level of groundwater development is 6 per cent in Orisssa, 19 per cent in Bihar, 24 per cent in West Bengal and 34 per cent in Western Uttar Pradesh. Since, the depth of groundwater in these states varies from 5 to 10 metres, hence use of groundwater can be comparatively easier here.

Management of Water Resources in India

Management of water resources means: "Formulate such programmes which ensure sufficient supply of good quality water for different uses without causing damage to any water resource or water reservoir." For water conservation, effort should be made keeping the following things in mind:

1. Water management means replenishing underground reservoirs and supplying water from excess water areas to water scarcity areas.
2. Replenishing of underground water is the most important aspect of water management. On mountains and hills, the water divide or watershed is covered with vegetation. Rain water easily infiltrates into the aquifer through grasses and vegetation of watershed.
3. In urban and rural areas, rain water, used water or water of domestic drains reaches pits or depressions. Flood water reaches the aquifer through such pits or spreads in pits or depressions in the fields.
4. By suitable treatment of domestic or urban waste, water can be made available for agriculture and industrial use. Pollutants, harmful bacteria and poisonous substances can be removed through treatment of waste.
5. Sea water may be de-salined. By use of solar energy saline water of sea can be distilled so that de-saline good quality water can be made available. The process of desalination of sea water has been adopted in India at some places like Bhavnagar in Gujarat and Churu in Rajasthan.
6. Use of excess quantity of water may be reduced. Reduction of use of excess water is most essential because using more water than is essential is a waste of this valuable resource which cannot be excused at any cost. In our country, a lot of water is wasted due to leakage from pipes or faults of plumbing. Similarly, use of excess water for irrigation has also to be checked.
7. Flood water or excess water in normal flow can be diverted to water scarcity areas. It will not only remove the problem of damages due to floods but also benefit the drought areas.

It has become evident from the above discussion that India being a large country, sufficient natural water resources are available but even then there is water crisis because of lack of proper use of water.

The government has been making efforts for its solution and proper utilization of water by framing the National Water Policy or National Water Grid for optimum use of this natural resource so as to face famines and droughts successfully and save people from sufferings for want of water.

National Water Policy, 2002

The National Water Resources Council sanctioned the National Water Policy with consensus on 1st April, 2002, which emphasized on institutional arrangements for total management and development of water resources along with constituting a 'River Basin Organization' for settlement of river water and land related disputes. The revised Water Policy, 2002, which replaced the Water Policy 1987, gave topmost priority to drinking water arrangements for all. The new revised Water Policy has kept international water disputes outside its purview. It has also emphasized on optimum and sufficient use and development and management of water resources in an integrated manner. Regarding implementation of the new Water Policy, all the state governments have also been asked to prepare water policies for their states within two years.

Priorities for use of water resources as per this Water Policy have been decided as under:

1. *First Priority:* Availability of drinking water for all citizens.
2. *Second Priority:* Arrangements for irrigation water.
3. *Third Priority:* Availability of water for electricity generation.
4. *Fourth Priority:* To maintain continuous flow of water in rivers up to a specific limit so as to maintain ecological balance.
5. *Fifth Priority:* Use of water for industries and navigation.

It has been also provided in the new Water Policy that the states can make some changes in the policy as per their own priorities.

Main Features of National Water Policy

While recognizing water as the most special natural resource, one of the basic needs of man and a priceless national asset, and with the main objective of ensuring planning and development of water resources, the 'National Water Policy' was announced in 1987 which was revised in 2002. As per the new policy, water has been considered as a resource which is not divided by borders of the states, different sources or areas but it is a part of the large ecological system. The main features and guiding principles of our new National Water Policy are as under:

1. Establish suitable types of organizations and units to ensure optimum and suitable use of available water resources in the country.
2. Consider recycling and re-use of water as essential aspects of development of water resources, and give them priority.
3. Manage suitable organizational arrangements at the national and state levels for ensuring security arrangement of water reservoirs/dams and other structures and arrange their review by experts in a definite time frame.
4. Lay emphasis on development of available water resources in the country in the national context, and preserve them taking all necessary steps for their suitable use and management.
5. Finalize projects for development of water resources after study of their possible effects on human life, human settlements, human occupations, as well as economic and social development etc., along with environment, ecological balance and local situations.
6. Ensure use of underground water in such a way and to such an extent that it provides for the possibility of its replenishment, so that it can be conserved for maximum period.
7. Evaluate water resources through scientific methods at definite intervals keeping in mind the quality of water and cost-benefit analysis.
8. Give priority to water resources development projects in tribal, Scheduled Caste and other weaker section dominated areas.
9. Plan and develop water resources development projects as multipurpose projects and give priority to arrangement of drinking water supply.
10. Make suitable arrangements for transfer of water to water scarcity areas from other areas, keeping in mind the needs of different areas.
11. Organize awareness campaigns in different parts of the country to ensure suitable use of water and to inform people regarding paucity of water resources. Take assistance of education, law, rewards and penalties to develop sense of conservation of water.
12. After identifying water areas in the country, classify them as per availability of water, and accordingly prepare and implement

plans for special economic activities like agriculture, industry and urban development.

13. Maintain priorities of drinking water, irrigation, electricity generation, river-water flow and industry and transportation etc. in the sequence given in the National Water Policy while planning and developing strategies for the use of water resources.
14. Manage regular review of quality of surface water and underground water and take stepwise programmes for improving quality of water.
15. Adopt a long-term organized training programme for development of water resources in the country.
16. Stress on preparing a 'Master Plan' for prevention and management of floods in flood-prone areas in different parts of the country so as to reduce the regular burden of flood relief assistance.

National Water Policy, 1987

The main objectives of the National Water Policy adopted in 1987 are that planning and development of water resources should be governed by national perspectives and the drainage basin should be considered as the basic unit of planning for development of water resources and their optimum utilization. The salient features of the policy are given below. Water being a state subject, necessary follow up actions on the National Water Policy (NWP) are being taken by all the states. State Water Policy has been framed by the governments of Kerala, Orissa, Tamil Nadu and Uttar Pradesh. The central government gives necessary assistance to states by way of technical guidance and financial support in this endeavor. The inter-state issues are resolved through mutual discussions amongst the states and over 130 inter-state agreements have been arrived at between states so far. Issues not settled through negotiations are referred to an inter-state water dispute tribunal which can be constituted under the Inter-State Water Disputes (ISWD) Act, 1956. Amendments to the Act are proposed on the recommendations of the Sarkaria Commission, agreed to by the Inter-State Council. The draft Policy Guidelines for Allocation of Water amongst the states prepared by the National Water Board to ensure fair and equitable distribution of the existing water resources has been referred to the Working Group of Ministers of all states/Union Territories constituted under the Chairmanship of Union Minister of Water Resources to achieve greater convergence and agreement on issues involving divergent views by the National Water Resources Council. This was stated by Smt. Bijoya Chakravarty in Rajya Sabha on 28 November, 2000.

The salient features of the NWP 1987 are as follows:

- Water is a precious national resource and its development should be governed by national perspectives.
- The available water resources, both surface and groundwater, should be made utilizable to the maximum extent.
- Planning for water resources should be on the basis of the hydrological unit such as a drainage basin of a sub-basin. Appropriate organizations should be established for the planned development and management of the river basins as a whole.
- Water should be made available to areas where there is a shortage by transfer from other areas including transfer from one river basin to another, after taking into account the requirements of the basins.
- Project planning for development of water resources should, as far as possible, be for multiple benefits based on an integrated and multi-disciplinary approach having regard to human and ecological aspects and the special needs of disadvantaged sections of the society.
- In the allocation of water ordinarily first priority should be for drinking water, with irrigation, hydro-power, industrial and other uses following in that order.
- The groundwater potential should be periodically reassessed and its exploitation regulated with reference to reuse possibilities and consideration of social equality.
- Due care should be taken right from the project planning stage to promote conjunctive use of surface and groundwater.
- Maintenance, modernization and safety of structures should be ensured through proper organizational arrangements.
- There should be close integration of water use and land use policies, and distribution of water should be with due regard to equity and social justice.
- Efficiency of utilization should be improved in all the diverse uses of water and conservation consciousness promoted through education, regulation, incentives and disincentive.

- Water rates should be such as to foster the motivation for economy in water use and should adequately cover the annual maintenance and operational charges and a part of the fixed cost.
- Farmers should be progressively involved in the management of irrigation system.
- There should be a Master Plan for flood control and management for each flood-prone basin. In flood control and management, the strategy should be to reduce the intensity of floods by sound watershed management and provision of adequate flood cushion in water storage projects wherever feasible, to facilitate better flood management of each flood-prone basin.
- Land erosion by sea or river should be minimized by suitable cost-effective measures. Indiscriminate occupation of, and illegal economic activities in coastal areas and flood plain zones should be regulated.
- Needs of drought-prone areas should be given priority in the planning of projects for development of water resources. These areas should be made less vulnerable through soil moisture conservation measures, water harvesting practices, the minimization of evaporation losses, the development of ground-water potential and transfer of surface water from surplus areas where feasible and appropriate. Pastures, forestry or other modes of development which are relatively less water demanding should be encouraged.
- A rational information system on water resources should be established with a network of data banks and databases integrating and strengthening the existing Central and State level agencies.
- Training and research efforts should be intensified as an integral part of water resources development programmes.

4

Groundwater

Water existing in the voids of the geological stratum below the surface of the earth is called groundwater. Groundwater is found in pores and fissures of rocks. It is regulated by the quantum and speed of rains, extent of vaporization at the time of rain, temperature, slope of land, dryness of air, porosity and permeability of rocks, vegetative cover and water absorbing capacity of the soil. Groundwater is 0.58 per cent of the total water resources available in nature and it is 22.21 per cent of fresh water part (2.6%) of total water reservoirs. It is located up to a depth of 4 kms of the earth's surface (Table 4.1). It is also called sub-surface water since it is found below the surface of earth.

Description of development of groundwater is available since ancient times. *The Old Testament* contains many details about groundwater in the form of springs and wells. Rollman has given description of underground water channels of Egypt and Syria in 80 B.C., while Greek and Roman philosophers have also described the principles of formation of waterfalls. Homer, Thales and Plato have explained creation of waterfalls from the water of seas. They said that sea water flows underground through water routes under the mountains. Seneca and Pilny also followed the Greek thinking. Shilpy Vitrviyas, while explaining infiltration, clearly told that in hilly areas it rains heavily due to which the water, after percolating into lower layers, appears in the form of waterfalls at lower levels of hills. French philosopher Bernard Pelisy (1510-1589) repeated the principle of infiltration in 1580.

Clear knowledge about the hydrological cycle became available in the 17th century when for the first time analysis was made on the basis of observations and quantitative figures. From this point of view, the contribution of Pire Perat (1608-1680), Edmen Meriote (1620-1689) and Edmund Hailey (1652-1742) remain praiseworthy. Perat measured rain for three years and found out the flow of the upper part of the basin of river Siene. Meriotte, after measurements of river Siene near Paris, clarified the work of Perat. Meenjar has called Marriott as 'Father of Hydrogeology'. English geographer and scientist Edmund Hailey, after measurement of the process of vaporization concluded that flow of all rivers and waterfalls depends on vaporization activity of seas. During the first decade of the nineteenth century, the first artesian wells were constructed in France. Thus, at present hydrogeology has become an independent branch in which various geographical aspects of groundwater are studied.

Origin of Groundwater

Total water existing on earth is 13,84,12,0000 cubic kms, out of which 8,00,0042 cubic kms is groundwater. Apart from this, 61,234 cubic kms is in form of soil moisture. Groundwater and soil moisture together constitute the sub-surface quantity of water. Groundwater is stored in different layers of earth by infiltration through pores and fissures of permeable rocks.

Groundwater mainly comes from three sources. They are, *first: 'Meteoric Water'*, which is the main source of groundwater and is received in the form of rain and snow. This water infiltrates from the surface through fissures, pores and joints of rocks till it is stored on non-permeable rocks in the form of groundwater; *Second: 'Connate Water'*, which exists in pores and cavities of sedimentary rocks of seas and lakes. It is also called sedimentary water. *Thirdly: 'Magmatic Water'* which converts into water after condensation of vapour as a result of volcanic action at the time of entering hot rocks.

The main source of groundwater is rainfall. It infiltrates through seepage slowly into the earth and collects there. It is also called 'plutonic water'. Groundwater is an important part of the water cycle, which also includes that part of surface and atmospheric water which goes underground through rainfall, rivers and lakes.

Sources of Groundwater

Water received on the surface of the earth from different sources becomes groundwater when it goes underground after infiltration through pores of permeable rocks. It is found from following sources:

(i) *Meteoric Water:* This is the main source of groundwater. This water is received in the form of rain and snow. Water from tanks, lakes, rivers and seas is again received by earth after vaporization. Water is received by melting of snow or rain, hence it is called 'meteoric' or 'shooting star water'. From the surface of the earth, this water infiltrates down below through rock joints, pores and fissures of rocks and is stored at the level of impermeable rocks in the form of groundwater. It originates in the atmosphere, falls as precipitation and percolates through the soil to become groundwater. You may have noticed the fluctuation of the water level in wells. During the rainy season the level goes up, while in the summers the level goes down. This is indicative of the fact that groundwater significantly depends upon water from the atmosphere.

Another way in which the groundwater may be derived directly from atmospheric moisture is condensation of water vapour from air circulating through the pores and interstices. This is also known as 'condensational water' and is the basic source of replenishment in the arid and semi-arid areas. During summers, the land is warmer than the air in the soil. This results in a difference of pressure between the water vapour in the atmosphere and the soil. The water vapour from the atmosphere penetrates into the rocks as the temperature of the water vapour drops in the cooler soil. A certain amount of water may accumulate this way. A third source is effluent seepage from lakes, rivers, oceans and also man-made channels, but the importance of this varies with the climate of the area concerned. In fact, in humid regions, the groundwater contributes to streamflow by means of effluent seepage, and the gradient of this saturated groundwater more often than not slopes towards the surface water bodies and the oceans.

(ii) *Connate Water:* Water contained in pores and cavities of sedimentary rocks under seas and lakes is called connate water. It is also called 'sediment water'. It is the second important source of groundwater. This is the water that is entrapped in the interstices of sedimentary and volanic rocks at the time of deposition. Connate water is highly mineralized and salty and does not mix readily with meteoric groundwater. Connate water is usually found deep down in the lower layers of the zone of saturation.

(iii) *Magmatic Water:* Hot magma enters rocks due to volcanic action after which its vapour drops are condensed and converted into water. This is called magmatic water. Apart from it, other sources are those in which groundwater becomes again available on the surface of the earth. They are mainly springs, wells, and geysers, which have been explained in detail in chapter 2 of this book. Such water is considered to have been generated in the interior of the earth. It has consequently travelled to the upper layers of the earth's surface for the first time; this is also known as magmatic water.

Box 4.1

Groundwater in Hydrological Cycle

Groundwater is an important part of water cycle on earth. In this water cycle, groundwater also acts as an organizing link. Group of rocks on earth filled with water work as pipeline and also as water reservoir. When surface water of the earth enters into any part of these rocks, slowly and slowly the moving water flow is again brought back on the surface of earth by man or by vegetation. 0.0001 per cent part (1,600 cubic kms) of the total water moving in the water cycle goes towards oceans from groundwater. Thus, in the water cycle, water is supplied to living organisms on earth. After that, infiltration of rain and river water into underground takes palce. The water then reaches oceans, from where it again evaporates and reaches the surface of the earth in the form of rain. From this point of view, groundwater is an important link in the water cycle.

Rock Structure and Groundwater

Below the surface of earth, availability of groundwater depends on composition of rocks. Water holding capacity and water yield depends on the composition of rocks and on such basis is decided vertical or horizontal distribution of water. It is clear that geology occupies an important place in hydrogeology. Springs and streams are included in special groundwater in lands permanently covered by snow.

Below the surface of the earth, groundwater remains in permeable group of rocks. Such group of rocks are called aquifers. From aquifers water can move towards springs or wells in sufficient quantity. Due to the composition of aquifers, sufficient water remains mobile in ordinary local conditions. Groundwater reservoirs and water-filled group of rocks (rock bed, strata or deposit) are synonyms. As against it, aquifers is such a non-permeable group of rocks which neither holds water nor it is permeable. Its composition is like solid granite.

Rocks without solid minerals or other such parts of soil can retain groundwater. Such spaces are known as voids, pores or fissures. Because such voids work as water pipes for groundwater, they are very important. Hence their shape, type, irregularity and distribution are specially studied in groundwater. Basic spaces are formed by those very geological reactions which form rock groups and they are found in igneous and sedimentary rocks. Interstices are formed after rock formation. Rock joints, fracture, solution holes and voids created by vegetation come in this category. According to shape or form, these spaces are classified as capillary, over-capillary, and sub-capillary. Capillary spaces are so small that water is held there by surface tension. Over-capillary spaces are those which are bigger than capillaries. Sub-capillary spaces are so small that water is held there by adhesive forces. Because of their linkage with other voids, they are called isolated ones.

Figure 4.1

Relation between Rock Spaces and Rock Structure with Porosity (D.K. Tod, 1984)

Figure 4.1 shows different types of spaces and their relation with porosity. From the viewpoint of recycling of groundwater, granular sedimentary deposits are of special importance. Porosity of such deposits depends on different types of granules and their arrangement, distribution as per size, degree of cementation and compaction.

Solution of mineral substances emanating from consolidated groups of rocks and the condition of broken rocks is also important. Due to above mentioned reasons, range of porosity varies from 0 to 50. Table 4.1 mentions main details of sedimentary materials.

Table 4.1

Porosity of Sedimentary Materials

Material	*Percentage*	*Porosity*
Soil	50	60
Clay	45	55
Silt	40	50
Mixed medium and thick sand	35	40
Uniform sand	30	40
Mixed fine and mixed sand	30	35
Gravel	30	40
Bajri and sand	20	35
Sandstone	10	20
Shale	1	10
Limestone	1	1

Source: Ground Water Hydrology by D.K. Tod.

Vertical Distribution of Groundwater

Underground availability of groundwater can be divided into 'Saturated Zone' and 'Zone of Aeration'. In spaces of saturated zone, compressed liquid water exists. In spaces of aerated zone, partly water and partly air exists. In most of the parts of the earth, above every single saturated zone there exists a single aerated zone which extends up to the surface of the earth. This has been shown in Figure 4.2.

The upper part of the saturated zone remains either up to the point of saturation or is bounded by impervious levels. The lower

Figure 4.2
Parts of Water under Land (D.K. Tod, 1984)

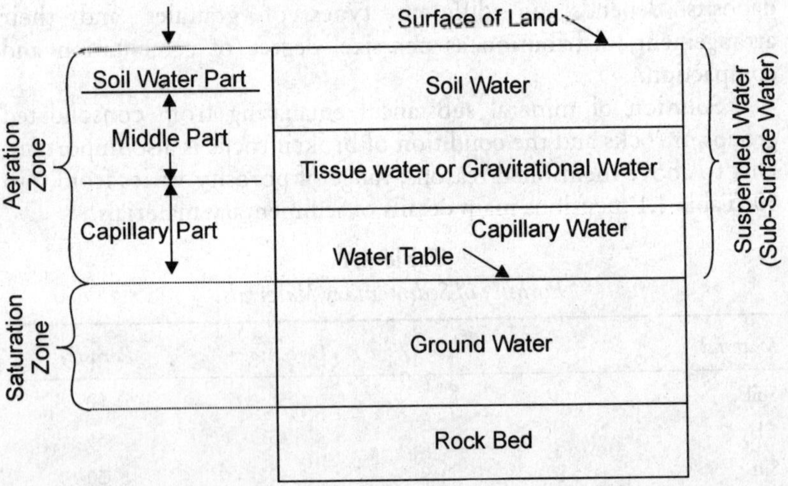

part of saturated zone extends to the underlying impermeable rocks like bed rock or clay. In the absence of overlying impermeable rocks, upper level of saturated zone is called water table or groundwater level. It has been termed as level of atmospheric pressure and after intrusion of level of the aquifer, it is found on the basis of height of water level in the well. In reality, saturation remains a little above the underground level due to capillary gravitation, even then water is available at lesser pressure than atmospheric pressure.

Normally, water existing in the saturation zone is considered as the form of underground water. In the aerated zone, suspended water or vadox water exists. This normal part can also be further sub-divided into soil water portion, middle portion and capillary portion (Figure 4.2). Expansion and distribution of water in each zone has been described in the coming paragraphs.

Inter-granular spaces vary widely in terms of size. Minute voids between the component particles of clay, shale and slate may feature on one end, while large spaces between the pebbles of well-sorted and unconsolidated valley gravel may feature on the other end of the spectrum.

Massive spaces are those that occur between large blocks of rocks such as fractures, joints and bedding planes sometimes enlarged by the process of solution.

Figure 4.3

Types of Interstices

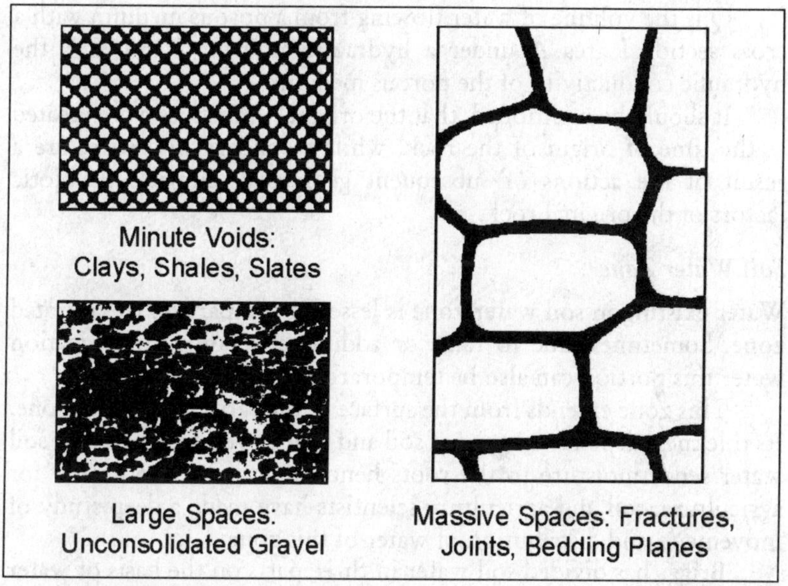

Capillary interstices or spaces are those that are small enough to hold surface tension forces (Fig. 4.3). They can be further classified into two types namely supercapillary and subcapillary. The former is large and may sometimes be as large as a limestone cave. The latter is very small and water is held in them mainly by molecular forces.

Here, let us briefly mention an important dimension of groundwater flow. The movement of groundwater is influenced by gravity just like surface water. Just as everything else has a low, a low was also formulated to express the relationship between capillary of laminar flow and the hydraulic gradient. This was first stated by Poiseuille and is referred to as Poiseuille's law in physics. But it was Darcy who confirmed the application of this law to the movement of groundwater through natural materials. Since then geologists know it as Darcy's law, which is expressed in the form of an equation:

$$q = KH/L$$

q is the velocity of groundwater flow H is the difference in head between the two points separated by the distance L

K is the hydraulic conductivity.

The amount of groundwater flow can be determined with the help of the equation

$$Q = KIA$$

Q is the volume of water flowing from a porous medium with a cross sectional area A under a hydraulic gradient I and K is the hydraulic conductivity of the porous medium

It should be mentioned, that the original interstices were created at the time of origin of the rock, while secondary interstices are a result of the actions of subsequent geological, climatic or biotic factors of the original rock.

Soil Water Zone

Water existing in soil water zone is lesser in comparison to saturated zone. Sometimes, due to rains or additional quantity of irrigation water this portion can also be temporarily saturated.

This zone extends from the surface of the earth to the root zone. Its thickness depends on type of soil and type of vegetation. Since, soil water sends moisture to the roots hence, due to its importance for agriculture, soil and agriculture scientists have made a deep study of movement, and distribution of water of this zone.

Briggs has divided soil water in three parts on the basis of water concentration. The first is *Hygroscopic Water:* This is the water drawn from air which forms thin films of moisture on level of granules. Due to excessive adhesive strength, this water does not become available to plants. The second is *Capillary Water,* which exists on all sides of soil granules in continuous layers. It remains held up due to back pressure strength and starts moving by capillary action. This water becomes available for plants. *Gravitational Water* is the additional soil water which flows away from soil due to gravitational force.

Hygroscopic co-efficient is the quantity of maximum moisture which pre-dried soil draws from 50 per cent moisture containing atmosphere at 250 cms. *Wilting point* is the quantity of water drops in which plants fade away permanently. It has been proved by experimentation that it has no definite standard, but it depends on plants, environment, root system and place of tested soil. *Field capacity* is the quantity of water in soil which exists even after the flowing away of excess water by gravitational force and infiltration of water underground. *Moisture equivalent* is the quantity of water which, after using 1,000 times centrifugal force of gravitation, remains in saturated soil after being decentralized. Field capacity of sand is more than moisture balancing figure but it is equal for loamy sand. Because field capacity and wilting point indicate the maximum and minimum limits of water respectively for development of plants, hence water required for development of vegetation is equal to the difference of these two. Water necessary to saturate all voids of soil is the maximum water content in it. This is called maximum water capacity.

Study of soil moisture has developed many methods of measuring soil moisture according to change of place and time. The most correct method is 'weight measurement method of soil samples, in which samples are weighted and again weighted after drying. In this method, gravitational suction block is used, which is penetrated into and taken out of soil. These porous parts develop moisture equalization with soil, due to which correlation of weight can be established with moisture.

Figure 4.4

Figure Showing Classes and Balancing Points of Soil Moisture

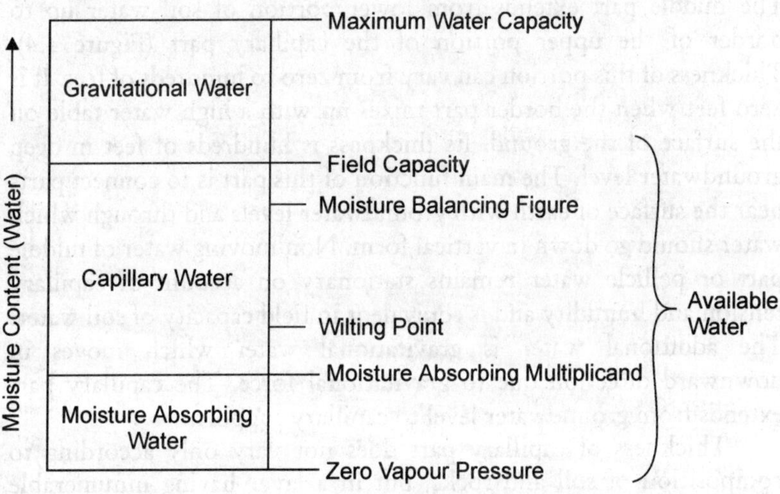

Richards and other persons told that for finding out tension measurement or capillary magnanimity, tension measurement zero (on saturation) to 0.85 remains limited to atmospheric pressure.

According to soil texture, more than half of the available range of water comes within this limit. Many instruments have been developed on the principle of electric resistance measurement from materials kept in soil, in which the relation between resistance and soil water content has been established. To amalgamate electrodes, various sucking materials are used, out of which some are Plaster of Paris, nylon and fiber glass. Because soil moisture is received from heat management of soil, hence based on this principle, different units have been developed when heating elements are buried under soil.

To determine soil moisture, 'Neutron Scattering' is a useful method. It is known that fast neutrons become dormant as compared to other organs while colliding with hydrogen. Hydrogen exists in most of the soils almost completely in the form of water. Using these facts, a fast neutron source is combined with a dormant neutron source and inserted into special type of joint hole in the soil. Counting of dormant neutrons due to hydrogen of the soil is measurement of soil moisture. After calibration, changes in soil moisture can be known easily according to time and depth in the well.

Intermediate Zone

The middle part extends from lower portion of soil water up to border of the upper portion of the capillary part (Figure 4.4). Thickness of this portion can vary from zero to hundreds of feet. It is zero feet when the border part mixes up with a high water table on the surface of the ground. Its thickness is hundreds of feet in deep groundwater level. The main function of this part is to connect parts near the surface of earth with groundwater levels and through which water should go down in vertical form. Non-moving water of middle part or pellicle water remains stationary on account of capillary tension and humidity and is equivalent in field capacity of soil water. The additional water is gravitational water which moves in downward direction due to gravitational force. The capillary part extends from groundwater level to capillary support.

Thickness of capillary part does not vary only according to composition of soil and rocks, but in a layer having innumerable voids, where size of voids is large, its border looks like inverted sides even with a microscope. Physically, with rising height, quantity of moisture reduces or capillary water remains existent in almost all voids above groundwater level. As height increases, water remains only in small voids and at still higher levels, only the smallest interconnected voids contain water, in which water above the groundwater level is present.

Saturated Zone

Since all spaces of the saturated zone are filled with groundwater, hence per unit porosity is directly measured with sphere of available water content. The total quantity of such water cannot be pumped out of the well or discharged from the land, because atomic and backpressure retains some part of water at the same place. Hence, retained water is that water which remains retained in spite of gravitational force. Special retention capacity of any soil or rock is the percentage ratio between water retained against gravitational force after saturation, and total water retaining capacity.

Thus, special product is a part of porosity of the aquifer. Its measure depends on size and shape of granules, distribution of voids and layer tolerance. For uniform sand, special product can be up to 30 per cent, but in alluvial aquifers its measure is between 10 to 20 per cent. To find out special product, Meinzer has suggested seven methods:

1. To saturate samples in laboratory and allow discharge of water from them.
2. Saturated water level and sufficient mass of material remain over the capillary portion and discharge natural water downwards.
3. Collect samples just above the capillary part after lowering down of groundwater level.
4. After pumping out water from the well in known quantity, determine the width of sediments throwing out water.
5. To determine width of sediments saturated by measured seepage of one or two streams of water.
6. To know special product indirectly after finding out moisture measure by centrifuging.
7. Determine special product or special hold after finding out porosity through mechanical analysis.

All these methods have limitations. Samples of laboratory can be disturbed or may have defective representation. In field experiments, control and measurement of variables is a difficult task and thus decided counting may lack truthfulness.

Figure 4.5
Confined and Unconfined Aquifers

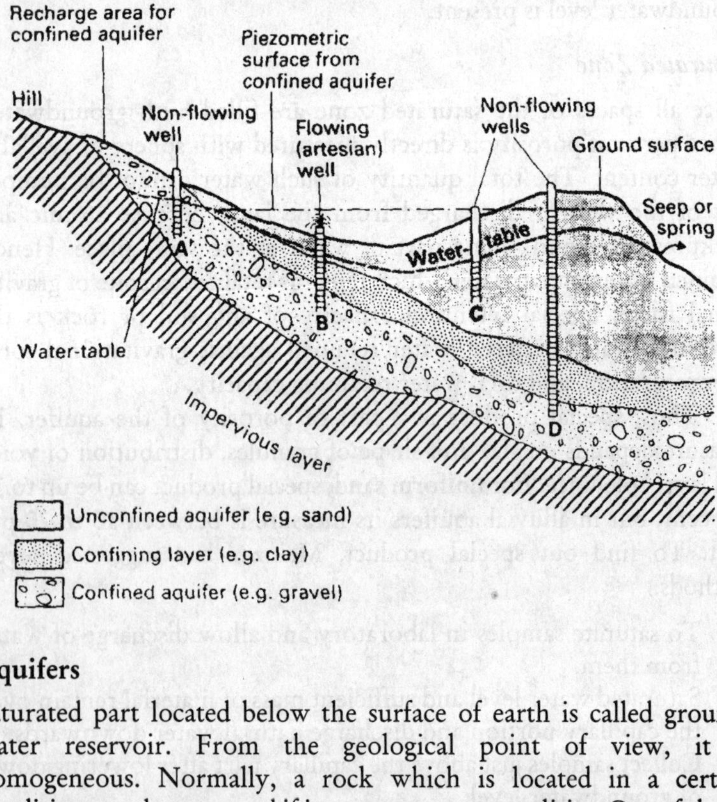

Aquifers

Saturated part located below the surface of earth is called groundwater reservoir. From the geological point of view, it is homogeneous. Normally, a rock which is located in a certain condition, can be saturated if it possesses water qualities. Out of them, some rocks can become aquifers. Water reservoirs are considered as carrying the same meaning as aquifers. The permeable rock group or the layer through which water can enter underground and which can recharge wells or springs in sufficient quantity, is called 'water bearing horizon' or 'water bearing layer'.

Aquifer is made of two words *'aqua'* and *'ferre'* from Latin language. *'Aqua'* means *water* and *'ferre'* means *'produce or bear'*. Thus, aquifer is a geological composition which is porous and permeable. There is storage and transmission of water in it, as well as yielding of water to wells and springs in sufficient quantity. On such basis, any rock group or material which can supply water in sufficient quantity is called aquifer. Almost 90 per cent part of all developed aquifers are unconsolidated rocks whose composition is from sand and gravel. Aquifers can be divided in four categories on the basis of water yielding capacity. First, 'Water Course', second, 'Left Out' or 'Inside Valleys', third, Plains, and fourth, Intermountain Valleys. The water route which forms the water channel, remains below the river bed. It is formed through alluvium of river. Wells located in most permeable rocks on river banks yield water in sufficient quantity because infiltration from the river maintains the quantity of groundwater. In internal valleys their formation streams remain absent. Replenishing of giant reservoirs takes place in those areas where infiltration of rain water and reach of contemporary streams is up to lower levels of the earth. Volcanic rocks can also form permeable aquifers. Basalt lava flow is more permeable as compared to limestone. Other permeable parts found in volcanic rocks are mainly permeable parts in flow breccias amidst lava layers, lava pipelines, fissures and joints are important. Big springs of the United States of America are mostly with basalt. Revolute has lesser permeability in comparison to basalt.

Types of Aquifers

From the point of view of geological structure and expansion, aquifers are of different types. Sufficiently wide aquifers are also called underground water reservoirs. Water enters these reservoirs by natural or artificial recharging and again due to gravitational force it comes on the surface of the earth. It is drawn out through wells and it comes automatically through artesian wells. From the geological angle, on the basis of existence or non-existence of water table, aquifers have been divided in following two categories:

1. Unconfined Aquifers
2. Confined Aquifers

1. *Unconfined Aquifers:* In such types of aquifers, groundwater level works like the upper layer of the zone of saturation. It is also called free or non-artesian groundwater. Wavy form and slope of groundwater is changeable, which depends on replenishing of groundwater, abandoned area and permeability. Fluctuation in the level of groundwater occurs according to changes in the stored groundwater in the space of the aquifer. On the basis of rise of water level in wells, a contour map of water level and sections can be prepared, with the help of which information about availability of water, and its distribution and transmission can be known.

Figure 4.6

Types of Aquifers and Perched Wel

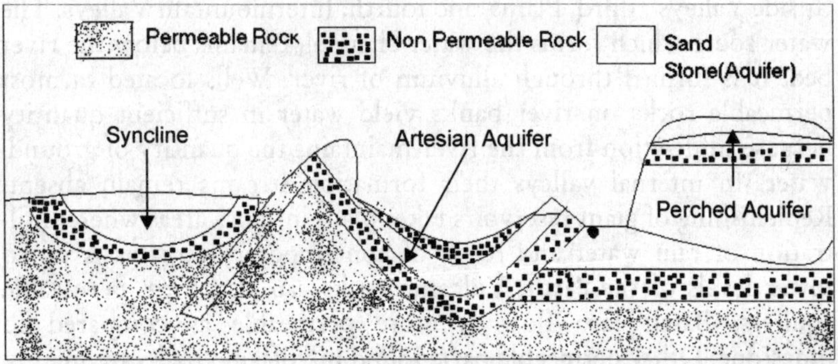

One special state of unconfined aquifers is 'perched aquifers' which is also called the sorrowful state. In this situation, groundwater part is situated at some distance from the main aquifer. It has comparatively non-permeable layers and is separated by the aeration zone above the main groundwater unit. Such aquifers are mainly found in clay parts of sedimentation deposits, which are flat in shape. From wells located in perched aquifers, water can be available in small quantity and for short periods.

2. *Confined Aquifers:* Confined aquifers are found at such places where pressure of groundwater is comparatively more than atmospheric pressure due to non-permeable layers. They are also called 'artesian aquifers'. Such type of aquifers are found above the base of confined beds of water level in punctured wells. Water enters confined aquifers from such areas where confined beds come up right up to the surface of the earth or disappear due to geological situations and aquifers become unconfined. The area supplying water to confined aquifers is called 'recharge area'. In punctured wells of confined aquifers, fluctuation in water level is more due to changes in pressure than due to change in quantity of stored water. Hence, there is hardly any change in storage water of confined aquifers and they mainly function as conduits for transmission of water from replenishing areas to natural or artificial final points. Pressure level of confined aquifer is imaginary, which is equal to the height of the hydrostatic level of the aquifer. Water level of a well boring confined aquifer shows its height at that point above piezometric surface. If pressure level is higher than the earth's surface, it becomes a flow well.

Figure 4.7

Perched Groundwater Resulting from Impervious Clay Layers in Alluvial

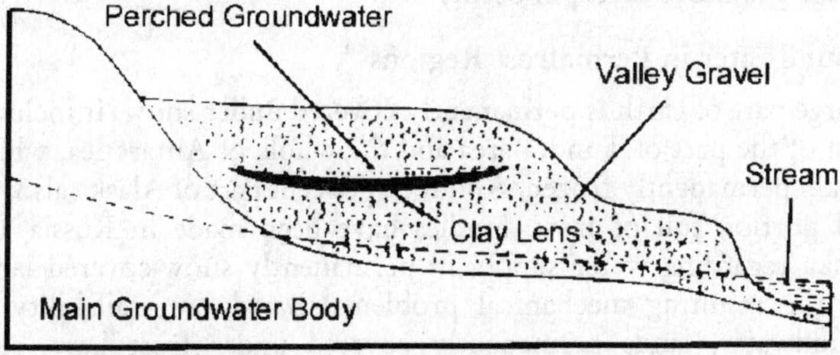

Groundwater that percolates through the soil into the ground exists as free, confined and perched groundwater.

The upper boundary of the zone of saturation determines whether the groundwater is free or confined. In fact it is this upper boundary of free groundwater which is known as the water table. C.F. Tolman defines this boundary as the "contact plane between free groundwater and the capillary fringe and this approximates to the level where the pore water pressure is equal to the atmospheric pressure".

The water table definitely depends on the nature of the groundwater bearing material. In fine textured clay, there will be no clear cut 'plane' surface dividing the zone of saturation from the lowest layer of the capillary fringe. In the case of granite, the water table may vary considerably in height over short distances and also may be interrupted in places. While in the case of an open textured rock such as well-jointed limestone, the groundwater will move through the interstices to form a more horizontal surface.

A piezometric surface may be defined as the imaginary surface that coincides with the static level of the water in the aquifer.

In the case of confined groundwater, the upper boundary of the water body is formed by an overlying confining stratum in which the spaces are so few and small that groundwater movement is extremely slow. The distinction between free and confined groundwater is made on the basis of hydraulic differences between the flow of groundwater under pressure of a confining layer and the flow of free unconfined groundwater.

Perched groundwater refers to a special case in which the confining layer is not continuous over a very large area and is situated at some height above the main confining zone. This confining zone is usually separated by open-textured strata in which the groundwater storage and movement can occur.

Groundwater in Permafrost Regions

A large part of earth is permanently covered under snow. It includes most of the part of Tundra area and the whole of Antarctica, which remain permanently frozen. About 60 per cent part of Alaska also has land portion full of frost. Studies have been made in Russia and Alaska regarding water supply in permanently snow-covered lands and the resulting mechanical problems. Based on availability of groundwater, lands permanently covered under frost have been categorized into supra-permafrost, intra-permafrost and sub-permafrost areas. Supra-permafrost groundwater being in the upper side of frost land creates a temporary supply of shallow water. Water located in supra-permafrost land is called artesian water or inter-frost groundwater. This water remains in open areas like near rivers, voids or fractures etc., of frost land.

Thus, water resources located under permanent snow can become optimistic sources, but there are difficulties like excessive expenditure, probable high salinity, and possible extension of frost land up to non-permeable rocks in its exploitation.

Groundwater Recharge

Nature takes care of most of this recharge. But that is not enough anymore. We need to build our awareness so that we may help nature to do its job.

The main components of natural groundwater recharge are:[1]

- infiltration and percolation of part of the total precipitation at the ground surface. The exact proportion actually reaching the water table however depends on the relationship between the rainfall intensity and the infiltration capacity and also the magnitude of the soil moisture deficit;
- influent seepage through the banks and bed of surface water bodies such as ditches, rivers, lakes and oceans; and
- the underground leakage of water from the adjacent aquicludes and aquifers.

But sadly the role of artificial recharge in terms of the total amount of groundwater storage is insignificant but it has grown is some areas of water shortage. The main methods normally used are:

- water spreading over relatively flat prepared surface incorporating a system of hollow basins or ridge and furrows. In the semi-arid areas the natural stream channels may be used during the dry season.
- the pumping of recylced water down the wells and boreholes, and, last but not the least are the various techniques that we are discussing under rain water harvesting so that the water table may be replenished in a few years.

Artificial Recharge Techniques and Designs

A wide spectrum of techniques are in vogue to recharge groundwater reservoir. Similar to the variations in hydrogeological framework, the artificial recharge techniques too vary widely. The artificial recharge techniques can be broadly categorized as follows:[2]

Artificial Recharge Structures

Ditch and Furrow Method

In areas with irregular topography, shallow, flat bottomed and closely-spaced ditches or furrows provide maximum water contact area for recharge water from source stream or canal. This technique requires soil preparation the recharge basins and is less sensitive to silting. Generally, three patterns of ditch and furrow system are adopted.

(i) Lateral Ditch Pattern

The water from streams is diverted to the feeder canal/ditch from which smaller ditches are made at right angles. The rate of flow of water from the feeder canal to these ditches is controlled by gate valves. The furrow depth is kept according to the topography and also with the aim that maximum wetted surface is available along with maintenance of uniform velocity. The excess water is routed to the main stream through a return canal along with the residual silt.

(ii) Dendritic Pattern

The water from streams can be diverted from the main canal to a series of smaller ditches spread in a dendritic pattern. The bifurcation of ditches continues until practically all the water is infiltrated in the ground.

(iii) Contour Pattern

The ditches are excavated following the ground surface contour of the area. When the ditch comes closer to the stream a switch back is made and thus the ditch is made to meander back and forth to spread water repeatedly. At the lowest point down the stream, the ditch joins the main stream, thus returning the excess water to it.

(iv) Site Characteristics and Design Guidelines

(a) Although this method is adaptable to irregular terrain, the water contact area seldom exceeds 10 per cent of the total recharge area.
(b) Ditches should have a slope to maintain flow velocity and minimum deposition of sediments.
(c) Ditches should be shallow, flat-bottomed, and closely-spaced to obtain maximum water contact area. Width of 0. 3 to 1. 8 inches are typical.
(d) A collecting ditch to convey the excess water back to the main stream channel should be provided.

Percolation Tanks (PT)/Spreading Basin

These are the most prevalent structures in India as a measure to recharge the groundwater reservoir, both in alluvial as well as hard rock formations. The efficacy and feasibility of these structures is more in hard rock formation where the rocks are highly fractured and weathered.

(i) Important Aspects of Percolation Tanks

(a) A detailed analysis of rainfall pattern, number of rainy days, dry spells, evaporation rate and detailed hydrogeogical studies are required to demarcate suitable percolation tank sites.
(b) Percolation tanks should be normally constructed on second to third order streams since the catchment and also the submergence area would be smaller.

(c) The submergence area should be in uncultivable land as far as possible.
(d) Percolation tank should be located on highly fractured and weathered rock for speedy recharge. In case of alluvium, the boundary formations are ideal for locating percolation tanks.
(e) The aquifer to be recharged should have sufficient thickness of permeable vadose zone to accommodate recharge.
(f) The benefitted area should have sufficient number of wells and cultivable land to develop the recharge water.
(g) Detailed hydrological studies for run-off assessment should be done and design capacity should not normally be more than 50 per cent of the total quantum of rainfall in the catchment.
(h) Waste weir or spillway should be suitably designed to allow flow of surplus water based on single day maximum rainfall, after the tank is filled to its maximum capacity.
(i) Cut off trench should be provided to minimize seepage losses both below and above the nalla bed.
(j) To avoid erosion of embankment due to ripple action, stone pitching should be provided upstream up to HFL.
(k) Monitoring mechanism in benefitted as well as catchment areas using observation well and staff gauges should be provided to assess the impact and benefits of percolation tank.

Check Dams, Cement Plug, Nalla Bunds

A series of small bunds or weirs are made across selected nalla sections such that the flow of surface water in the stream channel is impeded and water is retained on pervious soil/rock surface for a longer period. Nalla bunds are constructed across bigger nallas of second order streams in areas having gentler slopes. A nalla bund acts like a mini percolation tank.

Site Characteristic and Design Guidelines: For selecting a site for check dams/nalla bunds the following conditions may be observed:

1. The total catchment of the nalla should normally be between 40 to 100 hectares, though the local situations can be the guiding factor in this.
2. The rainfall in the catchment should be less than 1,000 mm/annum.

3. The width of the nalla bed should be at least 5 metres and should not exceed 15 metres. The depth of the bed should not be less than one metre.
4. The soil downstream of the bund should not be prone to water-logging and should have pH between 6.5 to 8.
5. The lands downstream of check dam/bund should have irrigable land under well irrigation.
6. The nalla bunds should be preferably located in areas where contour or graded bunding or lands have been carried out.
7. The rock strata exposed in the ponded area should be adequately permeable to cause groundwater recharge through ponded water.
8. Nalla bund is generally a small earthen dam, with a cut off core wall of brickwork, though cement bunds/plugs are now prevalent.
9. For the foundation for core wall a trench is dug 0.6 metres wide in hard rock or 1.2 metres in soft rock of impervious nature. A core brick cement wall is erected 0.6 metres wide to stand at least 2.5 metres above the nalla bed and the remaining portion of trench is back filled on the upstream side by impervious clay. The core wall is buttressed on both sides by a bund made up of clay and on the upstream face stone pitching is done.
10. Normally, the final dimensions of the nalla bund are, length 10 to 15 metres, height 2 to 3 metres and width 1 to 3 metres, generally constructed in a trapezoidal form. If the bed rock is highly fractured, cement grouting is done to make the foundation leakage-free.

Gabion Structure

This is a kind of check dam being commonly constructed across small streams to conserve stream flows having practically no submergence beyond stream course. The boulders locally available are stored in a steel wire. This is put up across the stream's mesh to make it as a small dam by anchoring it to the streamside (Figure 4.6). The height of such structures is around 0.5 metres and is normally used in the streams with width of about 10 to 15 metres. The cost of such structures is around Rs 10,000 to Rs 15,000 The excess water overflows this structure, storing some water to serve as a source of recharge. The silt content of stream water in due course is deposited in the interstices of the boulders to make it more impermeable.

Figure 4.8
Recharge Through Dug-out Pit

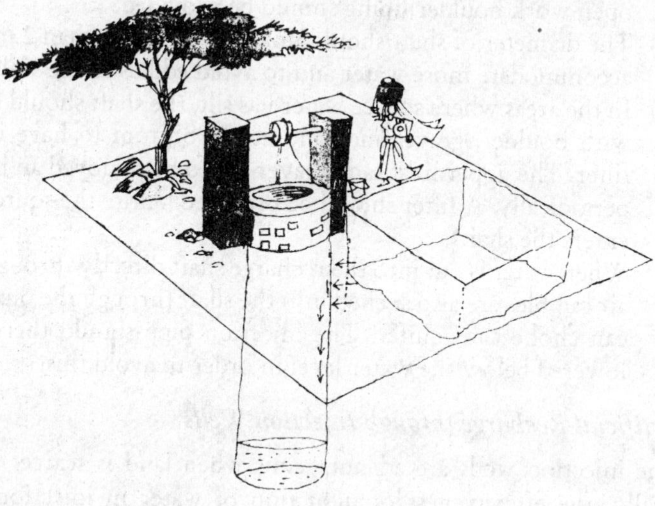

Dug Well Recharge

In alluvial as well as hard rock areas, there are thousands of dug wells which have either gone dry or the water levels have declined considerably. These dug wells can be used as structures to recharge (Figures 4.7 and 4.8) the groundwater reservior. Stream water, tank water, canal water etc. can be diverted into these structures to directly recharge the dried aquifer. By doing so the soil moisture losses during the normal process of artificial recharge are reduced. The recharged water is guided through a pipe to the bottom of the well, below the water level to avoid scoring of bottom and entrapment of air-bubbles in the aquifer. The quality of source water including the silt content should be such that the quality of groundwater reservior does not deteriorate. Schematic diagrams of dug well recharge are given in Figure 4.8.

Recharge Shaft

These are the most efficient and cost-effective structures to recharge the aquifer directly. In the areas where source of water is available either for some time or perennially, e.g., base flow, springs etc., the recharge shaft can be constructed (Figure 4.8). Following are the site characteristics and design guidelines:

(i) To be dug manually if the strata is of non-caving nature.
(ii) If the strata is caving, proper permeable lining in the form of open work boulder lining should be provided.
(iii) The diameter of shaft should normally be more than 2 inches to accommodate more water and to avoid eddies in the well.
(iv) In the areas where source water has silt, the shaft should be filled with boulder, gravel and sand at the bottom to have inverted filter. The uppermost sandy layer has to be removed and cleaned periodically. A filter should be provided before the source water enters the shaft.
(v) When water is put into the recharge shaft directly through pipes, air bubbles are also sucked into the shaft through the pipe which can choke the aquifer. The injection pipe should therefore be lowered below the water level in order to avoid this.

Artificial Recharge through Injection Wells

The injection wells are advantageous when land is scarce. Hydraulically, the effectiveness of induction of water in injection well is determined by:

(a) Pumping rate
(b) Permeability of aquifer
(c) Distance from stream
(d) Natural groundwater gradient
(e) Type of well

In alluvial areas injection well recharging a single aquifer or multiple aquifers can be changed to normal gravel packed pumping well. An injection pipe with opening against the aquifer to be recharged may be sufficient. However, in case of number of permeable zones separated by impervious rocks, properly designed injection well with inlet pipe against each aquifer to be recharged needs to be constructed. Injection wells as a means of artificial recharge are comparatively costlier and require specialized technique of tubewell constructed and supported by operation and maintenance to protect the recharge well from clogging.

Groundwater Dams or Sub-Surface Dykes or Underground Bandharas (UGB)

A groundwater dam is a sub-surface barrier across a stream which retards the natural groundwater flow of the system and store

Recharge Trench

- Suitable for buildings having roof area of 200-300 sq m and where permeable strata is available at a shallow depth.
- It may be 0.5 to 1 m wide, 1 to 1.5 m deep and 10 to 20 m long depending upon availability of water to be recharged.
- These are back-filled with boulders, gravel and coarse sand from bottom upward.
- Top layer of sand should be cleaned periodically to maintain recharge rate.

Notes

1. *Geography and You*, Vol. 2, No. 7, 2002.
2. *Artificial Recharge to Ground Water*, Courtesy, Central Ground Water Board, Western Region Jaipur, (Eds.) S.K. Gupta, L.N. Mathur and G.K. Sharma.

5

Hydrological Cycle

Water is a basic resource on the earth, which maintains its prominence among all natural resources. As a companion of nature, water provides the basis for all living organisms on the earth. Its distribution on the earth is found in many forms at various places. Its form keeps changing according to location and climate. It remains in gaseous form in atmosphere as water vapour, in solid form as micro snow granules and in liquid form as drops of water. All these forms keep changing because of change in temperature and they affect weather pattern. On the earth living community needs regular supply of water. This water should be balanced and qualitative, which is mainly available in rivers, lakes and aquifers. Sea water is not suitable for human consumption because of being saline. Most of the fresh water is frozen in the form of snow which is sufficiently beyond the reach of man and its use is also normally possible.

Hydrosphere, atmosphere and lithosphere of water are regularly included in the water cycle or hydrological cycle. Water reaches in the atmosphere through evaporation from seas, lakes, rivers, soil moisture and transpiration from plants etc., and according to changing climatic conditions condenses in the form of clouds and again reaches the hydrosphere and lithosphere. This cyclical movement of water in different forms is called hydrological cycle. In the water cycle, movement of water takes place independently within different realms. Among these, vertical and horizontal movement of air transferring moisture from one place to other in the atmosphere, movement of water through sea streams in the hydrosphere and movement of water through rivers and glaciers towards the sea in the lithosphere are included. Similarly, vaporized water from soil and evapotranspirated water from plants reaches the groundwater through infiltration. Thus, three main drive processes participate in the water cycle actively (Box 5.1).

> **Box 5.1**
>
> *Main Drive Processes of Hydrological Cycle*
>
> *Evaporation and Evapo-transpiration:* The process through which water reaches the atmosphere from ground level.
>
> *Precipitation:* Through this process, water reaches back on the ground through rain.
>
> *Air Movement:* This includes air and climatic system through which redistribution of water is made possible from one place to the another through atmosphere.

Table 5.1

Annual Estimated Movement of Water in Hydrological Cycle

Water	*Volume (km^2)*	*Percentage of Total Water*
Vaporization		
From oceans	3,50,000	0.026
From ground level	70,000	0.005
Total	*4,20,000*	*0.031*
Precipitation		
On oceans	3,20,000	0.024
On ground level	1,00,000	0.007
Total	*4,20,000*	*0.031*
– Flow from Rivers and snow tops towards seas	38,000	0.003
– Flow from ground level to seas	1,600	0.0001
Total	*39,600*	*0.0031*

Source: R.L. Nace, U.S. Geological Survey Circular 536,1967.

Every year, only one per cent of the water available on earth goes in the water cycle. Total quantity of water found in the water cycle has been estimated to be about 1,384 million kms^3. This cycle is quite complicated from the point of view of different types of branches and water reservoirs. Major portion of the water as companion of the water cycle remains in seas and oceans (Table 5.1). Only 2 per cent of water is pure while the remaining part is frozen in the form of snow.

Rivers play an important role in the universal water cycle because they allow movement of water from the ground towards seas and oceans. They flow water with fast speed and are not able to store it, whereas lakes store more water as compared to rivers. There are about 250 vast lakes on the earth. Water moves in different storage places but its speed is calculated in the water cycle at different rates.

Table 5.2

Permanency of Water in Different Conditions in Nature

Place	*Size (km^3)*	*Duration of Stay*
Plants and living organisms	700	1 week
Atmosphere	13,000	8 to 10 days
Rivers	1,700	2 weeks
Soil	65,000	2 weeks to 1 year
Lakes, reservoirs, moist land	1,25,000	For years
Rocks (underground water)	70,00,000	Few days to thousands of years
Snow	26,00,000	For thousands of years
Oceans	1,37,00,00,000	For thousands of years

Source: After Park, 1997.

Water remains stored in oceans, snow caps and rocks for a very long time whereas, it remains for a very short period in rivers as well as atmosphere, though they have a great importance in the water cycle.

Mechanism of Hydrological Cycle

There are many agencies which affect movement of water cycle on the earth. Condition and position of moisture are continuously changing relatively and the moisture which is received by the atmosphere can again be received by the ground or oceans in the form of water, dew, snow, frost etc. Hence, due to changes in condition and position, obstacles are created in the process of the hydrological cycle.

Figure 5.1
Hydrological Cycle

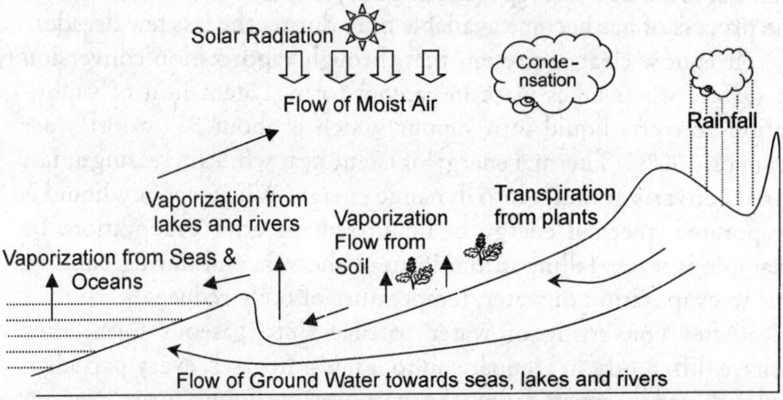

Water of oceans takes the form of vapour due to solar radiation and enters the atmosphere. Air flowing in oceans towards the ground gives speed to this vapour and transfers it from one place to other. After this, when water vapour condenses, it rains on the surface of the earth and the water received on the earth from precipitation flows on the surface in the form of drains and rivers and ultimately enters the oceans. Solar energy increases the speed of water cycle. Some part of the water received from rain is lost through transpiration by plants and some part again reaches the atmosphere through evaporation of water from rivers, ponds and lakes etc. Some part of precipitation on ground level goes underground through infiltration. This water reservoir in soil is called 'soil water reservoir' from where transpiration through plants continues, but some part again comes to the ground level in the form of water sources. Some part of soil water reservoir moves downwards. The portion of water stored below the surface is called underground water. Thus, in universal water cycle, the following processes are important:

Evaporation

As a result of evaporation, visible water converts into vapour. The process of conversion of water into vapour due to solar radiation is called evaporation. Through this process oceanic water reaches into the atmosphere. Evaporation of water from oceans takes place more from continents. Its main reason is sufficient availability of water.

People knew about the process of evaporation about 2,500 years back as Greek scholars have given description about loss of water from water sources due to evaporation, but systematic information about the process of has become available only during the last few decades.

It is now clear at present that through vaporization conversion of surface water takes place in gaseous form. Latent heat of vaporization converts liquid into vapour which is about 539 calories per gram on 100°C. Thermal energy of latent heat while increasing at fast speed converts granules into dynamic energy. Whenever any liquid is evaporated, thermal energy of liquid reduces after evaporation. Its example is water falling in the form of showers and during bathing, due to evaporation of water, temperature of body reduces.

After converting of water particles into gaseous form, they behave differently in comparison to liquids. In gases, every particle is isolated and is away from the particles in liquid form. Vapour particles located above a water lake on a summer day will be 40 times farther from the water level at the time of relative humid air.

During evaporation average mutual distance between particles increases. Water expands on a large scale. At normal temperature, one gram liquid water is one millilitre whereas, at 25°C temperature one gram saturated water vapour is approximately 42,000 millilitre. This expansion is about 42,000 times. In this state, vapour particles are at quite a distance. This fact is also important that up to about 60 per cent, water vapour is so heavy like other atmospheric gases that it inclines air to rise above water level. Besides such presence of high thermal energy in vapour, it also possesses some normal gaseous qualities. It can be compressed and can also expand. Its particles can become pressure free, which it is called 'partial pressure'. An English chemical scientist, John Dalton had established this rule of partial pressure in 1801 after different experiments.

At certain temperature and pressure, quantity of water vapour existing in air of a certain size is called humidity. Relative humidity of saturated air is 100 per cent. Less than 100 per cent air shows that it has still some capacity. For example, at a temperature of 44° Celsius, if temperature of 100 per cent relative humidity air is made 15.5°C , then its relative humidity will become 50 per cent because with increase in temperature, the capacity of absorbing humidity increases.

Thus, it is clear that fog existing in the morning disappears as soon as sun rises because the heat in the air vaporises the fog away.

Factors Affecting Evaporation

In nature, evaporation continues till water gets heat. Rate of evaporation of water continues changing according to environment. Thermal energy is essential for evaporation and it can be obtained from the radiation of the sun. Relative humidity of air and movement of air above water surface also affect the rate of vaporization. Near the equator where sea and air both are hot, water is taken away in large quantity through evaporation. Tropical climate is typically humid. There exist such conditions in the eastern continental group, Amazon basin and Central Africa. Cool air prevailing near the Poles cannot absorb humidity because of minimum temperature. Hence, Arctic climate is typically dry.

Effect of Water Quality: Quality of water is the main reason affecting the rate of evaporation of water. Increase of one per cent salinity in water reduces one per cent rate of evaporation of water. Due to this reason, as compared to saline sea water of oceans there is more evaporation from clean water portions.

Effect of Water Depth: Water has the capacity to absorb thermal heat. Hence, bigger water reservoirs can store more thermal energy. One deep lake, will absorb more thermal energy as compared to a shallow lake which will create difference in rate of evaporation. It also happens with change of seasons. In cold winter climate, temperature of a shallow lake would be near 4°C (39°F) and when its surface starts freezing, its temperature would reach 0°C (32°F).

Measuring Evaporation

Theoretically speaking, evaporation can be measured on the basis of rate of vaporization over any water level. This is possible only in laboratories. This function is very difficult in open water area because nature cannot be controlled so minutely. John Dalton discovered the 'Rule of Partial Pressure' in 1801. He tried to measure evaporation indirectly by measuring air on water level and pressure of vapour by using an equation which was later on revised by many researchers. His method is called 'Mass Transfer Method' of measuring evaporation.

As per one estimate, 4,55,000 cubic kms water from the sea reachs in the atmosphere every year through evaporation. Similarly,

62,000 cubic kms reaches the atmosphere through evaporation from water sources situated on the earth. As against it, when 98,000 cubic kms water through evaporation enters into the atmosphere, it is absorbed in it. Water vapour can be absorbed in atmosphere in a definite quantity. As soon as the relative humidity becomes 100 per cent, air is called saturated. It has been seen that hardly one per cent of the total water available on earth is used by the hydrological cycle. It is normally believed that nowhere is there equality between

Figure 5.2
Process of Transpiration

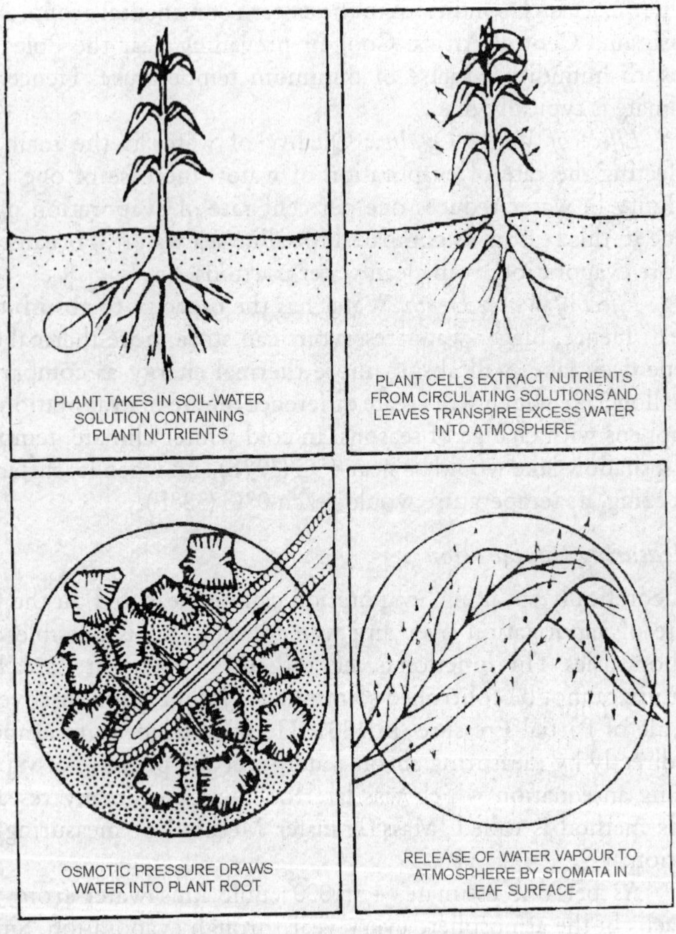

evaporation and precipitation. As estimated by M.I.Budyoko (1971), evaporation from oceans is more than rain. According to him, loss of 46,000 cubic kms water takes palce from oceans every year. On the other side, 62,000 cubic kms water evaporates every year from the land, but it receives 1,08,000 cubic kms water through rain. Hence, the land receives 46,000 cubic kms additional water every year. Such additional water again reaches oceans through surface flow. Thus, more rain is received by continents as compared to evaporation.

Transpiration

Transpiration is the process through which trees and plants, after absorbing moisture from their roots, releasc water into the atmosphere through leaves. This process is performed by the stomata existing in leaves. There occurs considerable loss of groundwater in this process. Such loss takes place mainly during the day time. Capacity of transpiration is different in different plants. Transpiration has an important contribution in the water cycle. In desert areas due to lesser vegetation, evaporation is also lesser, whereas in areas having dense vegetative cover, evaporation is faster, hence the process of hydrological cycle is also faster. The direct effect of such a process of water cycle is visible on the quantity of groundwater. Soil composition and structure also affect the quantity of evaporation, because water reaching plant roots reach underground through infiltration. This water infiltrates on the basis of nature of soil.

Precipitation

Rate of rain at a particular place is known as heaviness or the rate at which it is raining at a particular place per hour or per day. At places where rain air enters first of all, there remains a possibility of heavy rains. Heavy rain areas have more rains with speed whereas, in regions having scanty rainfall, speedy rain occurs occasionally. Speedy rain has more disadvantages as compared to advantages. Water received during speedy rain cannot infiltrate with such speed due to which underground water cycle is affected. Development of vegetation and crops in any area depends on the total rainfall in it.

Along with quantity of rainfall, the number of days of rainfall are also important. For balanced water cycle, it is essential that total rainfall at any place is distributed over majority of days of a year. Drought takes place due to scanty rainfall and flood occurs due to heavy rainfall. Thus, nature of rainfall affects the process of hydrological cycle. Along with rain, snowfall, dew, frost etc. also regulate water. Relief of surface is also important along with quantity of rain. Nature of rainfall may be fast or slow, slope of land affects its flow, infiltration and evaporation.

Infiltration

As soon as rain water reaches the surface of the earth, some part of it is absorbed by land. Rate of infiltration of water depends on the structure and composition of soil. The process of absorption of water by earth is called infiltration. Some part of rain flows away with the slope. Thus, is called surface flow. Nature of soil directly affects infiltration. Rate of infiltration is minimum in clay soil whereas, it is fast in sandy soil. Similarly, if voids of land are bigger, infiltration would be faster. Along with structure of land, organic substances found inside the soil would also increase the rate of infiltration. Through infiltration, rain water, after entering voids of lands, exists in form of groundwater. Ordinarily air exists in voids of soil. As soon as surface water enters inside the land, it replaces air. State of reduction of air and increase in quantity of water is called state of waterlogging. It is a saturated state of land as thereafter it cannot absorb groundwater any more. Water exists in the soil in three forms. First, the water which remains in big voids, called 'Gravitational water'. This water cannot be used by plants. Second, 'Capillary Water', which clings on all sides of soil granules in the form of a film by mutual gravitation. This water remains in capillaries and pores. This water is mainly used by plants. Third, 'Hygroscopic Water', which remains on the surface of soil granules in the form of a thin layer.

Due to high temperature or absence of rain for many days, water exhausts from small pores. After that plants start withering and such a state is called 'wilting point'. Wilting point comes faster in sandy soil as compared to clayey soil.

Surface Flow

That part of water received from rain which cannot reach underground, flows on the surface in different directions according to the slope. This is called surface flow. Speed of the flow depends on the slope. If slope is steep, flow will be faster and if slope is less, flow will be slow and infiltration will be more. In steep slope, infiltration is lesser. Due to infiltration, sub-surface flow starts. This drainage happens through big voids and its speed is comparatively slow. Surface flow is also harmful. Due to this flow, ablation of land water occurs and soil particles start flowing away. Speed of flow determines its quantity. If flow is slow, only fine particles flow away, but with faster flow, even pieces of big rocks start flowing. Surface drainage is affected by quantity, duration and rate of rainfall. With fast rain drainage is also fast and with lesser rainfall, drainage is slow. Fastness of rainfall affects both quantity and rate of surface flow. Similarly, area of watershed also affects the surface flow. Slope of the surface of the earth affects the flow. In parts having steep slope, lower areas are affected very soon when it rains on higher plane. Thus, flow and quantity of surface water both control the hydrological cycle.

Underground water plays an important role in balancing the movement of the water cycle. Rain water collects in different sources on the surface of the earth. From there it reaches underground by infiltration, from where it becomes available for use of plants and animals. Plants make its use in natural way through their roots, whereas man uses groundwater by boring artificial wells. This water goes underground by infiltration from rivers, lakes, ponds and direct rainfall. Plants absorb water through roots and transpirate through leaves, after which it is included in the water cycle. Similarly, during biological use also water reaches in the atmosphere through direct evaporation. Water also evaporates directly from the soil.

The water which goes underground out of surface flow is called underground water. This water moves towards soils, vegetation, rivers and oceans. Thus, directly or indirectly this water is released for hydrological cycle. Plants release underground water through transpiration. These processes complete the water cycle in such a manner that loss of water caused to water reservoirs (oceans) by evaporation, returns to the same source. This cycle is also affected by climate. In arid climate, it acts without distortion but in cold climate implementation of water cycle becomes very limited. As a result of weak implementation of water cycle, the quantity of rainfall would also be lesser.

Table 5.3

Rainfall, Evaporation and Flow according to Horizontal Zones on Earth

Horizontal Zone (N and S Hemisphere)	*Precipitation (mm)*	*Evaporation (mm)*	*Flow (mm)*
00-300 latitude	1180	1307	–127
300-600 latitude	908	812	96
600-900 latitude	380	176	202

Source: Baumgartner and Reichel, 1975.

Different Phases of Water in Hydrological Cycle

Water is found in nature in different forms, either in the form of liquid in a vast area or in the form of snow cover. Water found in atmosphere is in the form of vapour. A large part of water in the water sphere, i.e., 97.3 per cent remains in vast oceans. About 2.01 per cent part of the total water resources is found in the form of ice caps, icebergs and glaciers.

Figure 5.3

Conversion of Water to Different Forms
Three Phases of Water: Gas (Water Vapour), Solid (Snow) and Liquid (Water) (Park. C. 1997)

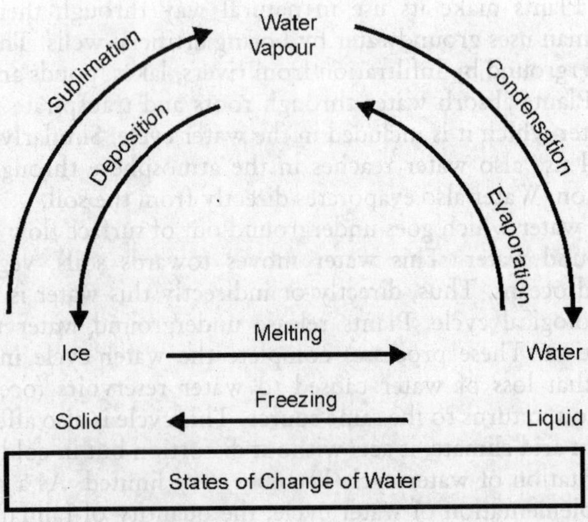

Water is found in the atmosphere in the following three forms:

Solid: in the form of micro snow granules.
Liquid: in the form of drops of water.
Gas: in the form of water vapour.

Due to change in temperature, the different forms of water also change, which control the monsoon image. Among the above three forms, there occur six changes in series, which are as under:

Solid to liquid: By melting
Liquid to Solid: By freezing

Different Forms of Water

Liquid to Gas: By evaporation
Gas to liquid: By condensation
Solid to gas: By melting and evaporation
Gas to solid: By sublimation when gas becomes cool

Sub-cycles of Hydrological Cycle

Water cycle always remains moving. Sun's heat is available for its smooth movement and the above described water cycle is rarely visible because of its of complicated nature. The Hydrological cycle is completed by combination of water sub-cycles. They are as follow:

(1) Micro Sub-cycle

Micro sub-cycle completes in a short period which can again be divided into two parts:

(i) *Sub-Micro Cycle:* While it is raining, evaporation of drops of water is taking place simultaneously. After evaporation, the same water again enters the atmosphere after condensation due to temperature and falls on ground through rain and completes the micro sub-cycle.

(ii) *Micro Cycle:* After evaporation of water from oceans, it moves as vapour and after condensation through friction, it again reaches the oceans. Thus water of oceans to atmosphere and from atmosphere again to oceans, completes the micro-cycle.

Figure 5.4
Sub-cycles of Hydrological Cycle

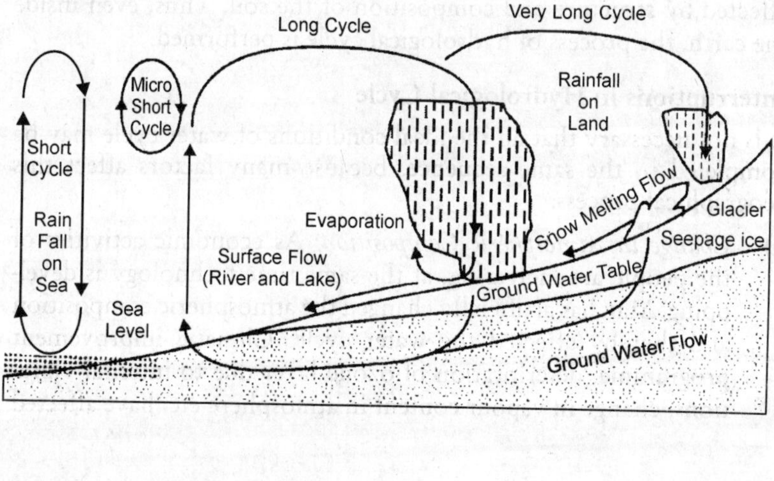

Macro Sub-cycle

This cycle takes comparatively more time as compared to the micro-sub-cycle. It is also divided in two parts:

(i) *Long Macro Sub-cycle:* When water falls on the surface of ground at the time of rain, some part of water (out of flow water) evaporates and reaches the atmosphere and after friction falls on the surface of the earth and completes the long macro sub-circle.

(ii) *Very Long Minor Macro Sub-cycle:* After rainfall, some part of the flowing water oscillates on earth, which is absorbed by roots of plants. Through transpiration and evaporation, this water again reaches the atmosphere and again comes to earth in the form of rainfall. Thus, this macro sub-cycle takes a very long time to complete.

An ideal water cycle is found on oceans. On the ground level, from rivers, lakes and other minor water storage places, before water reaches the oceans, some part of the water evaporates and water cycles get completed.

Inland Stage of Hydrological Cycle

Inland stage of hydrological cycle starts right from the seepage of rain water inside the surface of the ground. It is known that about 25 to 50 per cent of average rainfall flows away on the surface of the ground. A large portion of surface water evaporates and the remaining part goes underground. Inland water changes its flow according to flow routes as per geological structure of the relevant area. Some rocks are aquifers, others are aquiclude. Flow of water inside the ground is affected by structure and composition of the soil. Thus, even inside the earth, the process of hydrological cycle is performed.

Interruptions in Hydrological Cycle

It is not necessary that all the ideal conditions of water cycle may be completed in the same sequence, because many factors affect this geographical process:

(i) *Change in Atmospheric Composition:* As economic activities of the people are increasing, at the same time technology is developing. Man has artificially changed the atmospheric composition and thereby affected the water cycle. Climatic improvement programme, deterioration of ozone layer due to internal vibrations, change in vapour content in atmosphere etc. have affected possibilities of climatic system changes. As a result of such change, there can be change in movement of evaporation from ground as well as oceans and its direct effect is on quantity of rainfall.

(ii) *Due to Increasing Industrialization:* Quantity of air particles is changing because granules emitted by industries work as humidity centers for formation of drops of rain. Thus, solid particles in atmosphere, effect of pollution and formation of clouds affect the quantity and image of rainfall.

(iii) *Process of Cloud Formation* also affect the hydrological cycle.

(iv) Man has affected water cycle by disturbing the ecological balance of nature through *deforestation.*

(v) *Urbanization* has also affected environment, for example air pollution, deforestation for urban expansion etc., also disturb the hydrological cycle.

(vi) Hydrological cycle is also affected by internal movement of water. Deforestation reduces it, whereas irrigation, tree plantation, dam construction and construction of water reservoirs increase it and evaporation also increases. Man affects the water cycle by impacting on surface flow of water in many ways.

(vii) Different types of pollution created on surface of the earth also hinder the movement of water cycle. Due to the economic race, man has done mining by unscientific methods, and thereby disturbed the balance of nature and the functioning of the hydrological cycle.

Importance of Hydrological Cycle in Nature

Hydrological cycle is important on the earth for various organic activities because hydrological balance would be disturbed without water communication and any type of life would become impossible. Evaporation of water and its storage in atmosphere is a very important process. Seasonal variations depend on it. Performance of rainfall on the earth is completed only by the hydrological cycle. Biogeochemical cycle is completed on the earth with water cycle. Among different biogeochemical cycles on the earth, vegetation is an effective medium for movement of sediments and chemical substances. Thus, biogeochemical cycles on the earth are possible only through movement of water.

Figure 5.5
Effect of Urbanization on Hydrological Cycle

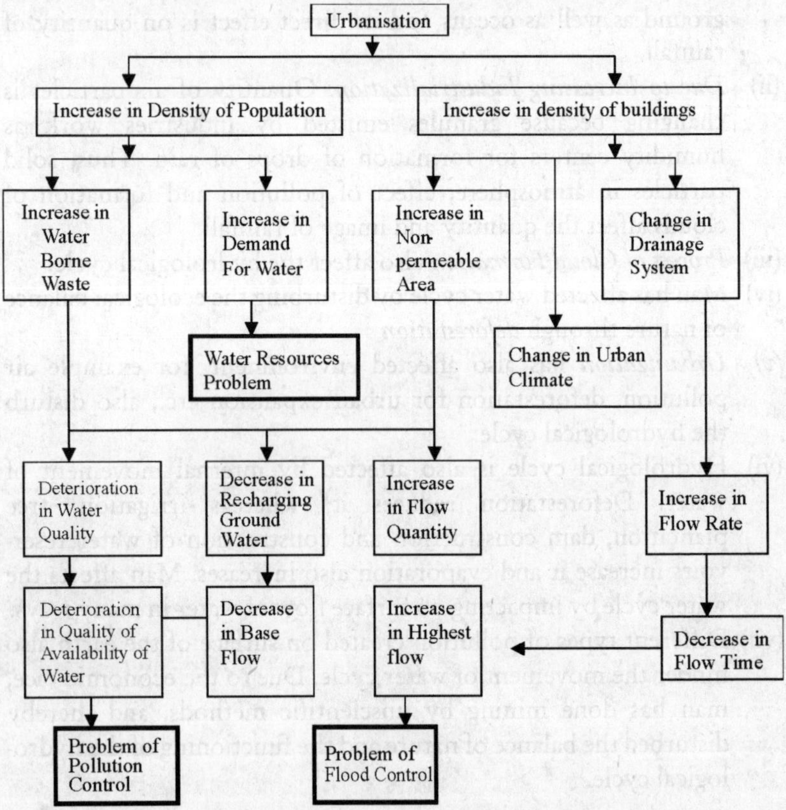

Balance in Hydrological Cycle

In nature, universal hydrological cycle is a closed ecological system because hydrological cycle on the earth does not exchange energy in the universe, it only gets energy. On the earth, it does not get humidity from outside, and the companion water in hydrological cycle is stable. It does not increase or decrease. There is only seasonal variation but on the world level there is no change in its quantity.

Total evaporation from oceans is 86 per cent and 14 per cent from ground level whereas, rainfall is 79 per cent on oceans and 21 per cent on the ground level. This additional 7 per cent moisture moves from seas towards the ground through clouds and again transfer of water from ground towards seas through rivers and groundwater

Figure 5.6

State of Balance in Universal Hydrological Balance

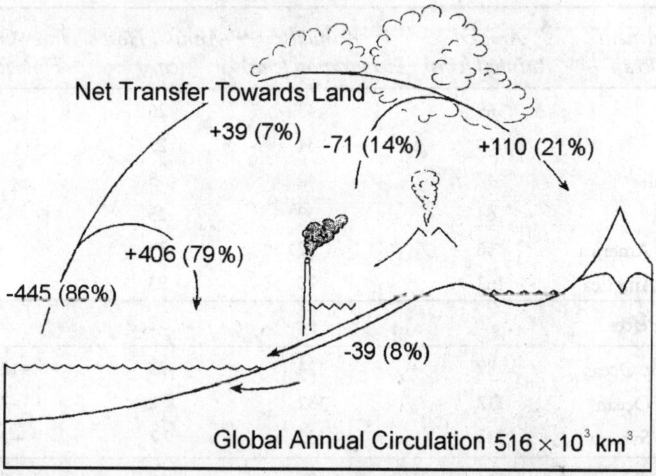

maintains the balance. In water balancing, water moves from one place to another. For example, evaporation from oceans is more than rainfall over oceans, but on ground level (except arid and desert areas) this situation is reverse. On the whole, universal evaporation balances rainfall. If evaporation is lesser, there would be increase or decrease in quantum of humidity in atmosphere from time to time. Flow of rivers on the earth balances the evaporation and rainfall. On this basis, normal equation of water balancing is as under:

Precipitation/Rainfall = Evaporation +/– Flow

Water and ground both combined make the balance on universal level though its local form may differ, which is mainly affected by variations in climate (Legates and Mather, 1992). For example, in dry and hot Australia, rainfall is lesser than evaporation, whereas in tropical south American areas, in spite of more evaporation, there is high flow because rainfall is sufficient (Park, 1997). Oceans play an important role in universal hydrological cycle because they contain maximum portion of water in them. Maximum rainfall occurs in the Pacific Ocean. Thus, it is clear that in the hydrological cycle in nature, existing water only moves in different forms and in different places. Its actual quantity remains stable. It is called universal water balance and the hydrological cycle is important for balancing quantity of water sphere.

Table 5.4

Average Water Balance of Continents and Oceans on Earth

Continent/ Ocean	*Annual Rainfall (cms)*	*Annual Evaporation (cms)*	*Annual Flow (cms)*	*Inter-Ocean Flow (cms)*
Africa	69	43	26	
Asia	60	31	29	
Australia	47	42	5	
Europe	64	39	25	
North America	66	32	34	
South America	163	70	93	
Total Places	73	42	31	
Atlantic Ocean	89	124	23	–12
Indian Ocean	117	132	8	–7
Pacific System	133	132	7	+7
World Oceans	114	126	12	0

Source: M.Budyko, 1974, pp. 227-228 and 238.

Water Balance (Budget) of the Earth's Surface

The word 'water balance' was first of all used in 1944 by the famous meteorological and climate scholar C. Warren Thornthwaite (1899-1963), who meant that it is a balance between precipitation,

Figure 5.7

Constituents of Water Cycle of a Drainage Basin in a Hilly Area

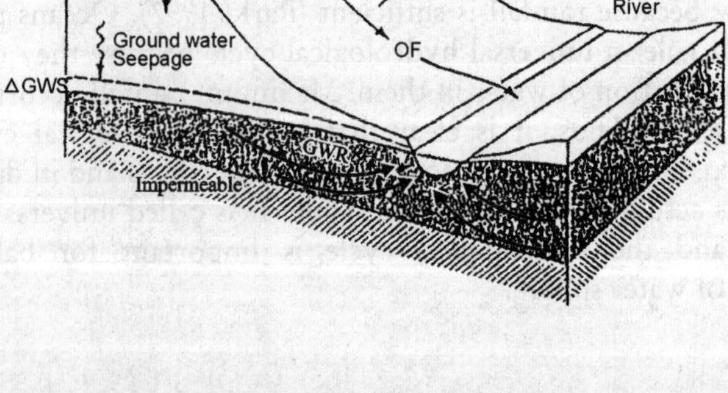

water obtained by melting of snow and evaporation, groundwater recharging and surface flow of water. For its assessment, statistics of soil moisture, transpiration, groundwater recharging and figures of flow of rivers are collected from meteorological department and of moisture and vegetation. Water balance is useful for assessment of demand for irrigation water and geographical as well as climatic image of soil moisture. Water balance study gives information about increase or decrease of water of lakes and water ponds as well as their salinity. For utilization of land and water resources and to assess possibilities of economic and ecological study, statistics regarding soil moisture, groundwater and river water flow can be obtained only on the basis of water balance.

Water balance also explains the effect of human activities on hydrological cycle. Changes in climate, vegetative cover etc. can be studied on the basis of water balance , hence water balance is an important tool for analyzing water problems in a geographical region.

Measurement of Water Balance

The following formula can assess measurement of water balance of a micro drainage area:

$$P = I + AET + OF + \blacktriangle SM + \blacktriangle GWS + GWR$$

It is an area of non-permeable rocks up to a certain depth.

Here, these symbols explain changes in precipitation, interception, transpiration, sub-soil surface flow, soil moisture, quantity and flow of groundwater. First of all, precipitation and interception are measured. According to the above formula, actual transpiration can also be assessed:

P= Precipitation, I=Interception, AET= Actual Evapotranspiration, OF= Over Land Flow ▲SM= Change in Soil Moisture, ▲GWS= Change in Ground Water Storage, GWR= Ground Water Runoff

Computation of Water Balance

The method first suggested by C. Warren Thornthwaite had suggested the method of assessing water balance which has been transformed many times for different objectives. Thornthwaite and Mather (1957) have given suggestions from time to time regarding changes in the formula related to water balance. For assessing water balance, analytical study of precipitation, interception, transpiration, evaporation, overland flow, changes in soil moisture, changes in groundwater storage and groundwater flow is made, the detailed description of which is as under:

Precipitation: It is the quantity of water which reaches the surface of earth after movement in different forms in the atmosphere.

Evaporation: Evaporation is that seasonal process in which water from independent ground level evaporates and reaches in the atmosphere in the form of vapour. This process takes place when temperature reaches boiling point of thermal energy. Factors like quantity of solar radiation on earth, temperature of air, quantity of water vapour existing in the atmosphere, nature of ground level, current air etc. determine the rate of evaporation. Quantity of evaporation is maximum in case of high temperature, fast flow of air and presence of vast water surface.

Evapo-transpiration: It is the total moisture which mixes with atmosphere as a result of evaporation and action of transpiration from plants. Rate of evapo-transpiration is mainly based on temperature. It is an important factor in climatic studies because it decides all the atmospheric conditions.

Potential evapo-transpiration: This process depends on the availability of water. If sufficient moisture is available to meet the needs of vegetation, the resulting evaporation is called potential evapo-transpiration.

Actual evapo-transpiration: Actual evapo-transpiration is found in special conditions. Its quantity is found out by adding precipitation and changing soil moisture.

$$AET = P + \blacktriangle SM$$

where AET = Actual Evapo-transpiration

P = Precipitation

▲SM = Changing soil moisture

Infiltration: Infiltration is the flow of water into the ground through the soil surface. After rain or irrigation, as soon as the water comes on the surface of the earth, it inclines to enter soil from the nearest point. Thus, entry of water into the soil is called infiltration.

Run-off: That part of water from rain or of melted snow which flows as drainage on the surface, is called 'run-off'.

Percolation: Percolation is downward movement of water in soil profile. Percolation is not the same as infiltration. It is the first stage of infiltration. Thus, till water does not percolate, its infiltration cannot take place.

Method of Assessment of Water Balance

1. First of all, assessment of potential evapo-transpiration is done.
2. P=PET is counted where P=Precipitation.
3. Loss of storage water is counted., which is added by reducing analysis method (P=PET). Only for dry season, moist climate is counted in the end.
4. Soil moisture (Soil Moisture – SM) is counted from figures of potential loss of storage water and soil and land utilization.
5. Changes in soil moisture (▲SM) is counted from soil moisture rules, i.e., difference between current month and next month.
6. Actual evapo-transpiration (AET) is counted Precipitation + changes in soil moisture. If rainfall is more than potential evapo-transpiration, then actual transpiration is equal to actual transpiration.
7. Loss of soil moisture (D) is counted by deducting actual evapo-transpiration from potential evapo-transpiration (PET-AET).
8. Excess soil moisture (S) is analyzed. Once the shortage of soil moisture is calculated, then in humid climate additional rainfall after collection of water soil moisture becomes excessive. Thus, more than required soil moisture is counted.
9. Additional moisture mixes with underground water sources. It cannot flow fully in the same month. Its remaining part flows in the next month. Its quantity differs according to texture and depth of soils of that basin. Thus, during any month, actually 50 per cent of water becomes available for flow.

6

Demand and Use of Water

Use of Water

Water resource is an important natural resource, which is the basis of all resources, and exploitation as well as conservation of other natural resources is possible mainly due to the presence of water. It is a renewable resource, which can be made re-usable after using it once through the process of purification. Water alone is the only resource whose regular supply is essential for us and we get it through rivers, lakes, ponds, groundwater and other traditional water conservation areas. Seawater can also be used, but it is necessary to remove its salinity. In present times, Libya, Qatar, Saudi Arabia, United Arab Emirates, Israel, Yemen and a few other countries are using oceanic water after purifying it.

The maximum use of water (70%) is made for irrigation whereas industries (23%) are at the second place. Only 7 per cent water is utilized for domestic and other purposes. People are using less than 10 per cent pure water available on the earth, but since this resource is not equitably distributed hence, water cannot be equally supplied at all places. Consequently, a few countries are facing grave water crisis at present. At many other places, a situation of water crisis is created in spite of availability of water because of continuous quality deterioration of available water resources.

Water resource can be utilized in the following areas:

1. In irrigation
2. In industries

3. For domestic purposes
4. For navigation
5. For canals
6. For hydropower

1. *In irrigation:* Maximum use of water is made for irrigation. Surface and groundwater is used for this purpose. Surface water is used through canals and tanks whereas, groundwater is used through dug wells and tubewells. One-fourth part of the world has arid conditions, which is fully dependent on irrigation. Russia, Turkey, West Asian region, Egyptian valley, European River Basin, Laplata Basin (South America), Nigeria, Murry Darling Basin of Australia etc., are under canal area irrigation. Even in Asia, besides rainfed areas, some areas depend on irrigation, which are important areas for crop production. Large-scale production of rice, wheat, sugarcane, cotton etc., takes place in such areas.

In some countries like China, India, Bangladesh, Southern Japan, Vietnam and Malaysia having greater burden of population, two to three crops of rice are grown, which need irrigation. Likewise irrigation is also needed for raising summer crops. At present, some countries, which have exploited more groundwater as compared to surface water, are facing acute water problem. United States of America has been using 25 per cent groundwater and 75 per cent surface water for irrigation, whereas in countries like India, groundwater is being blindly used for irrigation while the major portion of surface water flows away without use into oceans. Consequently, water crisis is assuming dangerous proportions in India.

2. *In industries:* About 23 per cent of pure water is utilized by industries. That is why most of the industries are established near water sources (bank of river or lake). Many industrial units also have their own water purifying plants so as to have regular supply of water. Water is used in industries for making steam, condensation of steam, solutions of chemicals, washing, dyeing, printing, temperature control through humidifiers and refrigerators, smelting of iron in steel industries, coal washing, leather processing and tanning, paper pulp making, and for preparing acids and alkalies. Presently, the increasing number of industries in the wake of industrialization has greatly deteriorated the quality of water due to which scarcity of water is being felt in spite of its availability.

3. *For domestic purpose:* Water is not equitably distributed in nature, but man has made adjustment with nature by developing different practices regarding use of water as per its availability. In arid regions, multipurpose use of water is made due to scarcity of water. Water is needed for drinking, cooking, bathing, washing clothes etc. in the domestic sector. Even in cities inhabited on riverbanks, problems have arisen in spite of availability of water because cities have contaminated these water sources. London on Thames River, St. Louis on Bainjar and Kanpur, Varanasi, on river Ganges, Kolkata on river Hoogly and Delhi and Agra on river Yamuna, are also facing problems of water supply. These cities are draining domestic wastewater into rivers.

4. *For navigation:* Surface water resources contained in rivers, canals and lakes are used for navigation. Direction of flow of water, quantum of water, seasonal effects, and length of rivers and canals play an important role in navigation. Rhine river of Europe is highly useful from the point of view of navigation. Besides it, rivers Nile, Danube, Yangtze, Volga, Mississippi, St. Lawrence, Siene, Elbe, Mekong, Iravadi, Parana, Amazon, and in India rivers Brahmaputra (up to Dibrugarh) Ganges (Farakka Barrage), Godavari etc. have navigation facility. The main problems faced in navigation are virulent flow of water, hurdles during route, presence of waterfalls and streams, and floods and changes in the route of the river.

5. *For canals:* Canals are constructed by the side of rivers in case of uneven surface of the soil. Canals are made for multipurpose uses of water like irrigation, navigation, hydropower, flood control etc. From the point of view of multipurpose uses of water, Grand Canal of China (1,900 km long) is most important. From the point of view of navigation, Eric Canal of United States of America, Seattle (on the Western Coast) 500 Canal are important. 500 Canal remains closed for four months during the winter season. Volga, Dnepr, Don and Danube rivers in Russia have been connected with canal system. Similarly, in southern India, transportation has developed between Vijaywada and Tidivanam on Coromandal coast through Kombaumur canal and Buckingham canal.

6. *For hydropower:* Water resource is a renewable resource which can never end. Hence, as an alternative to depleting energy resources, power is being produced in different forms from water. The equatorial regions receive maximum rainfall on earth, but due to its

unevenness, possible hydropower has not been produced. In Africa, 23 per cent of hydroelectric power of the world exists but developed hydropower is only one per cent of the world's potential. In Asia also, 23 per cent of hydropower of the world exists but developed hydropower is only eleven per cent. Similarly, in South America, as against the potential of 17 per cent hydroelectric power, developed hydropower is only four per cent.

Thus, we are expediting development by utilizing available water reserves of the water zone like oceanic water, surface and groundwater for different purposes. Many countries are utilizing seawater for drinking and irrigation purposes after purifying it. Oceans are being mainly used for navigation. Besides, oceans are reservoirs of energy for the future. Tidal power is being produced from ocean tides. Oceans are also food reservoirs where sufficient quantity of food is found in the form of plankton fish. Many countries prepare food items from various creatures of the sea.

Role of Water in Plantation

Water plays an important role in the life of trees and plants directly and indirectly in two ways. Firstly, by changing soil properties and secondly, by its functions in different physiological activities.

1. Water is considered as an important component in the protoplasm of cells of trees and plants. There exists 60 to 90 per cent water in the body of plants. The direct effect of water is visible in metabolic reactions in protoplasms.
2. Water serves as a solvent by carrying minerals and gases in large amounts to plants. They move from one cell to other as nutrients in plant body.
3. Water facilitates the process of photosynthesis, which provides food for plants, this water also works as an agent for hydrolysis reactions.
4. Water is an essential ingredient for division of cells, for their extension and cell wall formation in plants.
5. Water keeps plant cells healthy, due to which pores of plants remain open for exchange of gases. It is also essential for photosynthesis.

Role of Water in Soil

1. Water plays an important role in the process of soil formation and expediting eluviations. It also assists in the soil base formation.
2. Water affects physical qualities of the soil.
3. Water plays an important role in soil erosion.
4. Water being an ideal solvent, strengthens the process of leaching, chemical reactions etc.·
5. H_2, which is essential for trees and plants, is supplied only through water. In photosynthesis, glucose is formed by combination of H_2 and CO_2 in the presence of light.

Effect of water balancing on different conditions of crops: Interrelation among rain and water due to melting of snow, underground or surface flow and process of evaporation on any surface basin of earth is called water balancing. This balance affects different conditions of plants. Shoot growth of plants is weakened by scarcity of water but not in proportion to balancing by photosynthesis. In case of acute shortage of water, plant leaves start falling down. Such a condition is more in the beginning stage. In case of fruits and vegetables also, such an effect is visible. For example, growth and freshness of fruits diminishes. In many fast growing vegetables, production decreases faster compared to normal growing crops. In the early life of crops, fewer flowers grow in flowering plants. In fruit plants, production of fruits is lesser. In sugarcane, oilseeds, rubber etc. protein, oil and rubber production also falls down due to water shortage.

Necessary quantum of water: The assessment of required water quantity for plants is done between wilting point and capacity of area. This quantity of water stays mainly in pores of cells. Hence, it is similar to capillary water.

This state of water is very important for proper growth of crops. Necessary quantum of water depends on composition of soil and living organisms etc. (Table 6.1).

On the surface of the earth, soil is of different composition. Silt and loamy soil requires lesser water whereas, sandy soil needs more water. On the other side, sandy soil has greater water retention capacity, but due to higher wilting point, the necessary water quantity reduces.

Table 6.1
Wilting Point, Area Capacity and Necessary Water Retention capacity in Soils of Different Composition

(in %)

Soil Composition	*Wilting Point*	*Area Capacity*	*Necessary Water Retention Capacity*
Coarse Sand	1.7	6.8	5.1
Fine Sand	2.3	8.5	6.2
Sandy Loam	3.4	11.3	7.9
Fine Sandy Loam	4.5	14.7	10.2
Loam	6.8	18.1	11.3
Silt Loam	7.9	19.8	11.9
Clay Loam	10.2	21.5	11.3
Clay	14.7	22.6	7.9

Due to presence of living organisms in the soil, soil composition improves and it increases its water retention capacity. Vishwas and Ali (1969) have studied effects of living organisms on soils at various places. These studies reveal that soils having more living organisms retain more water content on area capacity. In other samples of the same soil, almost the same quantity of water was noticed. Hence, it can be said that water quantum increases with the increase in living organisms in the soil.

Necessary Quantity of Water and Growth of Plants

The effect of necessary quantity of water for different plants is directly evident on growth of plants. Its assessment is made on the basis of nature of plants (Table 6.2). According to Wihamiya Per and Handrix (1950), all plants need the same water, but in 1952, Richard and Bowleg, through their study, said that it was wrong. They explained that as soon as necessary water quantity as per area capacity reduces, availability of remaining water is comparatively lesser. Hence, before reaching fading point, there is adverse effect on growth of plants because of shortage of necessary quantum of water.

Table 6.2

Necessary Demand of Water for Different Plants

(in cms.)

Crop	*Demand of Water*
Wheat	40 to 50
Maize	46 to 60
Sugarcane	100 to 140
Tobacco	60 to 65
Cotton	50 to 65
Rice	130 to 160
Potato	40 to 60

Necessary quantity of water develops plant roots and affects related growth of plants. Lesser development in roots of any plant, which spread over very far, maybe because growth of such plants is affected by necessary quantity of water. In climatic conditions, evaporation, transpiration, rains, quantity of temperature etc. are important. In short, necessary quantity of water directly affects the growth of plants. Vegetative growth of plants reduces with scarcity of water. The extent of this increase and decrease affects the particular status of these plants.

Demand of Water for Different Crops

Demand of water means such quantity of water that is required for normal growth of a crop for a fixed period, as per local conditions. Except climatic factors (evaporation and transpiration) and water necessary for physical development of plants, it also includes water, which is used in leakage, flowing, field preparation, planting, transmitting salts up to roots down below, etc.

Demand of water can be found out by the following formulae:

(i) Demand of water = Transpiration + Evaporation + Losses during irrigation + Special needs

(ii) Demand of water = Need of irrigation water + Effective rains + Collected humidity

Demand for quantity of water in crops is calculated for the whole lifetime of a particular plant. Thereafter, weight is taken after ripening and drying of fruits and flowers, taking out from field along with roots, and drying the crop.

Factors Affecting Demand for Water

Demand of water for crops is affected by the following factors:

1. *Nature of Crops:* Different nature of crops needs different quantity of water. For example, more quantity of water is needed for rice and sugarcane, lesser quantity for bajra and jowar. Duration of crop and stage of growth also determines the demand of water. Long duration crops need more water, whereas shorter duration crops need lesser water. Demand for water also depends on growth time of plants. For example, in earlier stages, plants need less water whereas, more water is required in their growth period. Structure of plant roots also affects demand for water. Deeper the roots, the more water they retain.

2. *Nature of Soil:* Water retention capacity depends on composition and compaction of soil. Crops sown in sandy soils need more water because of more infiltration rate. Likewise, surface soils need more water as compared to deeper soils. Soil fertility directly decides necessary quantity of water for the plants. In soils having lower fertility, evaporation is also more.

Water discharge in different types of soil also affects its status. In waterlogged soils, plants utilize lesser water due to lesser evaporation. Thus, in cases of faulty discharge of water, plants need lesser water.

3. *Effect of Climatic Factors:* The particular climate for sowing crops also affects it. Crops sown in summer, needs more water as compared to crops sown in winter because there is more evaporation during summer. Similarly, extent of humidity also affects demand for water. Demand for water decreases during rains. Winds provide speed to evaporation. Lesser winds will reduce demand. Duration of the day and direct or indirect sunrays also affect demand of water. Higher temperature will increase evaporation resulting in more demand for water.

4. *Effect of Nature of Irrigation:* Irrigation system directly affects demand for water. Water leakage in flow system of drains requires more water. Drop irrigation or spray system require 60 to 70 per cent less water. Supply system of water from water source to fields also affects demand for water. Leakage and seepage in drains cause heavy loss of water. If such drains were constructed *pucca*, there would be lesser use of water.

5. *Attack of Pests and Diseases:* Various crop diseases also affect the demand for water because growth does not take place in diseased plants whereas, they need regular supply of water. Quality of irrigation water used in crops, agricultural practices, and soil

preparation methods also affect demand for water. Thus, admixture of type of agriculture and sources of irrigation directly affect the demand for water.

Absorption of Water by the Plants

Plants absorb water for development. Only a few xerophyte plants exist in arid climate which can grow without suction of water. In many plants, transpiration is equal to available quantity of water. Kremer (1969), through one study explained that a maize plant transpires 2 to 4 litre water (double its weight) on a dry and hot day. If such transpired quantity of water were not re-supplied, the plants would either wither away or die. Transpiration mainly depends on physical factors.

Absorption of water in soil cells is in the direction of decreasing water potential gradient. Hence, such absorption is of two types:

1. *Passive Absorption:* Due to water losses because of transpiration of water from plants, when water potential falls in cells of plant leaves, water potential falls to low level in the plant leaves.
2. *Active Absorption:* In dry and humid soil as well as moist environment, transpiration is minimum. In such conditions, absorption is called active absorption. According to Kremer (1969), active absorption is less important in water economy of plants. It happens in the following manner:
   (i) Absorption due to seepage activeness of main cells.
   (ii) Absorption due to ordinary osmotic reactions.

Transpiration of water is affected by the following factors:

1. *Development of Roots:* Plants absorb water through their roots. Plants having deeper and extended roots can absorb water as per their requirement whereas plants having less deep roots can absorb lesser water.
2. *Availability of Soil Water:* Quantum of water absorption capacity of plants is determined by existence of water in the soil. Necessary water is decided by interrelation between wilting point and area capacity but its quantity depends on type and composition of soils found in different areas. Soils having different structure and composition have different water retaining capacity. Sandy soils retain less water whereas loamy soils retain more water.

3. *Soil Temperature:* Soil temperature has direct as well as indirect effect on water absorption. Root growth stops with lower temperature and hindrances are created in the water flowing towards roots, which affects active absorption. Higher temperature has also adverse effect. Good aeration arrangement in soil provides active support in absorption of water.

Interrelation between Availability of Water and Growth of Crops

Scarcity of water as well as excessive water both are harmful for crops. Scarcity of water refers to the condition when water does not become available for absorption and plants start drying. Scarcity of water does

Figure 6.1

Mutual Relationship Between Soil Water and Plants (quoted from Surajbhan 1983, Regards to Dakshin Murty and Saathi)

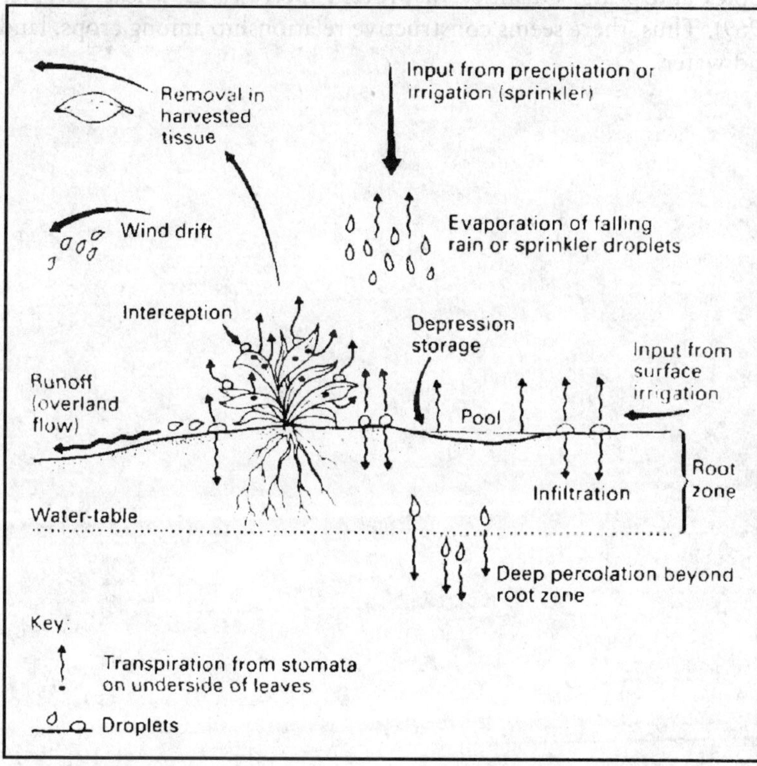

not equally affect all the living processes of plants. The effect on blossoming buds of plants and short crops is not the same as in old leaves and developed plants. Normally, all crops need sufficient quantity of water at the flowering and fruit bearing stage and in such situation even a little shortage of water causes serious effects, for example at the time of coming of ears of in corn or wheat. Transpiration of water from plants also causes adverse effect on crops. Due to water scarcity, constitution of cells cannot develop properly. Its direct effect is visible on photosynthesis and respiratory processes. As compared to the respiratory processes, photosynthesis is highly affected, as a result of which plants utilize sufficient food material for necessary energy. Scarcity of water in cells also affects nature and quantity of carbohydrate. Quantity of starch also reduces if shortage of water continues for a longer duration.

Scarcity of water is also sometimes advantageous. Shortage of water improves the quality of many crops and plants as in the case of apples and pears. Quantity of protein increases in wheat (Kremer 1969). Thus, there seems constructive relationship among crops, land and water.

7

Irrigation

Man has been using water for irrigation since ancient times. Important civilizations of the world have developed on the basis of irrigation management. Normally, groundwater and surface water are used for irrigation and when water available in these sources is taken away artificially by flowing it for supplying water in required quantity to crops, it is called irrigation. This is done at such places where rainfall is insufficient. Man has discovered many methods to fight with nature and irrigation is one such means. According to Peter Wales, "It is an artificial means of watering the crops or plants or an art of supplying water to the crops".

According to Dudley Stamp, "In the absence of rainfall, supply of water to land by artificial means for growth of plants and for increasing productivity is irrigation."

For optimum development of crops, balanced irrigation has been considered essential, because if irrigation is done beyond a limited quantity, the water would go downwards through seepage due to gravitational force of the earth. This leads to not only leaching of nutritive elements of soil but a state of waterlogging is also created in the soil. Similarly, less than necessary quantity of irrigation, is also harmful because in that case water does not reach the lithosphere. Through balanced irrigation, quantity of water received by plants from the soil is properly recharged. Plants from their root zone, lose that water through evaporation and transpiration. Instead of irrigating very deeply and for a long time, it should be done less but at short intervals.

While irrigating through different methods of irrigation, all the water does not reach the soil but some water goes waste by infiltration, transpiration, flow etc., and some water is filled in drains and lower parts of the fields. Efficiency of irrigation depends on soil texture, method of irrigation and quantity of water stream. Eficiency of irrigation is possible through the following formula (Surajbhan 1993):

$$\text{Efficiency of irrigation} = \frac{\begin{array}{c}\text{Actual loss of water available in soil}\\ \text{which was recharged by irrigation}\end{array}}{\text{Total quantity of water given by irrigation}} \times 100$$

After estimation of efficiency of irrigation, the interval between irrigation is most important. This is affected by season and nature of crops.

Need for Irrigation

In any geographical region, equal distribution of natural substances is not necessary. Somewhere there is dry climate and at some other places, it is humid. Similarly, on the surface of the earth also, supply of water is not in the same proportion. In such circumstances, human beings have to arrange supply of water for crops through artificial means for fulfilling their primary needs. Thus, irrigation becomes necessary on account of the following reasons:

1. *Nature of Rainfall:* Due to irregular, uncertain and limited rainfall, scarcity of water is caused. Consequently, the need for irrigation arises. Sometimes, it can rain heavily and for a long duration, whereas sometimes it can be light and for a short duration. If there is torrential rain, it is harmful for agriculture whereas at other places it might rain at short intervals in a balanced way.
2. *Nature of Soil:* There is greater need of irrigation in sandy soil whereas in clay or alluvial soil the need for irrigation is lower.
3. *Probability of Drought:* At places having irregular rainfall, since there is probability of drought, irrigation is needed.
4. *Need of Irrigation in Dry Areas:* In dry areas, where rainfall ranges between 40 to 50 cm per year, the need arises for irrigation.
5. *More Need of Water for Special Crops:* Crops like rice, jute, sugarcane, etc. need more water, which can be fulfilled only through irrigation.

6. *More Need of Water for Improved Seed Varieties:* New and high-yielding seeds need additional water through irrigation for higher productivity. This was felt after the green revolution.
7. *For Pasture Development:* Along with agriculture, it is essential to develop pastures for cattle development and dairy development, which also need water.
8. *Population Increase:* Population of the world is multiplying fast, and it needs additional food production . This can be possible only through irrigation.

Factors Affecting Irrigation Development

Different systems of irrigation are found in different geographical contexts. Factors affecting irrigation development are as follows:

1. *Topography and Slope of Soil:* Irrigation development is done according to structure of the ground level. If ground level is plain, canals or wells would be constructed, whereas in stony areas, the tanks would be suitable. Similarly, slope of ground level also affects irrigation development along with agriculture. Canal system was developed according to the slope. Ground level plays an important role in construction of wells. If ground level is hard and rocky, deeper wells cannot be constructed.
2. *Structure, Depth and Permeability of Soil:* Water can be absorbed in the soil according to the structure of the soil. If there is lesser depth, it affects permeability and it can create a condition of waterlogging.
3. *Depth of Groundwater:* Quantity of groundwater and its depth also affects irrigation development.
4. *Climate:* Quantity and duration of rainfall affect irrigation development. On this basis, rain water flow and seepage is affected. Evaporation and transpiration occurs according to temperature. Speed of air affects evaporation. Thus, climate exercises an important effect on irrigation development.
5. *Social Factors:* Social reasons also affect irrigation development. Labour is an important factor. Scattered fields or their consolidation directly affect the irrigation development.
6. *Economic Factors:* Among economic factors, means of transport apart from capital also affect irrigation development.
7. *Resettlement:* Problems of resettlement also affect some big or small irrigation development projects.

Means of Irrigation

On account of different physical conditions on the earth, different sources of irrigation are found in different places. Differences also occur due to ground level latitude, for example, canal irrigation is developed in plain areas, whereas on stony ground level, tanks, lakes or groundwater are used as means of irrigation. Surface water is used through canals, water reservoirs, tanks, lakes, ponds, while groundwater is used through wells, tubewells and artesian wells. Canal area development is done according to quantity of water of rivers, whereas development of wells is done according to quantity of groundwater, its depth and as per chemical qualities. Anicuts, check dams and loose stone check dams are used for irrigation on un-leveled ground level.

Canals

Canal irrigation is being done since ancient times, but due to the food grain requirements of the fast increasing population, development of modern canal system has become the necessity of the day. At the time of construction of canals, first of all a physiographic survey is done and thereafter the position of water is estimated. Physical and chemical qualities of soil affect canal irrigation development. Control over water plants and moss is essential for better development of canals. There should be control over weeds and grasses in earthen canals. Construction of canals is mainly affected by ground level and quantity of flow. Canal water is obtained from rivers and tanks, but

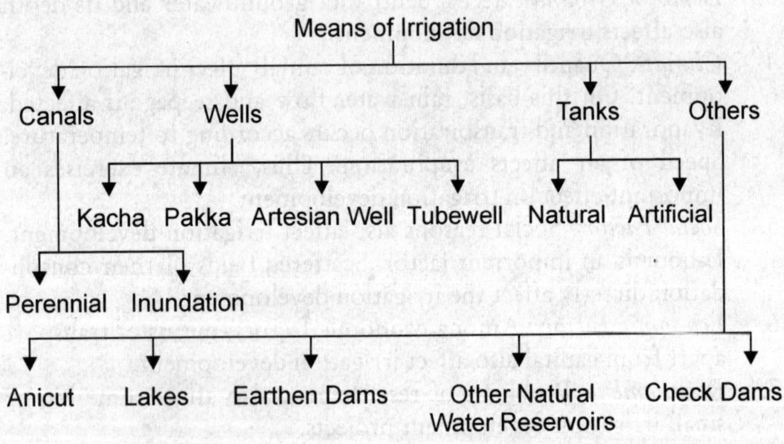

they take more water from rivers Which flow throughout the year. Whenever flood occurs in rivers, excess water is collected in reservoirs and diverted to canals. Canals are of two types:

(i) *Perennial Canals:* Perennial canals are constructed from such rivers in which water exists throughout the year. Water is checked in these rivers and nearby areas are irrigated by such canals.

(ii) *Inundational Canals:* Inundational canals are constructed where water does not flow in rivers throughout the year. Such rivers flow excess water during the rainy season which is checked by construction of dam. Only a single crop is taken in such areas. When irrigation water is not available in these canals, irrigation is done through wells.

Advantages of Canal Irrigation

1. Unirrigated wastelands can be developed by canal irrigation, which would increase the quantity of biomass in the area.
2. Economic development can be expedited by avoiding dangerous droughts. Dependence on rainfall can be minimized through canal development.
3. Canals are fed by rain water received by rivers, and the water is used for irrigation. Production of crops needing more water is also possible through canals. As compared to unirrigated soils, higher productivity per hectare is also possible due to canals.
4. Canal system is a permanent structure, hence only maintenance is required for getting its benefits for a long time.
5. Canals are multi-purpose where apart from irrigation hydro electricity generation, navigation, drinking water supply and fishery development is also done.
6. Groundwater level does not go down on account of canal irrigation, but on the contrary water level increases, which facilitates digging of wells.
7. Canals are also becoming a source of tourist attraction these days.

Disadvantages of Canal Irrigation

1. Due to imbalance in distribution of canal water, a situation of scarcity somewhere and waterlogging in other areas is caused due to collection of water there. It makes the soil unproductive as

harmful underground salts and alkalies come to the surface level due to waterlogging. Land can also become marshy there.

2. Many diseases are caused due to spread of mosquitoes, worms and insects on account of stationary water in canals.
3. Sometimes efficient canal management results in excessive production of crops, due to which the farmers are not able to get suitable price for their product in the market.
4. Due to shortage of water in inundational canals, crops are destroyed for want of water for irrigation.
5. Regular maintenance of canals is not done, due to which sediments are collected resulting in reduction of capacity of canals.
6. Due to excessive economic investment, it is not practicable to provide canal irrigation to all areas. Construction of canals also takes more time.
7. Many types of social evils are generated in canal areas. Sometimes, disputes arise for water distribution etc. resulting in murders and throwing of dead bodies by the bank of a canal.

Wells

Digging a vertical hole in the land to obtain groundwater from the earth is called a well. Wells are also constructed for artificial recharging of water, under ground investigations and inspection, disposal of industrial waste and disposal of flowing refuse. Wells are considered as the simplest traditional method of groundwater exploitation for different uses. During the present times, main use of wells is to get water for irrigation. According to hydrogeology experts, the well is the name of such water yielding structure through which maximum quantity of groundwater can be extracted on minimum cost according to the capacity of the aquifer. While constructing a well, depth of groundwater level, geological conditions, object of construction of well and economic factors etc. are kept in mind.

After the green revolution, hybrid seeds yielding higher production in agriculture are being used which need comparatively more water. Hence, during present times, a tendency for construction of wells and thereby exploiting water has increased, resulting in converting dark zones in many areas due to overexploitation of groundwater. It is necessary to control it for bringing the situation back to normal.

While constructing a well, the distance between pumping wells is significant because aquifers inside the earth are located at a definite distance and of definite size. The effective area of a well is dependent on duration and rate of groundwater exploitation, permeability multiplier of aquifer, saturated thickness etc. In case of difference in any of these factors, the effective area of a well changes. Even on economic basis, it is thought necessary to keep distance between wells. There should be also distance between irrigated areas because some areas may be repeated in the effective area of a well which may make the expenditure on well as unproductive.

For construction of a well, the following geographical conditions are essential:

1. Ground level should not be stony and hard. In stony and hard rocks, digging of a well is not only a difficult task but it is more expensive also.
2. Availability of groundwater reserves in sufficient quantity is necessary. It is useless to excavate a well without sufficient availability of groundwater.
3. Along with saturated aquifers, there should also be percolation of surface water.
4. There should also be recharge in proportion to exploitation of groundwater.
5. Work of construction of well is also affected by economic factors.
6. Groundwater should not be situated at great depth.

Types of Wells

Different type of wells are constructed in different areas on the basis of structure of the ground level, quantity of groundwater and its depth, demand of water for irrigation and economic factors. The quantum of groundwater mainly depends on qualities of aquifer supplying water to the well. Hence, construction of wells should be according to qualities of aquifer. Mainly, depth and diameter are kept in mind while constructing a well.

Depth of well is decided on the basis of distance of aquifer from ground level, size of aquifer, quality of water, and groundwater level. A well can be constructed to the extent of availability of depth of water in aquifer. Similarly, diameter of the well is kept on the basis of rate of exploitation of water and duration. This radius is not affected by natural reasons but it is decided on the basis of artificial technique. Depth of well and its radius is also affected by economic reasons apart from the above factors. Three types of wells have been constructed by the man. They are:

(1) *Kachha Well:* Such wells are constructed in the ground level having hard structure. Their depth is comparatively lesser. Internal wall around all sides of the well is made of clay. Its construction is not possible in case groundwater level is deeper.

(2) *Pakka Well:* Their construction is possible up to the depth of the water. Their internal walls are *pakka*. Their construction is possible due to deeper groundwater. They have greater capacity and their construction is possible even in weaker structure ground level.

(3) *Tubewell:* The construction by which digging of hole in the earth and bringing water to the ground level through pumps is done, is called a tubewell. Construction of tubewells is possible in areas where groundwater is found at a depth of more than 15 metres. They are constructed at places where canal water cannot reach. The following geological conditions are necessary for construction of tubewells:
   - (i) There should be sufficient groundwater below the surface of earth.
   - (ii) There should be intense demand for irrigation in the area.
   - (iii) Cheap electricity should be available in sufficient quantity for running the pumpset.
   - (iv) Soil should be productive so as to bear the expenditure of pumpset.
   - (v) Water level should not be at a greater depth than the water surface.
   - (vi) Need for irrigation exists round the year.

Tubewells are developed when the above conditions are favourıble. They are also called open wells. After the tubewell revolution, overexploitation of groundwater has resulted. Open wells are deep and their depth is 6-8 metres, more than the depth of groundwater level. Their internal walls are pakka. While constructing a tubewell, its size, depth and formation is kept in mind. Their construction is made by excavating equipments in fouı ways:

1. Digging by percussion
2. Core drilling
3. Through cyclical system
4. By blasting of rocks

Thus, tubewells are constructed according to ground level forms and suitable methods.

Advantages of Irrigation by Wells

1. Irrigation by wells is possible where canal irrigation cannot be done on ground level.
2. Problems created by canal irrigation like waterlogging, increase in salinity etc. are not faced in case of irrigation by wells.
3. Irrigation by groundwater makes soil fertile because many salts like nitrate, soda, chloride and sulphate are dissolved in groundwater.
4. Well water is cheaper as compared to canal water.
5. Water in well remains available throughout the year, whereas many canals are seasonal.
6. Wells are constructed privately, hence water can be obtained as per requirement, whereas canals are constructed for public use, creating problems of water distribution.

Disadvantages from Well Irrigation

1. Continuous exploitation of water from wells results in shortage of groundwater because recharge is not possible in the same proportion.
2. As compared to canals, well irrigation requires more labour.
3. Wells can irrigate only a limited area whereas, canals provide irrigation facility in a wide area.
4. Groundwater is available in the form of aquifers at various places in different quantities and different chemical qualities. If it possesses salts in greater quantity, it may not be suitable for irrigation.
5. Every farmer cannot construct a well but every farmer can get canal water.
6. After a certain time if a well becomes dry, it is costlier to dig another well.

Tanks

In plateau surface of the ground, irrigation is done by tanks where water can be retained on the ground for a long time. In such areas rivers do not flow for the whole year and water is available only during monsoon season through rainfall. Here, water springs are temporary and canals and wells cannot be constructed due to the hilly region. Rainwater needs to be stored for irrigation in such areas.

Tanks are created automatically by storage of rain water in a wide pit on the surface of the earth. These pits are deeper than the nearby ground level. They are excavated from time to time and for increasing their capacity, banks are raised to make a boundary. Tanks excavated artificially for meeting local needs are called artificial tanks. At some places where it is difficult to dig a well because groundwater is very deep, construction of tank is considered necesary.

Advantages of Irrigation by Tanks

1. It is a cheap and popular method of irrigation.
2. It maintains level of groundwater.
3. Where construction of wells is very costly due to hard rock surface, irrigation by tank is cheaper.
4. Expenditure is much less as compared to canals.
5. Use of additional water from rainfall is possible through tanks.

Disadvantages of Irrigation by Tanks

1. Source of water in tanks is only from rainfall. If there is variation in rainfall, it has adverse effects on agriculture.
2. Rain water flowing towards tanks carries sediments also which reduces the depth of tanks. It requires de-silting from time to time, which is very costly.
3. Taking water from tanks for irrigation is very costly.
4. Tanks occupy large space because many natural tanks are very extensive.

Other sources of water for irrigation include lakes, earthen check dams, anicuts, check dams etc. In hilly areas, for conservation of rain water, anicuts and check dams are constructed. Similarly, natural and artificial lakes and earthen check dams are mainly constructed for local needs of water.

Criteria

Criteria are observed for qualitative or quantitative data which predict the chance or magnitude of the effect of irrigation water on soil and plant growth under specific environmental circumstances.

Needs for Criteria

There are two basic needs for environmental criteria:

1. To assess the long and short-term consequences of soluble salts and other specific chemical substances in any concentration in

the irrigation water. This information is vital to control agencies if realistic standards are to be set, so that adequate environmental protection is achieved through prevention of salination in the irrigated soils.

2. To provide scientific information for soil and water usages and for environmental planning. This information is necessary if the environmental consequences of men's actions are to be adequately assessed. In addition, the systematic description of exposure-effect relationship between irrigation water on one hand and irrigated soil and plant growth on the other hand, will reveal areas where little or no information exists and allow future research needs to be defined.

Essential Elements for a Complete Criterion

Criteria are subject to many qualifying factors that need to be carefully identified in order that the final criteria provide a complete, accurate and useful presentation of the available evidence. It is important to have some idea as to what has not been investigated and to know the results that are available. For example, natural water mainly consists of inorganic salts in soluble form and they may be polluted due to external sources. It may also contain organic agents and/or heavy metals, toxic elements etc. The criteria for assessing irrigation water suitability must include information about chemical, physical and biological characteristics of soils, water and crops, which must be quantified since they may have to be controlled for specific uses of irrigation water.

Requirement of Criteria

Irrigation water quality criteria are required by:

1. Geologists and water resource engineers who are responsible for the development of water resources for irrigation.
2. Irrigation authorities who are responsible to the farmer for the quality of water delivered to them for irrigating their crops.
3. Soil scientists who are responsible for the protection of the soil from becoming saline and/or alkali (sodic).
4. Agronomists who have to decide the agronomic and cultural practices for obtaining optimum crop production.

5. Irrigation engineers who have to decide the suitable irrigation practices.
6. Plant scientists who have to decide the crops and their variety to be grown.
7. Drainage engineers who are responsible for the utilization of drainage waters.
8. Water pollution control authorities who must regulate the discharge of sewage or industrial waste to the rivers, lakes, estuaries and groundwater basins.
9. Environmental planning and protection authorities who are responsible for the conservation of the environment.

Criteria and Standards

Criteria are the scientific requirements on which a decision or judgement may be based concerning the ability of water of a given quality to support the irrigation use. On the other hand, standards are legally enforced levels that have been arrived at after consideration of water quality criteria, and the economic, social and political consequences of possible regulatory action. Thus, a compendium of water quality criteria presented herein, may be used as a guide by any agency that desires to set water quality standards. The tendency for criteria to become rigid and perhaps develop into standards and must at all times be avoided. The word 'criterion' should not be used interchangeably with, or as a synonym for the word 'standard'.

Quality of Irrigation Water

In every method of irrigation, many solvent salts are mixed in the supply of water. Production of these salts like sodium chloride, gypsum, sodium bicarbonate and sodium nitrate takes place during weathering of minerals from nature. Quantity of salts in groundwater depends on the nature of the source of water. When surface water infiltrates, minerals keep dissolving with it. Chemical deposition in water depends on source and pH value of water, nature of solvent gases, rate of weathering of rocks, soil geology, nature of soil profile, existence of quantity of salts, solution, water planning, hydrolysis, evaporation and rainfall etc. In areas where productivity is continuously falling down, use of saline water is resorted to in irrigation. Sodium, magnesium, calcium, chloride, sulphate and bicarbonate are the main ingredients of irrigation water. Apart from these, boron, lithium, silicon, iron, manganese and fluorine are also found in small quantities at some places. They are harmful for growth of boron plants.

Characteristics Determining Water Quality

The characteristics of irrigation water that appear to be most important in determining its quality, which however depend on climatic conditions, irrigation practices, soil-water retentivity characteristics, crop tolerance, depth to water table and agronomic practices, etc., are the following:

1. Salinity hazard (total concentration of soluble salts); Electrical Conductivity (EC).
2. Specific ion toxicity hazard (Ionic composition):
   (a) Major constituents (Sodium, Magnesium, Chloride, Bicarbonate, Carbonate, Silica, Nitrate)
   (b) Minor constituents (Boron, Lithium, Fluorine and other microtoxicants)
3. Sodicity hazard (Relative proportion of sodium to other actions); Sodium Adsorption Ratio (SAR).
4. Alkalinity hazard (Bicarbonate concentration as related to the concentration of calcium plus magnesium); Residual Sodium Carbonate (RSC).

In addition to the above individual parameters, combined evaluation of the two parameters given below are also of practical importance:

5. EC and SAR
6. SAR and RSC

Salinity Hazard

The total concentration of soluble salts is the single most important criterion which has been used conventionally for determining the quality of irrigation water. It is measured quantitatively in terms of 'Electrical Conductivity (EC), because this is very closely related with the sum of major actions (or anions) determined by chemical analysis, and because this correlates well with the value of total dissolved solids, as well as osmotic potential. The EC of a water sample is determined by measuring the electrical resistance between two parallel electrodes immersed in the solution. The EC is also called 'specific conductivity' or the conductance per unit cross-sectional area and across unit distance.

Specification Toxicity Hazard

Major Constituents

Sodium: Sodium is the most abundant chemical in all natural waters whereas, its presence is least desirable. Excess of sodium ions characterises that the water is saline or alkaline depending upon its occurrence in association with chloride/sulphate or carbonate/bicarbonate ions, and accordingly the irrigated soil may develop salinity or sodicity problem. For some time in the past, the quality of irrigation water used to be evaluated with respect to sodium on the basis of Soluble Sodium Percentage (SSP).

Soils and plants are adversely affected by high sodium in irrigation water. Sodium soils are relatively impermeable to air and water. They are hard when dry, difficult to tilt, and plastic and sticky when wet. These adverse physical conditions retard or prevent germination and are generally unfavourable for plant growth.

Magnesium: Magnesium is the second most abundant anion usually found in high salinity waters. However, in low salinity waters, calcium dominates over magnesium ions. Obviously, it can be stated that with an increase in EC of water, Mg Ca ratio tends to increase.

Magnesium is known to affect plant growth mainly by reducing calcium uptake and causing calcium deficiency. However, when high concentrations of magnesium occur together with high concentrations of calcium, there is no specific ion toxicity effect due to magnesium (Meiri and Shalhavet, 1973).

Chlorides: The occurrence of chloride ions in natural irrigation water increases with an increase in EC and sodium ions. Therefore, these ions are most dominant in very high salinity waters. Unlike sodium ions, neither do the chloride ions have any effect on the physical properties of the soil nor are they absorbed by the soil.

Bicarbonates and Carbonates: Bicarbonates occur in low salinity waters and their concentration usually decreases with an increase in EC. The proportion of bicarbonate ions is higher than calcium ions and have been considered to be undesirable because after evaporation of irrigation water, bicarbonate ions tend to precipitate calcium ions.

Table 7.1

Relative Plant Growth as Affected by Chlorides and Sulphates

Ece (dSm-1)		*Relative Plant Growth (%)*
Chlorides	*Sulphates*	
4	6	100
4-8	6-12	80-90
8-12	12-20	50-70
12-20	20-40	20-50
20	40	0

Silica: Silicon is the more abundant element in igneous rocks than some other types of deposits. Most silica in water is probably derived from the decomposition or metamorphism of silicate minerals rather than from a solution of quartz, as quartz is one of the rock minerals most resistant to attack by water. Many waters contain less than 10 mgL^{-1} of silica; those that drain deposits high in silicate minerals, particularly feldspars, often contain up to 60 mgL^{-1} and are not commonly found. It is believed that most silica is present in a non-ionsed form, but ionized silicate(s) is undoubtedly present in some waters.

Nitrates: Although nitrates are highly soluble, natural waters contain only small quantities, usually less than 10 mgL^{-1}. In Rajasthan and Uttar Pradesh, groundwater in some districts of Rajasthan has been shown in Table 7.2. Legumes take nitrogen from the atmosphere and fix it in the soil as nitrate. Nitrates in water may result from leaching of soils and rocks, from fertilisers, normal decomposition of plants and animals, sewage, industrial effluents, and living organisms. The nitrates occurring in saline waters from 2 to 5 $meqL^{-1}$ have been observed to cause beneficial effect on the plant growth and partly help in the concentration of boron ranges usually from traces to 5 mgL^{-1} in natural groundwaters. However, in very high salinity waters sometimes it can occur up to 10 mgL^{-1} (Gajbhiye et al., 1973). It increases with an increase in EC and SAR of the waters. In sea water, it occurs up to the extent of 4.8 mgL^{-1} In salt affected soils, boron may occur up to 10 mgL^{-1} (Bhargava et al., 1974). It differs from the other constituents of water in that it occurs as unionized boric acid. It is essential to plant growth but may prove exceedingly toxic at concentrations slightly above optimum.

Table 7.2

Distribution (%) of Nitrates in Groundwater of some Districts in Western Rajasthan

NO3 (mg L-1)	*Barmer*	*Jaisalmer*	*Bikaner*	*Average*
<20	18	50	66	45
20-50	8	19	27	18
50-100	63	22	2	28

Source: Gopal et al. and Reddy & Reddy (1995).

Lithium: Lithium is present in some minerals but is not abundant in nature. The more common minerals which contain from 2 to 10 per cent lithium oxide are petalite, spodumene, lepidolite and amblygonite. It is generally found in soils developed from mica minerals and is more frequent in arid regions. Although surface waters generally do not contain lithium more than 0.3 mgL^{-1}, lake brines have been reported to contain Li up to 16 mgL^{-1}.

Fluorine: Unlike chlorides, fluorides are only sparingly soluble and are present in most natural waters in only small amounts. Calcium fluoride (fluorite) is the principal source of fluoride, but there are some other complex fluoride rich areas that are underlain by crystalline rocks at places known to contain fluorspar and apatite, the main sources of fluoride in groundwater. The element is often characteristic of waters from deep strata and is frequently found in salt water from oil wells and in water from areas that have been subjected to recent volcanism. The concentration of fluoride ranges from traces to more than 10 mgL^{-1} in natural waters, although surface waters generally do not exceed 0.3 mgL^{-1} unless they are polluted from other sources.

Other Microtoxicants: Apart from boron, lithium and fluorine, there are some more trace elements, which if present in irrigation waters may also cause a toxic effect. Maximum concentrations of all trace elements in irrigation waters including B, Li and F, as recommended by National Academy of Science and National Academy of Engineering (1972), are shown in Table 7.2. These concentrations are based on the protection of soils for plant production under long continued use of water. Criteria for short-term is also suggested for soils that have high capacities to inactivate these trace elements.

Sodicity Hazard

Sodium Adsorption Ratio (SAR)

Apart from the total salinity, the next important consideration is whether the use of irrigation water of a given quality will cause sodicity in the soil. Water which might be considered suitable for irrigation on the basis of EC, may not be suitable if sodium predominates. In the past, sodic (alkali) soil was defined as one in which the exchangeable sodium percentage was greater than 15 and pHs (pH of the saturated soil paste) values usually ranged between 8.5 and 10. It implies that the quality of irrigation water should be such that the irrigated soil does not exceed ESP value of 15 and pHs value of 8.5 However, now the sodic soil is defined as one in which SAR is greater than 15. Obviously, the irrigation water which does not increase SAR of the soil beyond 15, is considered to be of good quality.

Sodium to Calcium Activity Ratio (SCAR)

The application of SAR to the group of waters which have EC higher than 5 dSm^{-1} and Mg/Ca ratio higher than 1, is obviously questionable. About 35 per cent of the groundwater occurring in 11 arid districts of Rajasthan have EC greater than 5 dSm^{-1} and nearly all waters are dominated by magnesium over calcium. Although some waters have Mg/Ca ratio as high as 16, the majority falls between 2 and 4.

Gupta and Abichandani (1970) examined field condition, the effect of irrigation with high salinity water having Mg/Ca ratio as 5.4 and SAR 8, on a deep sandy to sandy loamy soil, containing $CaCO_2$ from 1 to 2 per cent. Although SAR of the irrigated soil was low and in line with SAR of the irrigation water, ESP values were much higher than expected, than the SAR of irrigation water or irrigated soil.

Alkalinity Hazard

Residual Sodium Carbonate

Eaton (1950) gave the concept of residual sodium carbonate (RSC) and pointed out that development of alkali soils (saline or non-saline) may be expected when irrigation water containing CO_3+HCO_3 higher than Ca+Mg is used for irrigation, provided that it is used so sparingly that little leaching occurs. In water containing high concentrations of biocarbonate ion, there is tendency for calcium and to some extent for magnesium to precipitate as carbonates as the soil solution becomes more concentrated. On the other hand, if sodium is not too greatly in excess of calcium + magnesium and enough of such water is used that the drainage water continues to contain a substantial proportion of Ca and Mg, there may be no ill effects.

Reddy and Raddy suggested new classification of irrigation waters.

In the classification of irrigation waters, it is presumed that the water will be used under average conditions with respect to several factors which influence its quality in relation to its use. Whether the salts dissolved in irrigation water will accumulate in the soil in injurious amounts and affect plant growth, depends on one or more of the several parameters discussed. Large deviations from the average for one or more of the variable may make it unsafe to use what under average conditions would be a good water; or may make it safe to use what under average conditions would be a water of doubtful quality. This relationship to average conditions has therefore, to be kept in consideration with the use of any general method for the classification of irrigation waters.

Salinity (EC)

The classification of irrigation waters with respect to salinity hazard on the basis of EC is based primarily on the development of salinity in the soil to the extent that the yield is affected adversely. On the basis of electrical conductivity the irrigation water may be classified into six salinity classes:

1. *C-0, Non-saline waters* (*EC* < 0.2 *dSm*$^{-1}$) may create a severe permeability problem in the soil because infiltration rate into the soil is adversely affected due to lack of salts in the water to such an extent that the crop is not adequately supplied with water, and yield is reduced.
2. *C-1, Normal water* (*EC 0.2-1.5 dSm*$^{-1}$) can be used for irrigation of most crops and of most soils with little likelihood that soil salinity will develop. Large areas of citrus trees (very sensitive plants) have developed. They have been irrigated for many years with waters having EC 1.5 dSm^{-1}. These waters do not have any leaching requirement. These waters are generally relatively high

in Ca and low in chlorides. There is seldom any problem of high SAR/SCAR or of RSC/RSBC and toxic elements for water falling under this group.

3. *C-2, Low salinity waters (EC 1.5-3.0 dSm^{-1})* can be used if a moderate amount of leaching occurs due to the current irrigation practices. Most of the crops except sensitive ones comprising some horticultural and leguminous plants can be grown on all soils except very heavy textured ones, with impeded drainage. These waters do not have any significant problem of SAR/SCAR and toxic elements but may have some problem due to RSC/RSBC.
4. *C-3, Medium salinity waters (EC 3.0-5.0 dSm^{-1})* can be used of soils provided with good drainage such that leaching fraction is not less than 0.3. Most semi-tolerant and tolerant crops can be grown under good management. These waters may have a moderate problem of high SAR/SCAR and RSC/RSBC but seldom a significant problem of toxic elements. Interestingly, there is evidence to show that optimum yields of many tolerant crops are obtained under irrigation with waters belonging to this EC group.
5. *C-4, High salinity waters (EC 5.0-10.0 dSm^{-1})* are suitable for irrigation of soils provided with very good drainage so that leaching fraction is always greater than 0.3 and ECe does not exceed EC_{iw}. Only salt tolerant crops can be grown under good management. Higher the total and seasonal rainfall during the crop growth period, better the usability. These waters generally may have a significant problem of medium to high SAR/SCAR but no problem due to RSC/RSBC. However, high SAR (>30) values disqualify these waters form satisfactory use for prolonged periods. Summer fallowing is sometimes necessary for effecting desalinisation, and amendments may also have to be used to take care of hazards due to sodium and/or toxic elements.
6. *C-5, Very high salinity waters (EC> 10 dSm^{-1})*. These waters are directly not suitable for irrigation but may be used in cycle or in conjunction with low salinity waters, for example, canal waters.

Sodicity (SAR/SCAR)

Among the soluble salts or irrigation water, sodium is considered to be the most hazardous. The water which might be considered as suitable according to salinity may not be suitable if sodium predominates. The classification of irrigation waters with respect to sodic hazard on the basis of SAR/SCAR is established primarily on the increase of exchangeable sodium and its effect on the physical condition of the soil. Since, ESP of the soil is equal to SAR of the soil in equilibrium in turn with SAR of the water when EC_{iw} and EC_e are less than 5 dSm^{-1}, Mg to Ca ratio is less than one, SAR of the irrigation water directly reveals whether it is excessive for any crop and if it will deteriorate the physical properties of the soil. When EC_{iw} is higher (5-10 dSm^{-1}) and Mg/Ca ratio is also higher (2-4) then SCAR will give a better idea than SAR if ESP is to be depended on for soil appraisal and/or crop performance.

On the basis of SAR/SCAR, the irrigation waters may be classified in six classes:

1. *S-0, Non-sodic waters (SAR/SCAR < 5)* can be used for irrigation of almost all soils, for all crops even those sensitive to sodium such as stone-fruit trees or wood trees.
2. *S-1, Normal waters (SAR/SCAR 5-10)* can be used for irrigation on almost all soils with little danger of development of harmful levels of exchangeable sodium for growing all crops except some stone-fruit trees or wood trees which are specially sensitive to sodium. These waters do not have any requirement for leaching and/or amendments.
3. *S-2, Low sodicity waters (SAR/SCAR 10-20)* can be used for crops which are semi-tolerant or tolerant to sodium on almost all soils so that LF is around 0.3. If soils contain gypsum or calcium carbonate, these waters can be used more successfully. EC should not be less than 1 to 2 dSm^{-1} lest permeability problem develops.
4. *S-3, Medium sodicity waters (SAR/SCAR 20-30)* can be used only for crops which are tolerant to sodium on soils provided with good drainage such that LF is always greater than 0.3. If soils contain gypsum or calcium/carbonate and if rainfall is appreciable and effective, these waters can be used very successfully. EC should not be less than 2 to 3 dSm^{-1} lest permeability problem develops.
5. *S-4, High sodicity waters (SAR/SCAR 30-40).* These waters are directly unsuitable for irrigation without drastic treatment.

Alkalinity (RSC/RSBC)

The classification of irrigation waters with respect to bicarbonate or carbonate alkalinity hazard on the basis of RSC or RSBC is based primarily on the precipitation of CA/or Mg and pairing of residual CO_3 or HCO_3 with sodium and formation of Na_2CO_3 in the soil and increasing SAR/ESP, characterizing it as alkali soil. Since, appreciable amount of carbonates occurs only in the waters that have high pH(>8.5), RSC should be calculated for high pH waters and RSBC for low pH (<8.5) waters. These water, also cause a permeability problem in the soil.

On the basis of RSC/RSBC, the irrigation water may be classified in six classes:

1. *A-0, Non-alkaline waters (RSC/RsBC-ye)* can be used for irrigation on almost all soils, for all crops for indefinitely long periods without any problem.
2. *A-1, Normal water (RSC/RSBC 0 $meqL^{-1}$)* can be used for irrigation on almost all soils for all crops even those very sensitive to carbonates or bicarbonates. These waters may not create any permeability problem.
3. *A-2, Low-alkalinity water (RSC/RSBC tr-2.5 $meqL^{-1}$)* can be used for irrigation on almost all soils for all crops. These waters may not create any permeability problem unless drainage is impeded, rainfall is unduly low and evaporation is very high.
4. *A-3, Medium-alkalinity waters (RSC/RSBC 2.5-5.0 $meq\ L^{-1}$)* can be used for irrigation on almost all soils with little danger of the development of harmful levels of alkali for growing all crops except those which are specifically sensitive to carbonate or bicarbonates. Optimum yields of several alkali tolerant crops are obtained when RSBC is in this range.
5. *A-4, High-alkalinity waters (RSC/RSBC 5.0-10.0 $meqL^{-1}$)* can be used for irrigation on soils provided with good drainage such that LF is not less than 0.3, for growing semi-tolerant and tolerant crops to sodium. EC should be <3.0 dSm^{-1} and SAR<10. Rainfall should be appreciable and effective (>400 mm) and evaporation must be low (<2000 mm) for the prolonged successful utilization of such waters. If SAR is >10, use of gypsum may be required.
6. *A-5, Very high-alkalinity waters (RSC/RSBC> 10.0 $meqL^{-1}$)* These waters are directly not suitable for irrigation but may be used in

cycle or in conjunction with low alkalinity waters or with the use of amendments such as gypsum. High alkalinity with preponderance of carbonates associated with high EC and SAR/SCAR, prohibits the use of these waters for irrigation.

References

Anonymous (1983). Coordinated project on Management of Salt Affected Soils and Use of Saline Water in Agriculture. *Ann. Kep.* RBS College, Agra.

Ayers, R.S. and D.W. Westcot (1976). Water Quality for Agriculture. *Irrigation and Drainage Paper,* 29, FAO, Rome.

Ayers, R.S. and D.W. Westcot (1985). Water Quality for Agriculture. *Irrigation and Drainage Paper,* 29 Rev. 1, FAO, Rome.

Bajwa, M.S., G.S. Hira and N.T. Singh (1983). Effect of sodium and bicarbonate irrigation waters on sodium accumulation and on maize and wheat yields in northern India. Irrig. Sci. 4: 191-99.

Bandyopadhyay, A.K. (1974). Seasonal variation of boron in salinesodic soils. *Ann. Arid Zone* 13:125-28.

Banin, A., D. Shaked and Y. Chen (1972). Measurements of ionic activity in soils and soil extracts by means of ion selectivity electrodes. *Final Rep. Project* 05.8290, Faculty of Agriculture, Hebrew, Univ. Jerusalem, Rehovot.

Bhandari, L.M., A.D., Purohit, T.N. Bhargava and I.C. Gupta (1970). Potability classification of groundwaters in the arid zone of western Rajasthan, *Ann. Arid Zone* 9:221-28.

Bhargava, G.P., S.K. Singhla and I.C. Gupta (1974). Distribution of boron in salt affected and contiguous soils. *Ann. Arid Zone* 13: 32-36.

Bingham, F.T., A. Page and G.R. Bradford (1964). Tolerance of plants to lithium. *Calif. Agric.,* 18:6-7.

Bower, C.A., G. Ogata and Tucker, J.M. (1968). Sodium hazard of irrigation waters as influenced by leaching fraction and by precipitation or solution of calcium carbonate. *Soil Sci.* 106: 29-34.

Bresler, E., B.L., Mc-Neal and D.L. Carter (1982). *Saline and Sodic Soils* Springer-Verlag, New York, p. 238.

Doneen, L.D. (1963). Water Quality for Agriculture, In: *Plants in Saline Environments* A. Poljakoff-Mayber and J. Gale (eds.), Springer-Verlag, New York, pp. 56-76.

Eaton, F.M. (1935). Water quality for irrigated agriculture. In: *Plants in Saline Environments* A. Poljakoff-Mayber and J. Gale (eds.), Springer-Verlag, New York, pp. 56-76.

Eaton, F.M. (1935). Boron in soils and irrigation waters and its effect on plans with particular reference to the San Joaquin Valley of California. *USDA Tech. Bull.* 448.

Eaton, F.M. (1950). Significance of carbonates in irrigation waters. *Soil Sci,* 69: 123-33.

FAO/UNESCO (1967). International Source Book on Irrigation and Drainage of Arid Land in Relation to Salinity and Alkalinity. FAO, Rome.

8

Irrigation Methods

Methods for supply of water to crops are called methods of irrigation. Different methods of irrigation are adopted at different places according to nature of crops. The following factors are kept in mind while selecting any type of method of irrigation:

1. The soil in which irrigation is being done should be able to absorb irrigation water otherwise underground salts would collect over the soil through waterlogging.
2. It is essential that irrigation water must reach up to roots of the plants.
3. There should not be excessive flow of water through any special type of irrigation method.
4. It is also necessary to keep in mind the probable difficulties in use of special types of equipments in the selected irrigation method.
5. How does irrigation method adopted in any special area affect other activities.

The suitable irrigation method should be selected keeping the above facts in the mind. The following factors affect any selected method of irrigation:

1. *Topography:* Topography is the slope of the ground and how much uneven or leveled it is. The irrigation method is selected accordingly. If the slope is from 0.4 to 8 per cent, corrugation method of irrigation is suitable. If the slope is more than the above limit, then sprinkler method has been considered more

suitable for soils which are shallow and permeability is fast. Similarly, leveled surface flow method, check basin method, furrow irrigation method etc., are adopted. Apart from level differences, the structure and composition of soil also affects the selection of irrigation method. Deep soils have medium and fast permeability for which corrugated method of irrigation is suitable. Rate of seepage of water in the soil depends on its composition.

2. *Climate:* Ground level irrigation methods are directly affected by climate. The sprinkler system of irrigation is most suitable for such climate . Dryness, humidity and speed of air affect the methods of irrigation.
3. *Means of Irrigation:* The sources of irrigation and the chemical composition of the water also affect the irrigation method. At the time of irrigation by tubewells, the sprinkler method or drip irrigation method is suitable, but in canal irrigation, flood method is selected.
4. *Crops:* The irrigation method is selected according to type of crops and pattern of its sowing . Different crops require different quantities of water. Growth of plants and their height are also affected. In taller plants, check basin irrigation method is more suitable as compared to sprinkler method.
5. Texture of the soils crusting, cracking and infiltration characteristics of the surface soil; nature and depth of relatively impermeable layer(s) in the sub-soil, if any; the water storage capacity of the potential root zone; the magnitude and nature of land slope; the length and size of the field; surface drainage; the nature and extent of salts in the surface soil and the sub-soil, also decide the method of irrigation.
6. *Water Conservation:* Water conservation is the demand of the day when the whole world is facing water crisis. Hence, such a method of irrigation should be adopted which uses minimum water but provides maximum humidity to the plants. Sprinkler method and drip irrigation method are the best from the view point of water conservation.
7. *Economic Factors and Labour:* While selecting any method of irrigation, the economic condition must be kept in mind as many irrigation methods require heavy initial investment but lesser investment later on. Surface irrigation methods require continuous investment. Along with investment, availability of

labour is also an essential factor. As compared to sprinkler method, check basin method of irrigation requires more labour.

8. *General:* Outlook, management skills and financial resources of the farmer; nature of farm machinery used; availability and cost of labour; maintenance facilities and costs of irrigation equipment; availability of power supply should also be considered.

As far as possible, an irrigation method should not only provide a high level of water application efficiency but also ensure its economic viability, sustained soil productivity and wide adaptability to the prevalent features of the farm.

The irrigation method is selected keeping above factors in mind. Surface irrigation is done through the following methods:

1. *Traditional Irrigation Methods:*
   (i) Check Basin Method
   (ii) Furrow Irrigation Method
   (iii) Strip Irrigation Method
   (iv) Basin Irrigation Method
2. *Modern Irrigation Methods:*
   (v) Sprinkler Irrigation Method
   (vi) Drip Irrigation Method
   (vii) Pot Irrigation Method

Check Basin Method

In this method, the whole field is divided into basins according to the capacity of water. Basins are connected through a *'Dhora'* (A small drain type flow way), which has raised earthen walls on both sides. *'Dhora'* is of two types, one is the main *'Dhora'* and the other *'Dhora'* is connected to basins. Size of basins are made according to the inflow of water. These basins are surrounded by small furrows. Branch 'Dhora' flows towards the slope from the main 'Dhora'. If the slope of branch *'Dhora'* is steep, 'mooonja' or polythene is spread in it to prevent erosion of sides. The main source of water is located at the highest place in the field. The width of drains is affected by factors like flow of water, percentage, slope and structure of the ground etc. The length of *'Dhora'* is different depending on the basis of slope and formation of the fields.

This method is also prevalent in India as it does not cause any burden on the farmer. This method has the following advantages:

1. It is the best method of irrigation for levelled fields.

Figure 8.1
Check Basin Method

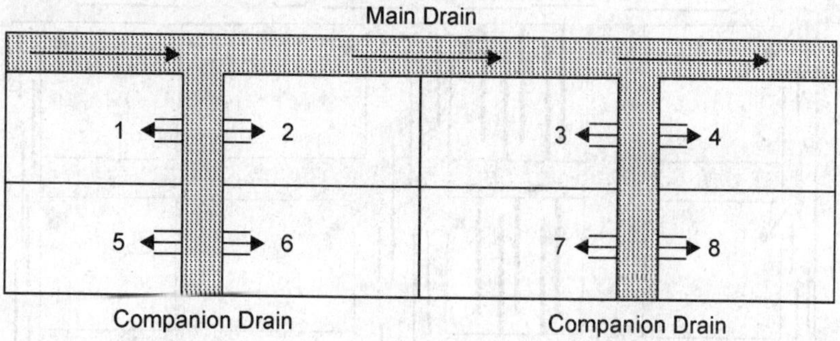

2. It does not require any technical knowledge.
3. This method is more useful in soils having lesser infiltration.
4. In this method, rain water stays in basins, hence soil erosion is not caused.
5. It has lesser economic investment.
6. It irrigates more area.
7. Crops gets sufficient water.

Following are disadvantages of check basin method:

1. Due to seepage in drains, wastage of water is caused.
2. Machines cannot be used in this method because during spray of insecticides or fertilizers, the earthen walls of basins are damaged.
3. There is imbalance in distribution of labour. After growth of crops, water reaches the basins in disproportionate quantity thereby causing wastage of water.
4. Creation of problem of waterlogging.

Furrow Irrigation Method

Furrow irrigation method is resorted to where crops are one grown in rows. Along the side of rows of crops, *'Dol'* is formed, and in between two such *'Dols'*, a furrow is formed in which water flows for irrigation. The quantity of flow of water depends on demand of water by plants and the rate of infiltration. In different situations, different

Figure 8.2

Furrow Irrigation Method

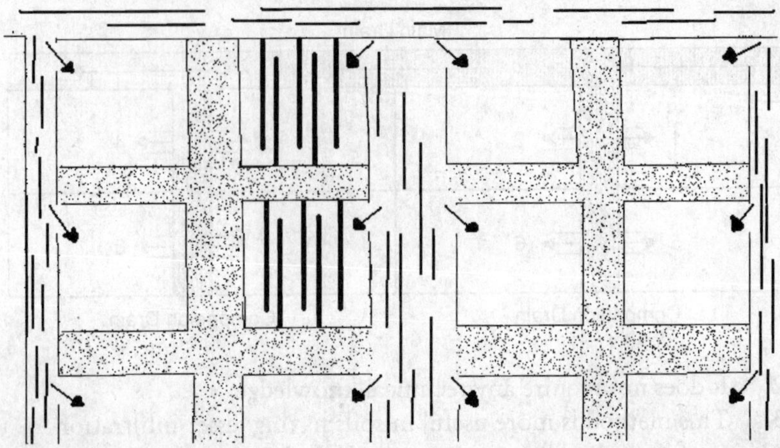

furrow methods are used (Surajbhan 1978). They are mainly of five types:

1. Slopy Furrow
2. Levelled Furrow
3. Contour Furrow
4. Serial Furrow
5. Corrugated Furrow

Advantages

1. Large areas can be irrigated at a time.
2. It saves labour since once the furrow is filled, it is not necessary to give water a second time.
3. It is a comparatively cheaper method.
4. Plants gets suitable quantity of water by this method.

Disadvantages

1. Due to imbalance in flow of water, wastage of water is caused in it.
2. It is not suitable in all types of crops.
3. Making *'Dol'* for drains requires more labour information.
4. Due to filling of excess water, there is risk of underground salts coming up to the surface layer.

Strip Irrigation Method

In strip irrigation method, fields are divided into strips of different size. A boundary called *'Med'* is formed to separate the strips. These strips are constructed according to the slope. The source of water is situated at the highest place in the field from where the whole field can get the flow of water. The width of strips is decided as per quantity of water. More wastage of water is caused if strips are wider. Length of strip is decided by the slope of land and its structure. Effect of soil composition is also visible on it.

Advantages

1. It is possible to irrigate more area at a lesser expenditure.
2. It requires less labour.
3. Method of irrigation is easy and it causes lesser erosion.

Disadvantages

1. It is not suitable for all types of crops.
2. It is not possible to get balanced supply of water.
3. It is not suitable for all soil compositions.

Basin Irrigation Method

This irrigation method is more suited for horticulture development. In this method, a raised platform called *'Thanvla'* is formed around trees or bushes and they are connected with each other through drains

Figure 8.3
Basin Irrigation Method

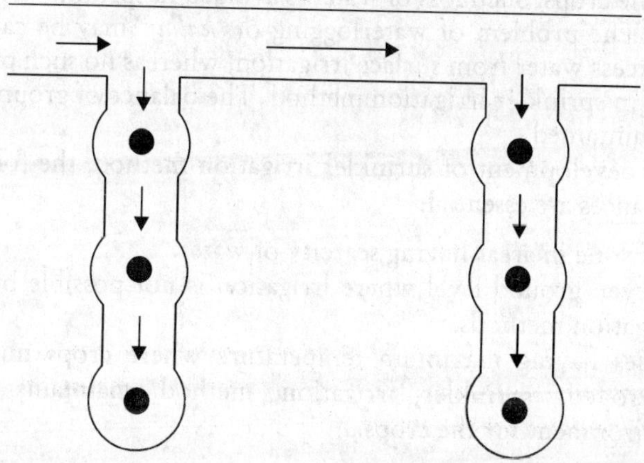

and the water reaches from one tree to the other. This method is not suitable for crops.

Advantages

1. It saves time. Once the water is opened, it reaches other trees automatically.
2. Its economic investment is less.
3. It is beneficial for more trees.

Disadvantages

1. It is not useful for all crops.
2. Wastage of water is caused in it.
3. Diseases spread in trees.

Sprinkler Irrigation Method

In present times, when water crisis is developing very fast every where, we should adopt improved techniques of irrigation to encourage suitable water management. Sprinkler irrigation method is an easy and simple method of irrigation in present times. The whole land becomes available for cultivation of crops, whereas in traditional irrigation methods, 15 to 20 per cent land remains vacant in depressions and boundaries. Modern equipments can also be used in it due to absence of depressions and boundaries. Rate of infiltration is higher in sandy soils where frequency of watering is more. Hence, sprinkler irrigation method is more suited to sandy soils. In sprinkler irrigation method, water is taken from source to the fields through pipes, whereas in surface irrigation methods only 30-45 per cent water reaches the crops. Such loss of water is avoided in sprinkler irrigation method. The problem of waterlogging or '*kallar*' may be caused in case of excess water from surface irrigation, whereas no such problem is caused in sprinkler irrigation method. The balance of groundwater is also maintained.

For development of sprinkler irrigation method, the following circumstances are essential:

1. It is done in areas having scarcity of water.
2. Uneven ground level where irrigation is not possible by other irrigation methods.
3. Places having maximum temperature where crops might get destroyed, sprinkler irrigation method maintains humid environment for the crops.

Figure 8.4

Sprinkler Irrigation Method

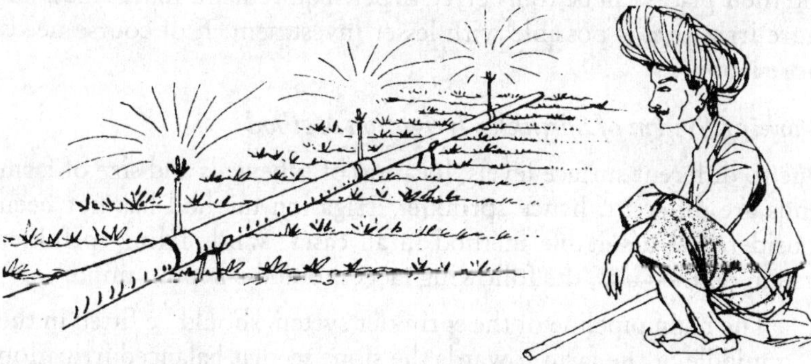

4. Where soil texture may be of different nature, for example, sandy soil at some places and stony soil at others places.
5. It requires lesser number of labourers hence, it can be developed even where there are less workers.
6. Irrigation may be required in large areas.
7. There should be average technical knowledge.

In areas where change in temperature of earth, environment and humidity is required for growth of crops, sprinkler irrigation method is possible to a certain extent. Due to continuous spray of water, there is improvement in physical conditions of earth and composition of soil. In *kallar* or *reh* soils, land can be improved by sprinkler irrigation, whereas surface irrigation needs much more water for it. Thus, it is a suitable irrigation method for sustainable development of water resources in present times. It is installed in fields by three methods:

1. *Permanent:* In this method, the main line and branch pipelines are permanently installed in the field. After that it is not possible to shift its place. In this system, labour involved in shifting of lines from time to time is saved but it is quite expensive. The pipeline remains safe being underground, whereas in case it is outside, the breakages are more by frequent changing. Such an arrangement is suitable for canal irrigated areas having *'barabandi'* where water becomes available for a very limited time.

2. *Semi-permanent:* In this method, the main pipeline is permanently fixed under the ground level but branch pipelines are kept

outside temporarily so that by changing their places, the whole of the land might be irrigated.

3. *Temporary:* In this case, the whole arrangement is temporary and their places can be transferred as per requirement. In this method, more irrigation is possible with lesser investment. It of course needs more labour.

Working System of Sprinkler Irrigation Method

Due to different surface levels, location of tubewells and size of farm lands are different, hence sprinkler irrigation method has not been considered as a suitable method in all cases. While fixing sprinkler system on the farm, the following facts should be kept in mind:

1. The main pipeline of the sprinkler system should be fitted in the middle of the farm towards the slope so that balanced irrigation is done on both the sides.
2. Branch pipeline should be fitted across the main line inversely to the slope.
3. Direction of the air should also be kept in view and it should be fitted at 90° of the direction of the air but in no case it should be lesser than 45°.
4. Irrigation of crops need to be done according to time hence, crops requiring water at similar time should be sown near each other, so that extra labour is saved.
5. Even after changing the branch lines, the number of nozzles should remain same.

In sprinkler system, parts needed are pipe, nozzle, risen, coupler, bend, reducer, foot baton and dot etc. While erecting this equipment in the farm, there are two types of expenditure, i.e., initial investment, and expenditure on maintenance of the set. Initial investment depends on location of the tubewell and area of the field. Maintenance is more beneficial in case of tubewell as compared to well and an engine. Initial expenditure is 55-60 per cent whereas maintenance expenditure ranges between 40-45 per cent.

The sprinkler set is kept according to the nature of the soil of the farm and on such basis is decided duration between one and the other irrigation. In sandy soil, sprinkler set is kept at one place for 3-4 hours and duration of 8-10 days is kept between the two irrigations.

As per soil, the time of keeping sprinkler at one place and duration between two irrigations should be as under:

Table 8.1
Time of Keeping Sprinkler According to Nature of Soil

Nature of soil	*Time of keeping Sprinkler Set at one place (Hours)*	*Duration between two irrigations (Days)*	*Rate of soaking by soil (MM/Hour)*
Sandy Soil	2 to 3 Hours	8 to 10	25 to 30
Sandy Loam	3 to 3.5 Hours	10 to 12	20 to 30
Loam	3.5 to 4.5 Hours	12 to 15	15 to 20
Clay Soil	8 to 12 Hours	15 to 18	Less than 8

Advantages

1. There is increase in production and compactness.
2. It is helpful in soil conservation and stabilization of sand dunes in desert areas.
3. Sprinkler system is considered more suitable in areas where slit is coagulated on surface of soil after rains, prevents growth of crop.
4. This system saves the crop from extreme frost or temperature.
5. Fertilizer application as well as insecticide spray can be done by sprinkler system.
6. Waste land can be improved by less water. Physical condition and composition of soil can be maintained in a balanced condition by continuous sprinkling.

Disadvantages/Defects

1. Sprinkler irrigation method is expensive.
2. It requires technical knowledge.
3. Sprinkler irrigation method cannot be used in all crops.
4. Crop is damaged by changing sprinkler system again and again.
5. Water to be used in sprinkler method should be clean.

In spite of the above defects, sprinkler irrigation method is being adopted with great speed due to increasing water crisis.

Drip Irrigation

A newly developed irrigation system known as drip irrigation or trickle irrigation, originally developed in Israel, is becoming popular in areas of water scarcity. In this irrigation system, a small amount of water is applied at frequent intervals in the form of water droplets through perforations in plastic pipes or through nozzles attached to tubes spread over the soil to irrigate a limited area around the plant. A precise amount of water equal to the daily consumptive use or the depleted soil water needs to be applied. The soil water can be maintained at the field capacity during the crop growing period. Deep percolation losses can be completely prevented and the evaporation loss is also reduced.

The application of water and piping systems needs to be designed according to the type of crops, topography and weather conditions typical to the geographical area.

The basic equipment for drip irrigation consists of a water supply head, a main pipe, lateral pipes and drippers. The water flow in the pipe system is controlled with control valves and fertilizers can be applied at the water source. As water passes through the very small outlets of drippers, it is filtered before it is distributed in the pipe system.

Structure of Drip Irrigation Method

The following are the main organs of drip irrigation method: water pump, main PVC, pipeline, branch PVC pipeline connected to main line, plastic pipes connected to branchline, drippers connected with plastic pipes, fertilizer tank for application of fertilizers, valve, water measure, pressure controller, filter etc. Internal radius of side pipe is from 10 to 32 mm. Side pipelines are fitted with drippers from where

Figure 8.5
Drip Irrigation Method

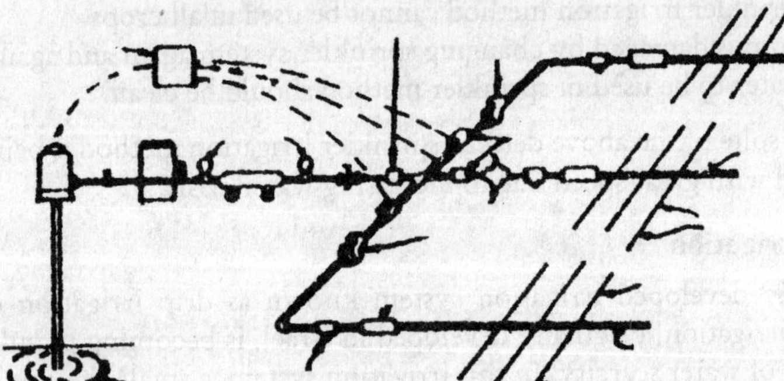

water falls in drops. Efficiency of drip irrigation method depends on suitable operation of drippers. Flow rate of drippers is 2 to 10 litres per hour.

This system of irrigation is established on the basis of type of crop, distance between plants, requirement of water for crops and distance of water source from the field.

Advantages

1. In this method, water directly reaches the roots of the plants, which take water to plants in balanced quantities.
2. Drip irrigation method saves 30 to 70 per cent water and it is possible to irrigate three times more area with the same amount of water.
3. In this method, weeds do not spread because water reaches only near plants and does not spread in the whole field.
4. Fertilizers and insecticides can also reach the plant directly by solution in the water and it saves 30 to 60 per cent chemical fertilizers as well as 40 to 50 per cent pesticides along with saving of water.
5. Even in case of uneven lands, drip irrigation method can do balanced irrigation.
6. Cultivation in saline and alkaline soil also can be done by this method of irrigation.
7. Crop production is higher by 20 to 40 per cent in drip irrigation method, because plants can get air and water in required quantities, resulting in regular growth of crops.
8. Lesser labourers are required for irrigation work.
9. Bacteria causing diseases in crops do not grow because of dry atmosphere near plants.

Disadvantages/Defects

1. Drip irrigation method is expensive.
2. It requires special technical knowledge for successful operation of this method.
3. In heavy soils, it creates problems of flow and water blockages.
4. Plants are able to get nutritive elements in a very limited area.
5. It is not suitable for every crop.
6. Utmost care has to be taken for holes of drippers, because soil may come along with water at any time, which will prevent water dripping smoothing from holes.

7. Animals may cause damage to branch pipelines and dripper pipelines.
8. Most of the drippers work on pressure. Wherever land is sloppy, pressure on valves increases by 50 to 10 per cent, which results in stoppage of working of valves on the upper side.

Pot Irrigation Method

Pot irrigation method is more suitable for areas having scanty rainfall. In saline areas where flow irrigation is not suited, pot irrigation method is successful. An earthen pitcher is used in this method. The pitcher is fixed in the ground up to neck. Holes are made in the pitcher and water is filled in it so that seepage of water through the holes keeps the nearby soil moist. Water is filled in these pitchers at regular intervals. This method can be considered as an alternative of drip irrigation method. Pot irrigation method can be adopted in the following conditions:

1. Unlevelled land which is uneven.
2. Area having maximum shortage of water.
3. Such difficult areas where supply of fruits and vegetables is difficult and they are costly.
4. Where there is saline water, making surface irrigation difficult.

In this method, distribution of humidity around sides of pitcher is affected by many factors, mainly size of the pitcher, seepage of water per unit of area and type of soil. Humidity is spread in the same proportion as the size of the pitcher. Distance of pitcher also affects the moist area. Normally, distance between two pitchers should be kept so much that the humid area between them does not overlap.

Advantages

1. In this method, only the area near the pot gets irrigated and not the whole area.
2. Evaporation of water is minimum in this method.
3. Water seepage below the ground is also in minimum quantity.
4. It is the best method for horticulture crops and vegetables.
5. Once the pitchers are fixed, irrigation can be done for six years, which reduces expenditure.
6. It needs minimum technical knowledge.

Defects

1. Irrigation in this method is possible in a limited area.

2. This method requires clean water because unclean water would cause blockage of minor holes, which would not be able to provide moisture any longer.
3. It is costly to draw out pitchers again and again and re-fix them.
4. It is not suitable for every crop.

New Methods of Watering

Some new methods are under trial in areas where water is scarce and costly, and water saving and economic returns are the main consideration. These methods are also found to be beneficial in forestry or fruit plantations, particularly in those drought-prone areas where survival and initial establishment is very poor. In these systems, there is practically no loss of water due to percolation and evaporation.

Sub-soil Irrigation

Sub-soil irrigation involves watering in the root zone up to about 22.5 cm below the ground. It results in huge savings on water, fertilizers and pesticides, and improvement in growth and yield has been clearly

Figure 8.6
Sub-soil Injector

evident. The instrument used for the purposes is known as a subsoil injector. It is an enlarged version of a medical syringe, operated by foot. It is fed from a 20-litre cane carried either on the operator's back or on a vehicle. It is easy to use. The needle is placed near the plant or seedling at an angle of about 45° and pushed into the ground in one movement. The pedal is then pressed with the foot. This operates a piston that forces out up to a litre of water at a depth of 22.5 cms, the ideal depth for optimum utilization (Figure 8.6).

In this system, evaporation losses are completely checked. It encourages deep roots, which act as a safeguard against subsequent droughts. This can also be used for injecting pesticides, fungicides and fertilizers. Besides, weeds are reduced as the surface is not wetted. This system is becoming increasingly popular for fruits and forestry plantation.

Adverse Effects of Irrigation

As growth and development of plants is prevented due to scarcity of water, similarly excessive water also has an adverse effect on growth of plants. Due to excess water in soil, voids are filled with water or water collects on the surface of soil. In such situation, balance of air and water is disturbed in pores and in the absence of air, plants are not able to breathe etc. Hence, neither irrigation should be more nor less but it should be as per requirement because excessive irrigation is dangerous. Excessive and irregular irrigation has the following effects:

1. *Increase in Saline and Alkaline Elements in Soil or Increase in Salinity:* When irrigation is done regularly with excessive water, groundwater level comes up and quantity of solvent salts increases. Due to evaporation, salts from saline groundwater collect on the surface of the soil. Seepage of canal water also results in more salts in irrigation water. As observed in Rajasthan, concentration of soluble salts salts is so much in these soils that it has an adverse effect on the growth of crops and the production is lesser. When concentration of soluble salts in the soil is more than 0.1, growth of plants stops.
2. *Problem of Waterlogging:* Excessive irrigation also causes problem of waterlogging as has happened in Indira Gandhi Canal area in Rajasthan. Soils and ground level of Indira Gandhi Canal area is very sensitive from the angle of irrigation as maximum soil there has bold granules and at a little depth below

the surface, there are hard rocks of calcium carbonate which do not allow excess water to infiltrate underground but keeps it filled above. Hence, the problem of waterlogging occurs there. Due to maximum dampness in Naurangdesar, Silvana, Lakhuwali and other areas near Indira Gandhi Main Canal, agriculture activity has almost completely stopped. Pilibanga and Badopal have also similar problem. Thus, 44,625 hectare land in the north of the canal has become waterlogged.

3. *Hindrance in Air Communication:* Excessive water replaces air in pores of the soil. Hence, roots of plants and beneficial organisms do not get sufficient air for breathing and living. Hence, growth of plants is stopped and diseases destroy crops.
4. *Reduction in Temperature of Soil:* Excess collected water becomes cold in the process of evaporation and it causes reduction in temperature of the soil. Seeds do not settle properly and growth of plants is affected. Activeness of organisms is also reduced.
5. *Shallowness of Roots:* Due to excess water and shortage of air, roots do not go deep and they remain on the upper surface. Plants are not able to get nutritive elements in sufficient quantity and shallow roots, because of weakness, are neither able to develop nor are they able to bear swift winds. Due to this the plants fall on the ground.
6. *Marshy Land:* Land becomes marshy due to stagnant water for a long duration and only forest grass can grow there. The land does not remain useful for agriculture.
7. *More Nitrate Formation:* Due to absence of oxygen in water-logged soil, process of de-nitrogenization starts. Organisms themselves start drawing available oxygen in nitrate, which convert nitrate into nitrus oxide, nitric oxide and nitrogen gases. Being in gas form, they are useless for plants. Such problems will not exist by balanced irrigation.
8. *Shortage of Soil Nutritive Elements and Decrease in Productivity:* Due to excessive irrigation, nutritive elements of soil flow to sub-soil, resulting in decrease in soil productivity and deficiency in crops.
9. *Acidity of Soil:* In excessively irrigated areas, H^+ replaces positively charged Ca^{2+}, Mg^{2+}, K^+, Na^+ stick to colloid granules of soil. These positively charged particles flow away to lower levels which cause decrease of basic positively charged particles

and there is an excess of colloidal granules. This results in excess of H^+ in soil water and soil becomes acidic.

10. Deformation comes in soil structure.
11. Attack of diseases and harmful insects.
12. It encourages the process of leaching.
13. Collection of other harmful salts and alkaline takes place on surface of the soil.
14. There is decrease in activities of living organisms.

Methods of Preventing Harmful Effects from Excessive Irrigation

1. Proper drainage of excess water.
2. Adopt new irrigation methods so that the soil is able to get limited water.
3. Implement water improvement management zone plan so that limited water is given, as is being done in areas of Naurangdesar, Silvana, Lakhuwali, Badopal and Pilibanga in Western Rajasthan.
4. Irrigation should be done according to crops. Different crops need different quantities of water. Hence, irrigation should be done according to the demand of the crop for water so that excess water does not collect there.
5. If the land of irrigated area is not levelled, it should be done so that equal distribution of water is possible.
6. Level of irrigated area should be lower than the source of water.
7. Right type of crops should be selected. In places where there is, irrigation management and rainfall is also more, such crops should be grown which need more water so that even after use of excess water, crops are prevented from harmful effects.
8. Irrigation should be done according to time. Different crops require different quantities of water that depend on irrigation at the time of critical stage of the plants. This depends on nature of crop, duration of maturity and climate.
9. Time of irrigation should be decided on the basis of climate.
10. Time of irrigation should be decided on the basis of soil water.
11. Irrigation may be done according to nature of soil. Water storage capacity of soil and availability of water depends on its physical qualities like soil granule composition etc.
12. Light sandy soil soaks more water as compared to heavy soils.
13. Soil humidity also affects quantity of water used by plants.

Essentials of Efficient Management of Irrigation Water

1. Proper repair/shaping and maintenance of delivery system before release of water.
2. Release of water in accordance with the requirement of cropping pattern/crops grown in the command.
3. When water is in short supply its release should be in accordance with the critical stages of major crops of the command.
4. Land levelling/shaping/grading should be done.
5. Adoption of efficient cropping pattern having crops of low water requirement and higher economic value.
6. Enforcement of *barabandi*.
7. Popularization of efficient irrigation systems like
   (i) Proper bed size of proper cut off ratio in flow irrigation system for various stream sizes in different types of soil
   (ii) Adoption of ridge and furrow methods of irrigation wherever feasible/possible.
   (iii) Adoption of sprinkler, drip and sub-surface irrigation wherever economically feasible/possible.
8. Over-irrigation is always detrimental. Hence, it should be essentially avoided.

9

Problems of Water Quality

Salinity and Alkalinity

Production of salts is caused by the process of weathering of original matter in the soil. Soil becomes saline by transfer of salts from underground water or from the solution of soil which comes from the lower layer to the upper layer. Thus, salinity develops from redistribution or transfer of salts in soil itself. In India, saline soils are found in an area of 71 lakh hectares, out of which 7.28 lakh hectares lie in Rajasthan. Moreover, such types of soil are mostly found in dry areas, because suitable arrangement for exit of water does not exist in this type of soil and evaporation is also more, due to which soluble salts collect in the form of a layer on the surface of the soil. The main salts are Sodium, Calcium, Magnesium, Chloride of Potassium, Sulphate, Carbonate, Bicarbonate etc. According to a salinity laboratory of United States of America, saline soils are of the following types:

1. *Saline Soil:* Soluble salts of Sodium, Calcium, Magnesium, Potassium of Chloride and Sulphate collect in saline soils, which cover the land in the form of a white layer. Gedroij has called it Solon-Chalk, which is also called as *reh*, *kallar* or barren land. Saline soil means such soil in which electro conductivity of saturation extract of soil is more than 4 desisimens per metre (ds/m), exchangeable sodium is less than 15 per cent and pH value of soil is less than 8.5.
2. *Non-saline Sodic Soils:* Soils in which electro conductivity of saturation extract of soil is less than 4 desisimens per metre

(ds/m), exchangeable sodium is more than 15 per cent and pH value of soil is more than 8.5 are called non-saline sodic soils. Due to sufficient sodium salt in these soils, quantity of exchangeable sodium is also more, as a result of which the physical condition of these soils is not good. There is lesser communication of water as well as air in non-saline soils. In the sub-surface soil, a layer of hard pebbles collects due to which water cannot percolate below it and the land becomes sticky. Some soil scientists have named it as black barren land and solonetz.

3. *Saline Sodic Soils:* Duc to more sodium salts in these lands, quantity of exchangeable sodium is also more, as a result of which physical condition of these soils becomes deformed and communication of water and air becomes lesser. It is also called barren land. In its main features, electro conductivity of saturated extract of soil is less than 4 ds/m, exchangeable sodium is more than 15 per cent and pH value of soil is more than 8.5. Such classification can be clarified as under:

Class	*Electro Conductivity at 250°C Temperature*	*Exchangeable Sodium*	*pH Value of Soil*
Saline Soil	More than 4	Less than 15	Less than 8.5
Saline Sodic Soil	More than 4	More than 15	More than 8.5
Non-Saline Sodic Soil	Less than 4	More than 15	More than 8.5

Effects

When soil is affected by salts, its productive capacity decreases and plants sown therein cannot develop in a good condition. In saline soils, sprouting of seeds does not take place in a proper way while raising crops. When land becomes saline, water and organisms from capillaries come back in the soluble soil, hence plants start withering and dying. The roots of plants are also not able to absorb water and nutritive elements in the required quantity. Even though humidity exists but plants feel absence of humidity.

Management of Saline and Alkaline Soils

Saline and alkaline soils are the main hindrance in water management. Without their management, proper water management cannot take a scientific form, whereas agriculture is strongly dependent on them. The following methods are adopted for their management:

1. Physical Measures
2. Chemical Measures
3. Soil Management Measures

Physical Measures

1. *Leaching:* When soil elements after dissolving with water go down from upper parts to the lower level, it is called leaching. In this method, saline elements of the soil are transferred to lower levels after dissolving with water. This water, along with the supply of water to crops, takes away salts after dissolving them. That is why more water is required in this method. This process is mostly adopted in dry season. The fields are divided into small fragments through bunding on the boundary so that water is conserved there, reducing the effect of salts.
2. *Bringing Salts out of Soil after Dissolving them:* Sometimes a layer of salts collects on the surface of the soil. Such salts are dissolved by spreading water and taken out. Though in this method there is a danger of loss of productive soil on the surface of soil also going out, but quantity of salts is reduced to a great extent and soil becomes fit for cultivation.
3. *Scratching:* Many times the salts are collected in a large quantity and it looks like a heap. Then it has to be scratched out from the field to reduce salinity of the soil. This method not practicable on a large scale.
4. *Water Discharge:* Discharge of additional water in excess of the required quantity in any field is called water discharge or exit. With discharge of additional water, concentration of salts does not take place. Drains may be constructed in the field so that excess water keeps on discharging.

Table 9.1

Extension of Waterlogged and Salinity Affected Areas in India

(in lakh hectates)

State	*Waterlogged*	*Salinity Affected Area*
Andhra Pradesh	2.66	3.70
Bihar and Jharkhand	3.63	3.20
Gujarat	0.89	12.10
Haryana	2.30	4.60
Karnataka	0.25	4.00
Kerala	0.12	-
Madhya Pradesh and Chhatisgarh	0.04	2.40
Maharashtra	0.60	6.0
Orissa	1.96	4.00
Punjab	2.00	5.20*
Rajasthan	1.80	9.90
Tamil Nadu	0.02	5.40**
Uttar Pradesh and Uttaranchal	0.35	13.0

*Alkaline soil

Source: Gupta and Tyagi (1991) and Sondhi, S.K.(1998).

Chemical Measures

In alkaline soils, exchangeable sodium exists in different quantities, hence leaching of sodium should be done. Elements assisting in removal of exchangeable sodium from surface of soil can be divided in three categories:

1. Solvent Calcium Salt: Gypsum and Phospho Gypsum
2. Little Soluble Calcium Salt: Lime Stone
3. Acid and Acid Producing: Sulphur, Acid of Sulphur, Pyrites etc.

The above reformers are helpful in improving such soils but some are not available at all places. Gypsum and Iron Pyrite are comparatively costly but they are easily available. Gypsum should be mixed up to 10 or 15 cm. depth. Calcium ion found in gypsum displaces exchangeable sodium ion and converts sodium chloride into calcium. Iron Pyrite contains 22 per cent sulphur. Due to sufficient humidity in soil, its oxidization takes place.

Soil Management Measures

Besides physical and chemical measures, soil improvement can also be done by soil management for which traditional knowledge of farmers is more important. If regular management of these soils is not done, they may again convert into saline soils. The following methods are important among them:

1. *Preparation of the Field:* Maintenance of fields is important for soil management. Fields should be kept levelled so that any soil improvement means can be easily adopted. Water distribution during irrigation is balanced in levelled field which prevents concentration of salts.
2. *Method of Preparation of Basin and Sowing:* In saline and alkaline soils, due to effect of salts and alkaline on seeds, sprouting is not easily possible. Hence, it is advised that sowing of crop should be done on the slope by following furrow and basin method. In basin irrigation, the salt collects near the basin, the adverse effect of which is visible on plants. Special attention should be given to irrigation.
3. *Make Crop Rotation Effective:* These soils should not be kept fallow but cultivation should be continuously done according to crop rotation. Crops should be sown which can bear salts, for example mainly beet, potato, barley, wheat, cotton etc.
4. *Irrigation:* Salinity affects sprouting of seeds. Hence, during first irrigation, more quantity of water should be supplied to neutralize the effect of salts. Thereafter, even during subsequent irrigation, along with reducing the quantity of water, even duration between two irrigations should be reduced so that soil does not get dry and dissolution of salts continues in the humid soil.
5. *Bring Reduction in Evaporation:* Saline and alkaline soils must be kept moist so that salts may not collect over the soil surface. Hence, prevention of evaporation is very essential. The cover of vegetation and grass etc. should be kept over the soil. Duration between crops should also be kept comparatively shorter.
6. *Use of Carbonic Material:* Biotic and carbonic materials, along with bringing improvement in physical qualities of soil, also generate carbon dioxide during their breaking up , which forms carbonic acid on contact with water. This reduces the quantity

of salinity. Use of green manure, cow dung manure, compost etc. also reduce salinity.

7. *Use of Fertilizers:* Use of fertilizers in somewhat greater quantity than normal in saline and alkaline soils is beneficial. Normally deficit of zinc, iron, manganese, and nitrogen etc. elements is found in such soils, hence productivity can be increased by use of these elements.
8. *Growing of Crops Bearing Salts and Alkaline:* In these soils, such crops may be sown which can bear salts. Such crops are mainly beet, rice, dhencha, mustard. Other normal crops are linseed, jowar, barley, jai, spinach, wheat, sugarcane, bajra, cotton, barseem, chari and rijka etc.

Overexploitation of Groundwater

Water crisis has been caused due to growing population and increase in demand of foodgrains and also subsequent Green Revolution in the late sixties. After 1970, groundwater has been exploited very fast by digging wells in every hectare and drawing out water by huge machines. It has definitely been a sorrowful and unscientific event. On the other hand, recharging of groundwater by rainfall has not been proportionate to its speed of exploitation, because quantity and duration of rainfall has continuously reduced due to deteriorating ecology. In the absence of forest cover, rain water drains away or it evaporates and vanishes, whereas previously groundwater was recharged more by blockage of rain water in the roots of trees. Dry leaves and branches of trees made a sponge on the surface of the earth which increased the capacity of the earth to absorb more water.At present it is not so possible due to destruction of forests.

A large part of the country has become prey to dark zone from the point of view of groundwater due to over exploitation, because, as compared to recharge of groundwater, its exploitation is on the higher side. To assess the development level of groundwater reservoir, the following accepted formula is used:

$$\text{Level of Groundwater Development} = \frac{\text{Annual Net Groundwater Use}}{\text{Net Groundwater Reservoir Available for Irrigation}} \times 100$$

When the quantity of groundwater available for irrigation is 85 per cent of total (annual) groundwater reservoir, it is called Grey

Zone. To assess the rate of groundwater development, average of total increase in use of groundwater during last 5 to 10 years is found out. This average is called rate of groundwater development.

By knowing the level of groundwater development, the possibility of quantity and direction of groundwater development for future can be assessed. To be saved from the stage of over exploitation of groundwater, it is divided into four categories on the basis of exploitation of groundwater:

Table 9.2
Level of Groundwater Development

Level of Groundwater Development	*Area*
Up to 65 per cent	White Area
65 to 85 per cent	Grey Area
85 to 100 per cent	Dark Area
More than 100 per cent	Overexploited Area

According to the above classification, areas where level of groundwater development is more than 85 per cent, groundwater exploitation should be completely banned. Similarly, in Grey Areas , groundwater use should be balanced. The object of balancing means that the stage of 85 per cent level of groundwater development should not reach through planned development. '*White Zones*' are those areas where the rate of groundwater exploitation is less than 65 per cent and groundwater is available in sufficient quantity. In '*Grey Zone*', the rate of groundwater exploitation ranges between 65 to 85 per cent and it is in transit zone. Areas where groundwater exploitation is more than 85 per cent, are '*Dark Zones*' where groundwater exploitation is playing with nature, because in a few days existing humidity of the earth would finish and the whole biomass would also reach at the brink of extinction.

Causes of Groundwater Depletion

Due to green revolution and rural electrification, construction of wells and tubewells was taken up at a very fast rate, resulting in cultivation of irrigation based crops. Thus began the uncontrolled over exploitation of groundwater, which has now reached the stage of crossing the limit of groundwater recharge capacity. Due to unlimited expansion of urban areas, increasing population and consequent increase in per capita consumption of water, overexploitation of groundwater has been resorted to in many areas. Industrial pollution has resulted in deterioration in quality of groundwater in many areas.

Main reasons for quantitative and qualitative decrease of groundwater are as follow:[1]

1. Scanty rainfall and dry climate in many areas.
2. Lack of surface water in most of the areas.
3. Problem of drinking water supply due to scattered hamlets and villages.
4. Continuous decrease in groundwater resources and groundwater level.
5. Decrease in water yielding capacity of tubewells and wells.
6. Deterioration in quality of falling groundwater level.
7. More consumption of electricity in exploitation of groundwater because of deep groundwater level.
8. Problem of waterlogging in canal irrigated areas.
9. More salinity in groundwater in various places.
10. Health-related problems due to excess fluoride in many areas.
11. Groundwater pollution due to urbanization and industrialization.
12. Challenge of fresh water supply for future generations.

Groundwater Conservation

Groundwater conservation can be done in many areas by conscious use of groundwater:

Agriculture Sector

1. Encourage sprinkler or drip irrigation method to save water in suitable quantity.
2. Grow such crops that require less water. Its knowledge should reach cultivators in the field. This can be done by units of agriculture department including agriculture extension officers and village assistants etc. They can organize demonstrations etc. for publicizing use of suitable quantity of chemical fertilizers for agriculture production and for adopting suitable but economical methods of irrigation.
3. Use of chemical fertilizers and pesticides in suitable quantities so that pure groundwater is saved from pollution.

4. Encourage installation of water pumps operated by solar energy and wind energy for groundwater exploitation.
5. Motivate cooperative irrigation so as to save water and electricity.

Industrial Sector

1. It is necessary for all industries to recycle 80 per cent of used water for re-use.
2. It should be compulsory for all industries to install joint pollution control equipment at a suitable place for effective prevention of groundwater pollution.

Domestic Sector

1. Water disposed after domestic use should be utilized for gardens etc.
2. Prevent uselessly flowing water from taps.
3. Use small size vessels for cooking and use water in reasonable quantity. Keep cooking vessels closed while cooking so as to save water loss by evaporation.
4. Impose control on cutting of trees and shrubs for cooking purposes so that average annual rainfall can increase and simultaneously soil conservation can also be done.
5. Make arrangement for rain water conservation in every house so that pressure is reduced for water used for domestic purposes.
6. Unnecessary flowing of water from public taps be should prevented.
7. Water should not be wasted through showers in houses and hotels.
8. In domestic uses like bathing, washing clothes etc. water should not flow on surface of nearby wells.
9. Organized sewerage system should be developed so that pollution caused by it can be prevented.
10. Prevent groundwater from pollution by suitable disposal of domestic waste.

Other Areas

1. Renovation of traditional water sources like tanks, *baoris*, wells, *tankas* etc. Desilting of tanks would not only increase storage of rain water but also increase groundwater recharge.

2. Take effective steps for minimizing rate of evaporation from tanks etc. containing surface water.
3. Public Health Engineering Department and other organizations should manage to prevent leakage from water supply schemes and pipelines with immediate effect.
4. Selection of suitable sites for groundwater sources.
5. Encourage hand pump construction. More water is wasted by piped water supply.
6. Implement compulsory registration of tubewells owned by boring machine and tubewell contractors or private persons with Central Ground Water Department.
7. Make prior permission of Central Ground Water Department compulsory for construction of wells/tubewell etc. in future.
8. Make every citizen aware about different methods of water conservation and their advantages. Various government and semi-government departments, non-government organizations (NGOs), voluntary agencies, environmentalists etc., should organize it as a people's campaign. Entrust responsibility to panchayats and other cooperative organizations for different works of groundwater management. Get public cooperation for implementation of surface water and groundwater development plans and repair works. Generate awareness about water conservation measures specially among women.

Groundwater Development: Artificial Recharging

1. Enforce compulsion for roof water storage in urban areas and artificial recharging of groundwater from surface rain water.
2. Ensure groundwater recharging through construction of check dams at suitable sites to prevent flowing rain water.
3. Prevent rain water in fields by contour bunding around fields.
4. Use unused wells, tubewells, handpumps etc. for artificial recharge plan of groundwater.
5. Optimum development of water conservation areas and maximum tree plantation.

Problem of Arsenic in Water

Arsenic is one among 10 important poisonous chemical substances: Lead, Mercury, Vinyl Chloride, Benzoic, Polycyclic, Aromatic Hydrocarbon, Polychlorinated Bylines Cadmium, Banzopyrin and Benz Florynathine. Arsenic is found in rocks, soil, water, air, plants and living organisms in natural form. It is a bright, grey and brittle substance having metallic and non-metallic qualities. Arsenic is a cumulative poison which is rarely found in form of an element of nature. It is mostly found in compound form with oxygen, chlorine, sulphur, carbon, hydrogen, lead, gold and iron. In compounds, arsenic is in trivalent and pentavalent position. Compounds made from carbon and hydrogen existing in plants and living organisms are called carbonic-arsenic compounds. Carbonic arsenic compounds are less poisonous as compared to non-carbonic arsenic compounds. Arsenic converts into other forms in the environment but it does not get destroyed, neither does it evaporate. By burning arsenic mixed things, it reaches in the air. Arsenic in the air stays in soil or dissolves in rain. Most of the compounds of arsenic dissolve in water.[2]

Different sources of concentration of arsenic in nature and human related conditions are as under:

Sources of Arsenic in Water

1. Geographical processes like rock slipping, mineral leakage and volcanic bursting
2. Flow of domestic and industrial effluent
3. Arsenic existing in air-from rain and dry concentrates
4. Soil

Sources of Arsenic in Air

1. Volcanic outburst
2. Fire in forests
3. Burning of fuel
4. Industrial processes like metal melting, heating, burning and other production processes
5. Use of pesticides in agriculture
6. Thermal power generation
7. Micro organic processes in soil

Sources of Arsenic in Soil

1. Use of pesticides in agriculture
2. Contact of air containing arsenic
3. Use of water containing arsenic in irrigation water
4. Volcanic outburst
5. Weathering of rocks containing arsenic
6. Mineral leakage

Source of Arsenic in Man

1. Use of arsenic polluted food
2. Use of arsenic polluted water
3. Breathing in arsenic polluted air
4. Working area for example working in industries and mines
5. Burning of arsenic painted fuel for safety and breathing in its fuel dust
6. Living near uncontrolled arsenic affected poisonous waste
7. Living near rocks containing higher quantity of arsenic

Arsenic moves mostly in conscious and sub-conscious forms of air, water and soil due to flexible and reactive reactions, where it makes different types of chemical and biochemical changes, e.g., oxidation, reduction, mithilation and de-mithilation. Finally, arsenic reaches the sea through air, water, soil, man etc.

Uses of Arsenic

Arsenic is harmful but it is useful also, as in United States of America 90 per cent arsenic is used in chemicals utilized for safety of wood and 10 per cent arsenic is used in manufacture of paints, drugs, dyeing, soap, metal, semi-conductors and other products. In the field of agriculture, arsenic is used for manufacture of insecticides. It is also used for manufacture of lead. Though arsenic is poisonous for man, but in the field of medicine also, it cures many diseases like skin disease, spirochetes infections, blood discrepancies etc.

Global Problem of Arsenic

In Novaskotia area of Canada, arsenic is found in a majority of rocks as well as groundwater. In New Mexico and Arizona regions of United States of America, local residents get drinking water from far off areaa by tanker because local water contains arsenic. Arsenic is affecting the health of native people in many areas due to its presence in large quantities in the groundwater in Argentina, China, Taiwan, Thailand, Mexico, America, Ghana, Hungary, England, Chile, New Zealand and Russia. In some areas of Argentina, concentration of arsenic in groundwater is 100 to 2,000 microgram per litre. According to health related tests, wherever presence of arsenic was more than 0.3 milligram per litre, problems from black spots on skin to skin cancer was found.

In China, deep handpumps in many areas were found to contain arsenic in high concentration. This problem was not found in less deeper hand pumps. Concentration of arsenic in deeper handpumps ranged from 0.6 to 1.2 milligram per litre.

In Thailand, the main reason for presence of arsenic in groundwater is the chemical composition of natural rocks and storage of arsenic affected waste in large quantities. The concentration of arsenic in less deep handpumps also was more than 0.5 milligram per litre, but in deeper handpumps there was low concentration. In Taiwan, reports regarding high concentration of arsenic are coming since 1968 and a large section of population is suffering from problems of skin diseases. In Taiwan, water of almost 19 per cent tubewells contains more than 50 microgram per litre concentration of arsenic, whereas in the rest of the tubewells, the concentration of arsenic was found to range from 10 to 1820 microgram per litre.

In the United States of America, the states of Nevada, Arizona and California have more than normal arsenic concentration in groundwater due to natural reasons. In many tubewells and five community water supply schemes of Nevada, arsenic concentration is more than the prescribed norm of 0.5 milligram per litre. About six crore population of America is affected because of this problem. At many places of North Mexico, a large proportion of population is suffering from skin diseases, skin cancer and problems of digestive system due to arsenic. In some areas, the average concentration of arsenic in groundwater is 0.4 milligram per litre.

In the South-eastern area of Hungary, the health of about four lakh citizens is being affected due to extreme use of insecticides containing arsenic. In the Cornwall area of England, the concentration of arsenic in groundwater was found to be between 10 to 50 micrograms. Aluminum hydroxide is used for curing such water. It neutralizes gray arsenic so that it does not reach water supply system.

In Chile, concentration of arsenic is more in surface water like rivers and lakes. In the water supply for Antofagasta city in Chile, average concentration of arsenic is 0.8 milligram per litre. It has been affecting skin of children there. In New Zealand and Japan also, water in some areas contain higher concentration of arsenic

In India, the water of six out of 17 districts of West Bengal contains arsenic concentration more than the permissible standard of 0.5 milligram per litre. In Bangladesh, 59 out of 64 districts contain arsenic concentrations in groundwater that is more than the permissible standard of 0.5 milligram per litre. Of the two crore and 40 lakh people residing there, 7,600 persons in all suffer from arsenocosis disease and one lakh from skin diseases. About 11.2 lakh tubewells contain water having higher concentration of arsenic. In America, 90 per cent arsenic affected water is utilized in production of chemicals for safety of wood. The remaining 10 per cent is used in paint, medicine, dye, soap, metal, semi conductor and other industries. For safety of wood, chromate copper and arsenate chemicals are used on a large scale. For the last few years, many organizations have been demanding ban on production of chemicals containing arsenic for domestic use. Ecology Security Agency of America categorizes arsenic as 'cancerous material'.

Though no concrete proof of evil effects of chromate copper arsenate on health has found, but three important American companies manufacturing it – Osmos, Archwood Protection and Chemical Specialties – had decided to stop its manufacture for domestic use up to the end of 2003. However, the production for occupational use was not stopped.

Effects of Arsenic on Human Health

Although arsenic affects human health but it depends how much, in what manner and in what form it enters the body and as a result of it, how the body responds to it.

Even a little quantity of non-carbonic arsenic harms the health of our body. It causes nausea, vomiting, leads to decrease in formation of red and white blood corpuscles, damage of blood vessel, and has the effect of paralysis in hands and feet. If arsenic concentrate water is used for a long time, it can also cause cancer of skin, lungs, liver, kidney and urinary tract. It also brings dryness of skin and change in colour of skin. However, water affected with arsenic does not cause any harm while washing hands, bathing and wearing clothes washed with such water in laundry because the water does not enter the skin.

As compared to normal persons, the maximum effect of arsenic is caused to infants in the womb, newly born infants, persons sick a since long time, AIDS patients, cancer patients and old and imbalanced persons taking little food.

Some health problems out of the above are also caused by diseases. Hence, the poisonous effect of arsenic does not become clear in the beginning. When people take water containing arsenic, the effects take place slowly. It becomes visible after a duration of 8 to 14 years. If its awareness is generated earlier, its effects can be minimized.

The level of effect of arsenic can be measured through examination of blood, urine, hair and nails. Arsenic stays in the body for a very short time, hence testing of the effect of arsenic should be done through urine examination as early as possible since it is more trustworthy. Extreme arsenic contact can be measured through testing of nails and hair within 6 to 12 months. Lesser quantity of arsenic contact cannot be measured by these tests and neither can these tests tell us about the harmful effects of arsenic on our health.

Problem of Arsenic in Drinking Water in India

In India, about 53.6 lakh persons in West Bengal are residing in arsenic affected areas and are compelled to drink water containing arsenic. The reason of this problem is geographical. According to a survey conducted by the Central Pollution Control Board, residents of these areas are suffering from diseases like Melanesia, Kurtosis and Gangrene and their normal problems are related to breathing.

Solution of the Arsenic Problem

Many techniques have been discovered for redemption of the problem of arsenic in India. They are, precipitation with the help of iron and aluminum hydroxide, ion exchange, reverse osmosis and suction of such materials on the surface. Some equipments based on such techniques are also working in West Bengal. According to a report "Arsenic Contamination in Ground Water and its Control" published by Central Pollution Control Board, whichever instruments based on absorption method are installed in West Bengal, they work properly only for some period. After some time, it is essential to clean the absorption medium, otherwise the instrument does not work properly. In arsenic affected areas of West Bengal, concentration of iron is excessive. Blocks of iron compounds are formed and they collect on the medium. They absorb comparatively more arsenic. Hence, if arsenic containing compounds are not removed from time to time, the medium would not work properly and its capacity will decrease.

A group of engineers under Nick P. Nicholedus of Connecticut University (USA) have prepared such a filter which separates arsenic from water and it can be directly fitted in a tubewell. In this filter, small pieces of iron and sand are filled and by rusting of iron, all the arsenic is absorbed from water.

Scientists of Canada have also discovered a new technique for removing arsenic which is not only cheaper and effective but can also be shifted from one place to another. Dust of Alumna is used in it. Atoms of arsenic are absorbed in alumna as sponge collects water in it. Thereafter a mixture of alumna and water goes to a micro filter which is based on a membrane. This water passes through filter whereas, alumna and arsenic come back in the reactor. This method can remove 99.9 per cent arsenic. When arsenic and alumna collect in sufficient quantity, then an acidic solution is mixed in it, which separates alumna from arsenic. This complete equipment is very small and can be taken anywhere. Hence, in view of scientists, it is very useful specially for developing countries.

Many techniques are available for removing arsenic from drinking water but in spite of it, crores of people are suffering from diseases on account of this problem. Hence, it is the responsibility of governments, local bodies and NGOs to make the public aware about the evil effects of arsenic and the available techniques for removeing it.

Notes

1. Ground Water Scenario in Rajasthan, Problems and Management, Central Ground Water Agency, Western Region, Jaipur.
2. Pandey, M. and Anand, C., *Geography and You*, July-August 2003, New Delhi, pp. 4-7.

10

Water Pollution

About 90 per cent of fresh water available in nature is frozen in the form of snow, whereas only 10 per cent of fresh water is available for use of human beings. If this whole water deposit remains at its original qualitative level, water crisis cannot arise, but even 10 per cent part of this fresh water (2.70% of total water in nature) is not free from pollution. Rivers, lakes, tanks and underground water are continuously becoming polluted. Oceanic water, in spite of being saline, necessarily has to remain free from pollution because the water is used not only for drinking purposes but it indirectly also maintains all living organisms. Hence, the whole water deposit in nature should remain free from pollution.

Human activities are extending day by day, due to which existing sources of water on the surface of the earth are getting polluted. Apart from oily pollution in oceans, disposal of effluents is made in large quantity every year in coastal areas. In important industrial countries of the world, including India, waste is disposed of by big cities habited on banks of rivers and at the same time waste from industrial plants is polluting river water. Two-third portion of the water of rivers Ganga, Yamuna, Godavari, Kaveri (Cauvery), Damodar, Hooghly etc. is not fit for human consumption due to disposal of polluting elements emitted by industrial plants located on the banks of these rivers in India. The Ganga enters in plain areas after Haridwar and thereafter up to Allahabad, its water is not fit for human consumption.

What is Water Pollution

At any place on the earth, deterioration of water due to inclusion of undesirable outside elements in water deposit, is called water pollution. Water is available from different sources in nature and its physical and chemical composition is also different. Oceanic water is saline which cannot be directly used for human consumption but it does not mean that oceanic water is polluted because quality of water is not only assessed from the point of view of human consumption but also from the angle of use by all living organisms. Hence, oceanic water is not polluted in spite of being saline. Saline water has its own ecological system which cannot be maintained by fresh water. Thus in the natural context, the water existing in its original form with its physical and chemical composition necessary for various biological communities, is free from pollution but when undesirable outside elements from natural and artificial sources mix in it, then it is called polluted water. Polluted water becomes poisonous and quantity of oxygen is reduced as compared to normal water. Therefore, the water becomes disease bearing rather than remaining nutritive.

Water pollution is change in physical, chemical and biological qualities of water due to human reasons and therefore such water becomes harmful for public health, since they become disease bearing, making them unfit for drinking by human beings and animals, and also various uses in industries, agriculture and other uses. While defining water pollution, C.S. Southwick (1976) has written that "the change in chemical, physical and biological qualities of water due to human activities and natural processes is called water pollution."

Normally, it can be said that interference by natural and human elements in quality of balanced water necessary for organisms living on the surface of the earth causes deterioration, which causes water pollution and therefore such water does not remain fit for use by those living organisms. Its effect is becoming visible on the surface level water sources and seas in present times.

Types of Water Pollution

Nature and forms of water pollution are related to many factors. Based on the factors which cause pollution of water, pollution of water can be of four types:

1. Sewerage Pollution

2. Domestic Waste Water Pollution
3. Industrial Waste Water Pollution
4. Solid Waste Water Pollution

Similarly based on its nature also, it is of four types:

1. *Physical Pollution:* This causes change in smell, temperature, colour, density, taste and warming qualities in the water.
2. *Chemical Pollution:* Due to chemical elements and layers mixing in water due to different industries and other sources water gets polluted.
3. *Physiological Pollution:* It is after such changes in qualities of water, harmful effect on human processes is caused.
4. *Biological Pollution:* Biological pollution is caused by harmful water organisms entering from various sources in water. Among such organisms, the main are Bacteria, Coliform, Algae and Virus etc. Normally, quality of water is decided on the basis of biological demand for oxygen and chemical oxygen, dissolved oxygen and pH value.

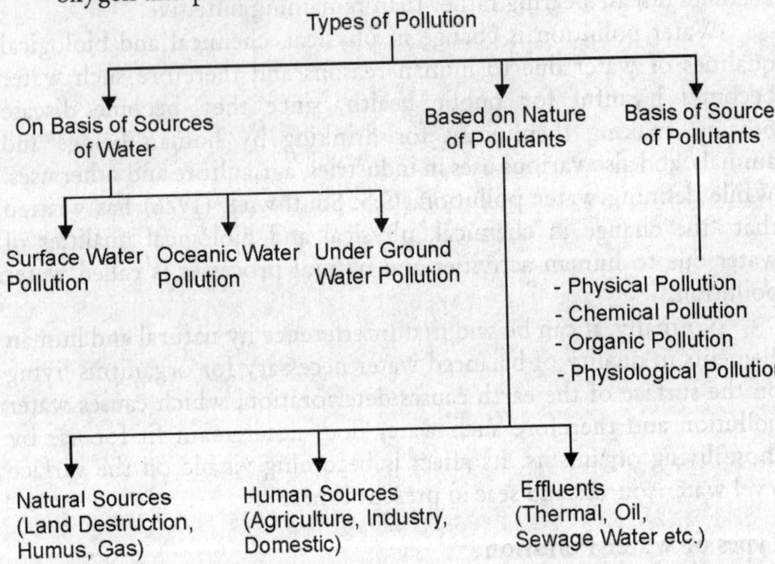

Classification of water pollution is normally made on the basis of sources of water as under:

1. *Surface Water Pollution:* Water is found in rivers, lakes, tanks and other micro places on the surface of the earth. On the surface level, sources of water are rainfall and melting of snow. Different ions

are found in rainwater. Other water sources contain sodium, chlorine, calcium, sulphate, magnesium and potassium ions. These ions from natural sources enter the atmosphere through oceanic water and they reappear on the earth with rain water.

Apart from them, surface water is also polluted by lead, zinc, mercury, cadmium, asbestos etc. Health, Education and Welfare Department of United States of America (HEWD) has divided surface water pollutants in the following eight categories:

1. Brewery and industrial waste
2. Plant nutritive elements and dissolved elements
3. Infectious factors (insects and virus)
4. Molecules and granules
5. Chemical elements
6. Effluented heat
7. Radioactive elements
8. Biochemical elements etc.

In every field different types of techniques are fast developing in the world. Due to industrialization and urbanization, 2,500 chemicals per year are damaging the pure surroundings resulting in pollution of surface water sources. The effect of this polluted water clearly is visible. Very soon adverse effects of pollution would appear, which would be a question mark on the existence of human beings and other living organisms. In efforts to grow more foodstuffs to maintain the population that is increasing at a fast rate, insecticides are being used in agriculture. Many chemical insecticides are also polluting surface water sources. Due to the activities of human beings, lakes are also becoming polluted along with rivers. They are also being affected by deforestation, since sedimentation is being caused in lakes because of soil erosion. In addition, many wastes reach lakes because of surface flows. Due to effluent from breweries and industries established in vast cities located on the banks of lakes, lake water is losing its quality. Many chemical substances disposed by agriculture also spread pollution in them.

2. *Oceanic Water Pollution:* Industries developed in coastal area of seas and cities developed there also spread pollution through disposal of waste. Brewery effluent and poisonous chemicals like mercury are the main waste materials among them. Unplanned disposal of waste in coastal cities is causing deterioration of quality of oceanic water. This directly affects the biotic life in seas. Coral and fish biotic life are struggling for their existence but man is ignoring

this problem because of his own economic development and self interest, which is having dangerous results.

Due to navigation in seas leakage of mineral oil from oil transporting ships is also causing pollution. Every year about 295 million gallon mineral oil is mixing with sea water due to oil leakage. Apart from this, leakage in sea oil wells and fire have also increased pollution. During wars, while using mineral oil for military purposes, some part mixes in sea water. Sometimes, damaged oil transporting ships also pollute water. Thus, a dangerous ecological crisis is arising from oceanic water pollution.

3. *Groundwater Pollution:* Groundwater pollution is considered as an important part of water pollution. It is caused by agricultural waste, effluent from chemical substances from industrial areas, and urban and rural waste disposal. All such disposed waste mixes with surface water and pollutes underground water. Similarly, agricultural waste percolates by leaching at the time of irrigation. The increasing population in urban as well as rural areas also disposes off their waste by unscientific methods and in an unplanned manner at any place they like. This goes underground after mixing with rain water. Besides this, the pollution in water deposits situated on the surface of the earth like lakes, tanks and baoris, pollutes underground water by infiltration. Nature of aquifers also affects pollution. Polluted water of upper layers stagnates in many aquifers, affecting the quality of groundwater in such areas.

Wells and hand pumps also pollute underground water. Poisonous chemical effluents disposed of by factories nearby wells pollute groundwater by infiltration. Similarly, in wells located near coastal areas of seas, due to over exploitation of groundwater some hollow spaces are created in which polluted oceanic water enters, polluting underground water.

Thus, groundwater is being polluted by different sources at a fast rate. Its effects are gradually becoming visible.

Sources of Water Pollution

Water as an invaluable element found in nature that possesses its own quality. Many undesirable elements, which we call as pollutants, deteriorate this quality. Pollutant materials are classified in categories like physical, chemical, corporal activities and biotic according to their nature. Factors causing deterioration in quality of water can be divided in two parts as joint compounds. One factor is effective on the basis of its availability and quantity, whereas in the second one, more than one factors are active. Water pollution is caused mainly as a result of human activities. To some extent, natural factors are also responsible for pollution. Sources of water pollution are indicated in the following chart:

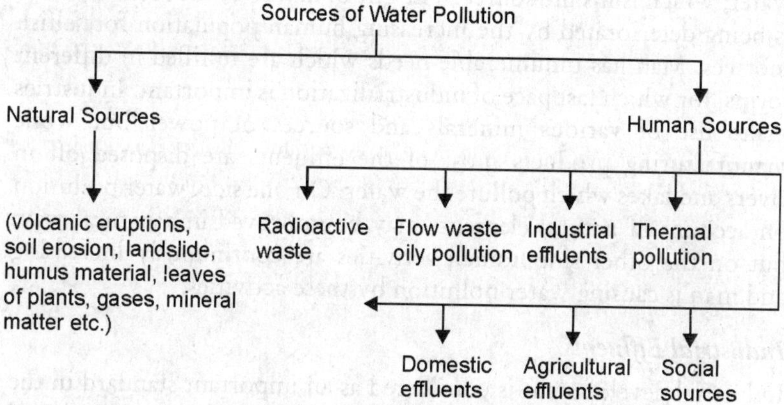

Natural Sources

Natural factors affecting water sources are volcanic eruptions, landslides, mineral matters, gases, soil, humus, animal excreta, leaves of plants and sometimes scarcity of rain also. Water moving on the surface of the earth flows in the form of tanks, rivers and springs. Wherever such water stores or stops, mineral matters situated in the soil mix with it. Other metals also mix with it. When the quantity of such materials becomes excessive, it starts its evil effects. These poisonous elements mainly include lead, mercury, cadmium and arsenic. The proportion of the concentration of mercury 0.1 mg/litre, lead, 0.01-0.05 mg/litre and cadmium, 1-10 mg/litre is normal. They become harmful as soon as they cross this limit.

Due to soil erosion, landslides and volcanic eruptions, sediment deposition takes place in fresh water, causing turbidity in water. Due to the refuse, the capacity of lakes etc. is reduced. Naini Lake of Nainital has reduced from 27.8 metres to 19.8 metres, resulting in the death of about one lakh fish every year. Reduction in depth of lakes is called *eutrophication* and such lakes are called eutrophic lakes. Decomposition of plants and animals is also a natural activity responsible for water pollution. Excreta of animals flowing in the water also causes water pollution. If a great quantity of vegetation is produced in water, it also affects water ecology.

Human Sources

In the present race of fast economic development, man is damaging water, which is his life source. The gift of nature in the form of water is being deteriorated by the increasing human population for selfish motives. Man has innumerable needs which are fulfilled in different forms, for which fast pace of industrialization is important. Industries make use of various minerals and sources of power but while manufacturing products most of the effluents are disposed off in rivers and lakes which pollute the water. On one side, water pollution on account of natural elements may be improved in a natural way, but on the other side human activities are continuously increasing and man is causing water pollution by these activities.

Industrial Effluents

Industrial development is considered as an important standard in the flow of modern development. The maximum population is employed in industries in developed countries. Those countries where industrial development is lesser, are in the category of developing countries. In industries, water is used in many forms. During the manufacturing process, used water is disposed off in rivers. Thus, industries located on banks of rivers or lakes or coasts of seas pollute them. If such water is not disposed in any water source, it goes underground and pollutes the groundwater. Water disposed off by industries contains many carbonic and non-carbonic elements dissolved in it and some of them are poisonous. Among developing industries, chemical industries are coming up in large numbers. They manufacture paint, cleaning equipments, artificial fibers, medicines, plastic etc. Similarly, sugar industries, breweries, paper industry, fertilizer industry and leather industry also dispose off effluents in fresh water sources.

Most of such disposed materials contain carbonic elements which neutralize through bacteria at a slow pace resulting in bad smell in water for a long time, change of taste in the water also occurs. During refining such water, other particles react with chlorine and produce chlorinated carbonic compounds. Effluents coming out of industries contain non-carbonic material, mainly calcium, sulphate, nitrate, ammonia chloride, potassium and sodium. Besides this, hot water coming out from industries increases the temperature of water in water sources whenever it flows into them. This also affects vegetation and biotic organisms in water. After mining, which is necessary for industries, the mining soil also flows with rain water into the water sources during rains. It then mixes in the water and the chemicals contained in it pollute the water. Mercury, arsenic, copper and cadmium are the main poisonous metals, which mix in water after disposal from industries. Among them, mercury is the most harmful metal, which mixes in water in liquid form at normal temperature. It evaporates in a poisonous state if temperature does not increase. Mercury is causing great harm to water at present. In the world, 5,000 tons of mercury in different forms is being disposed in natural surroundings. Its direct example is the tragedy of Minimata Gulf situated on the sea coast of Japan. In 1950, the inhabitants and fishermen of this Gulf suffered from blindness, mental diseases, weakening of muscles and paralysis. The reason was the consumption of polluted water of this gulf in which mercury effluent of a plastic factory, located in this area, were being disposed off.

Mercury compound converts into methyl mercury (H_3C, Hg, CH_3), which is a poisonous form. It then reaches the food chain and affects man and other living organisms.

Man is expanding industrialization at a fast rate for satisfying his needs, and also for a fast economic development. Disposed water flowing from industries pollutes lakes, rivers and ponds through ions present in it. All the vast flowing rivers in densely populated areas of the world have lost their original form. Now they have become dumping grounds of effluents disposed off by industries. Rivers of North America, Europe and Japan have lost their original form. In India also, water pollution through industries is increasing at a faster speed. In the capital region, 4.5 lakh litre DDT is released in the river Yamuna, every day. Similarly, on the banks of the sacred river Ganges especially in Kanpur, leather industry, fertilizer industry, paper industry, oil refinery and many other chemical plants have developed on both sides of the river to such an extent that the water in Kanpur has not remained fit for drinking.

In Tamil Nadu, the Palaeer river has become polluted because of leather industry units in Ranipet, Ayur and Baniyabadi cities. In West Bengal, river Hoogly is polluted due to jute and paper pulp industries and in Tamil Nadu, the river Kaveri (Cauvery) has lost its original form due to the presence of Bharat Heavy Electricals limited. Along with rivers, the water of lakes and groundwater is also becoming polluted by industrial effluents.

There may be hardly any river or lake which may be existing in the same form in the country as it was fifty years ago. Their area has shrunk and the capacity of holding water has reduced. These tanks are the main centers of drinking water, irrigation and fisheries development, but due to disposal of industrial effluents have all have become polluted. Chilka and Dal lakes are the direct examples.

Underground water is also being polluted by industrial effluents. In Rajasthan, poisonous water disposed by dyeing factories of Pali and Sanganer is destroying the quality of groundwater by infiltration under the ground. Industrial cities situated on the coast of oceans are polluting sea water. Thus, disposal of industrial wastes and effluent water in water sources is the main reason of water pollution, which is the result of wrong use of water by man because of his strong desire for economic benefit. Hence, only he can remove this problem of water pollution.

Domestic Effluents

Water is a basic need of the life of man. It is used for bathing, cleaning, cooking, air conditioning etc. After use in the above activities, waste water is flown as effluent in the surrounding areas. Such waste contains detergent for washing clothes, decomposed vegetables and fruits, ash of hearth, domestic refuse, pieces of cloth etc. Such waste water causes pollution by mixing with nearby water sources after passing through open drains.

Man requires 130 litre water per day on an average 70 to 80 per cent part of such water is flown away after being used. This water, accompanied by pollutants, flows into the water sources.

Due to fast increasing population, quantity of domestic effluents has also increased. With the increasing tendency of urbanization, problem of disposal of domestic effluents has increased. River Ganges is fast converting into a dirty drain due to disposal of domestic effluents, disposed in it by cities located on its banks. At present, pollutants of 48 'A' class cities and 66 'B' class cities situated on the banks of this river, are being disposed in large quantities every day.

In Delhi, 96 crore litre filthy water is poured in river Yamuna through 17 drains which results in a huge amount of pollution. Similarly, 87 per cent out of the total domestic waste in Mumbai and 89 per cent out of the total domestic waste in Kolkata is being disposed in rivers and lakes. It has been presumed since ancient times that river water is pure but it is no longer so. In developed countries, suitable method of treating domestic waste is being adopted but in developing countries, there has been no progress in this direction. In all cities, domestic refuse is thrown here and there, which gives birth to different epidemics.

Domestic waste also includes plastic. At present, the use of plastic bags is increasing. After being used they are flown away in the water. Since they neither decay, nor dissolve, hence their disposal has become difficult. Harmful bacteria develop when they are immersed in water which pollutes the water. There is no method for their disposal in India but developed countries have developed the technique for disposing them. Among all domestic wastes, the most harmful is water used for cleaning. This water is produced by cleaning of house, cleaning of utensils and washing of clothes. Ordinary soap does not fully clean the clothes, hence detergent soap is used to clean clothes, which is very harmful. Water is polluting fast because of the increasing use of detergent. It is difficult to refine detergent containing water, and the phosphorus produced by it develops algae which is a harmful vegetation. Growth of algae in the water of rivers and lakes makes the water unfit for drinking. If these detergents mix in any water source, they do not go away easily. Neither water treatment methods nor the bacteria in the water can affect them, and its permanent effect proves very harmful.

Sewage

Flowing sewage includes human excreta disposed from public and domestic latrines. It also contains carbonic as well as non-carbonic materials. Due to increasing population, people are becoming concentrated in small places in cities and towns, where in the absence of suitable sewage system for disposal of human excreta, it is disposed in nearby water sources. Some of this water also goes underground. Excreta disposed in water sources pollutes the surface water while the underground part pollutes groundwater. Hence, its successful disposal is the demand of the day. This problem did not exist in the past when the population was lesser. At that time people thought that water sources flowing in natural way would automatically get cleaned. But today this thought has become a curse for them.

Sewage of solid nature is mainly carbonic, which is nutritive for disease bearing micro saprophyte, and bacteria. Bacteria, virus, protozoa, algae, fungus etc. develop and multiply. Such sewage pouring in water sources without treatment would pollute the water. Besides, human beings as well as animals dispose excreta in open places which also flows with rain water and mixes in water sources. Thus, polluted water does not remain fit even for irrigation because such water would pollute groundwater through infiltration during irrigation. Apart from production of disease carrying bacteria, some big parasites also develop, systomomus being the main parasite among them. This parasite lays eggs after entering in the large intestine of human body with water. The eggs spread infectious diseases after being released with urine or excreta. Such diseases develop more in tropical regions.

Except sewage, dead persons and dead animals are also flown in water sources, which has become an important problem of the present world. Developed countries are producing manure and energy from sewage whereas, developing countries have yet to combat this problem. According to National Environment Engineering Research Institute (NEERI), Nagpur, only eight out of 3,119 cities of India have arrangement for fully developed sewerage system, whereas 209 cities have partial arrangement. Out of 142 metropolitan cities, 52 cities have partial arrangement and 82 cities have no arrangement.

On the banks of river Ganga, more than 50,000 people in 114 cities dispose untreated excreta in the river water. Waste material of Kolkata city is being disposed in Hoogly river through 361 drains, due to which the number of fish is continuously decreasing. In Egypt, fish production in Maryug lake near Alexandria city is in danger due to domestic sewage disposal in it. Tutee river in Brazil is being polluted due to disposal of 300 ton refuse of Sao Palo city.

Present world is facing the problem of searching methods for solving this problem caused due to increasing population. If disposal can be made for irrigation after making it fit for use in agriculture.

Agricultural Effluents

Agricultural effluents include the process of flowing of insecticides and other chemicals used in agriculture along with water. Soil erosion due to defective agricultural systems also supports it. At present, fertilizers, pesticides and insecticides and biotic killing chemicals are being used for increasing agricultural productivity. Their discovery and use are important for the present age. Many chemicals like sulphur, copper and lime have been in use for destroying harmful insects, for control of crop diseases since many years. Chemical materials for destroying insects include the following five types:

1. Fungicides
2. Weedicides
3. Insecticides
4. Nemoticides
5. Rodenticides

Different types of chemicals are used for making these materials which get mixed with irrigation water. Among them are poisonous elements like mercury, calcium, lead, arsenic, nickel, manganese, chlorine, boron etc. all of which have harmful effects on man as well as animals.

In the present times, sewerage farms are being developed from water mixed with urban effluents and vegetables are being cultivated there. Unfortunately, fungus, bacteria, harmful insects and non-carbonic elements are polluting the vegetables, which have a poisonous effect. Labourers working on sewerage farms also face health risk, because living amidst polluted water makes them suffer from skin diseases, breathing problems and digestive problems. During irrigation of sewerage farm, groundwater becomes polluted due to infiltration of water underground. This can have a harmful effect on trees, plants and biotic organisms. Excessive irrigation results in salinity in soil, while excess water brings existing salts and alkalinity from lower layer to surface and makes the soil sterile.

Fertilizers used in agriculture, mainly nitrates phosphates, urea etc. also cause pollution because of imbalanced use of these fertilizers flow away with water and pollute it . This water also develops algae. Useful bacteria in soil (Azotobacter, Azospirillum, Azolla) are greatly affected by groundwater pollution, resulting in deterioration of its natural productive capacity. In most of the chemicals used in agriculture, poisonous elements like mercury, arsenic, chlorine, fluorine, phosphorus etc. exist, which affect the biotic life and pollute water by agricultural effluents. In the beginning, fertilizers were used to the maximum extent but when their destructive effect was known, the developed countries started their balanced use. DDT insecticide being the most harmful, its used was banned. In India also, ban was imposed on 30 September 1992, on most harmful insecticides. The main among them were andrine, touxafan, haftachlor, aldrine nitrofan, tetradiphone, pentachlorophynol, chloroden, pentachloroniytobenzene (PCNB), hybromo-chloro-propane, methyl parathion and paraquait – diamethyl-sulphate.

During present times, some common chemical pesticides used in agriculture contain the following carbonic and non-carbonic elements:

- Organochlorine insecticide – aldrine, dia aldrine, BHC (Benzene Hexachloride), DDT (Diachloro-diaphenoyl- trichloromethyl methane), heftachlore, lidane
- Carbomate insecticide – Carbaril
- Diathaio carbamet fungicide – capton, furbace, meneb, menkozeb, propeneb, thyrum, jineb, jeerum
- Orgonophosphat insecticide – chlorophyrophos, diamythoate, fenaitrothion, melathion, foret

During the use of the above chemical insecticides, mercury, lead, arsenic and other poisonous elements spread. Disturbances are also caused in the ecological system by water pollution caused by agricultural effluents. Food chains are being disturbed. Hence, fertilizers and chemical insecticides should be used in agriculture, keeping in mind their overall effects.

Oil Pollution

When oil or petroleum from different sources mixes with water sources, it is known as oil pollution. This oil directly mixes with water sources from industrial sources and oily materials. Oil pollution mainly occurs in seas. Mixing of oil in water sources happens at the time of accident of oil transporting ships, discharging waste oil from ships and while filling or emptying oil from oil transporting ships. Apart from this, leakage from oil wells on the coast of seas also mixes oil with water. Oil pollution causes harm to human beings, creatures, fish, sea birds and vegetation.

According to one estimate, about 50 lakh to one crore oily substances from different sources mix in sea water every year. Oil also mixes in rivers and other water sources, e.g., the water of river Yamuna is being polluted by Mathura oil refinery. Oil exploration on coastal areas of seas also pollute sea water. In 1969, water pollution of a grave nature was caused by oil exploration on Santa Barbara coast in South California of United States of America, whose evil effects continued for years. Similarly, oil pollution is caused when of oil carrying ships meet with accidents. In 1976, a Greek oil carrying ship was lost near Waireyal in the sea. Its effect was seen in Indian coast for many days in the form of floating layer of oil. In 1976, due to accident of an oil carrying ship named Tory Kenyan, a vast sheet of oil was deposited over sea water. It had to be removed by spraying chemicals and such other detergents. About 10,000 tons of detergents were sprayed in it which created a layer of foam over the water and caused danger to life of conch-shell, oyster and snails.

Oil pollution is also caused in the world at the time of war. In 1991, oil was discharged in a large quantity in the Persian Gulf during the Gulf War, and fire was ignited in some oil wells, which not only destroyed sea creatures but also damaged water treatment plants situated in the gulf. It will take us years for bringing the Persian Gulf in normal position.

Apart from accidents of oil carrying ships, oil pollution is also caused by leakage from oil transporting ships. Every year in the world, more than about 12 crore tons oil is transported through seas and oil pollution is caused by leakage of oil. Control can be exercised over it to a large extent if precautions are taken to prevent this leakage. America is manufacturing oil transporting ships of such type. To prevent accidents also, planned operation of ships will have to be done.

Radioactive Wastes

Development of atomic power is being resorted to in the world at a fast rate for weapons as well as energy. Some metals like uranium and thorium have huge power hidden in the center of their atoms. This is called atomic power. While generating energy, many types of radioactive wastes are created and in the absence of their suitable disposal, most of these wastes are discharged in seas. This causes water pollution. Some part of the waste is buried underground, but their effect does not diminish. Sometimes the ships carrying radioactive elements meet with accidents in seas or they sink, which cause water pollution.

When atom bombs are exploded, innumerable radioactive granules spread out in far off areas. Their sub-division takes place very slowly. Due to this, radioactive elements mix with water sources. By entering the body of human beings and other creatures through water, they cause harm by becoming part of the food cycle. Among them strontium-90, radio-iodine, sisiag-137, carbon-14, sirium-144, zinc-65, iron-59 and cobalt-60 are the main ingredients. Due to present military competition, most of the developed countries are desirous of becoming atomic powers, hence some time back France carried out tests in Atlantic Ocean, while China tests in Pacific Ocean. Developing countries are also following this, which causes pollution through radioactive wastes.

Thermal Pollution

If temperature of water rises by inclusion of outside elements in it and the water becomes harmful for life, then it is called thermal pollution. Water from water sources is used by industries for removing of excess heat of reactors and this water is discharged by industries in water sources. This heats up water which is discharged directly in water sources. Sometimes, industries located on banks of lakes or rivers, discharge high temperature water industries or reactors situated on coastal areas release hot water in seas. Atomic power stations also cause thermal pollution as they use fresh water in huge quantity for cooling their condensers. Due to increasing demand for electricity generation, the number of atomic power stations is also increasing, which is resulting in dangerous dimensions of thermal pollution. As compared to normal or traditional power generation plants, atomic energy plants generate 40-50 per cent more energy.

Increase in water temperature is also caused due to atomic explosions in the military context and most of them are exploded in seas. Increase of temperature of sea water due to such explosions causes danger to sea creatures. Pacific Ocean is greatly affected by atomic tests since Australia, New Zealand, China and Japan have left maximum effect on it. Sometimes, thermal pollution is also caused by accidental alighting of space planes in water. Thermal pollution, along with extinguishing biotic variety of seas, also affects creatures and vegetation dependent on water sources on the earth.

Social Sources

Man pollutes water sources to a great extent on the occasion of cultural and religious festivals. These are called social sources of pollution. One example is the great Kumbh fair at Allahabad where 3 crore people collect every 12 years. Apart from this, people dispose dead bodies of human as well as animals in water sources. Natural water has the capacity to absorb natural pollutants in it. Hence, the maximum destructive reasons of water pollution are manmade ones.

Major Pollutants of Water

Natural form of water deteriorates due to many natural and human sources. Some pollutants enter into water from such sources and they badly affect other creatures and vegetation along with man. The extent of evil effect of that pollutant depends on the nature of the special polluting substance. Some pollutants leave their effect immediately whereas others act slowly. These pollutants are as follows:

Mercury: Mercury enters in water mainly from industrial and agricultural effluents. Some quantity mixes in water from natural elements also. Mercury is a proplasmic poisonous element of nature. After collecting in the body of human, it harms their liver, kidney, spleen and bones. Acidic mercury is more poisonous in the bodies. Harmful effect is caused to gums and face due to it.

Lead: The quantity of lead in water increases from industrial waste. Being a poisonous substance, it affects public health through different ailments.

Zinc: As a result of industrial processes, compounds of zinc, mainly zinc oxide and zinc chloride mixed in water, cause harmful effects. More concentration of zinc chloride in water spoils the lungs.

Fluoride: Quantity of fluoride should not exceed 1 PPM in water. Use of fluoride-affected water with its excessive quantity cause diseases of teeth, deformation of bodily limbs, curved bones, loss of blood, loss in weight etc. In India, maximum quantity of fluoride is found in groundwater of western Rajasthan. Many bodily deformations are visible due to it.

Table 10.1

Effect of Fluoride on Parts of Body

Quantity of Fluoride (Milligram/Per Litre)	*Medium*	*Effect*
1.0	Water	Reduction in space among teeth
More than 2	Water	Change of Colour of Teeth
8.0	Water	Bone Disease
20-80	Water and Air	Lameness
50	Food and Water	Stopping of Growth
125	Food and Water	Effect on Kidney
2.5 to 5 Gram	More Diet	Possibility of Death

Sodium: Sodium is found in the form of a compound due to its activeness. Hence, its effect is possible only due to positive ions.

Phenol: It is the most poisonous carbonic material which causes maximum effect on skin. It also affects organs of human body, specially kidney, liver, spleen and lungs. Touch of phenol for 15 to 20 minutes on skin, immediately shows effect on muscles, power of hearing and sight etc.

Nitride: By use of water having more nitride, babies suffer from 'Blue Baby' disease which causes vomiting and the colour of the skin becomes dark. Hence, it is a very dangerous compound.

Armenizum or Pathogenic Bacteria: After entering water, they cause fatal effect on human body. They spread diseases like plague, measles, pneumonia and influenza, which sometimes take a shape of plague.

Thus, the above pollutants cause harmful effects on man after mixing with water from different sources but their removal is also in the hands of man.

Harmful Effects of Water Pollution and Water-borne Diseases

Deterioration in the natural form of water is caused by mixing unwanted elements more than the specific limit. Use of such water causes various water-borne diseases. Pollutants deforming water enter from different sources on account of human activities and they cause harmful effects on man. Other creatures on earth and in water are also affected by it. Effects of water pollution can be described from the following view points:

Effects on Man

Water is a natural gift which has no alternative existing on earth. Hence, water has been termed as life. As per latest surveys, per capita requirement of water is 140 litre.

Man needs water for drinking, cooking, cleaning, irrigation etc., which should be fresh. Polluted water causes the following diseases:

1. *Water-borne diseases:* Water polluted by human and animal excreta, if drunk or used otherwise, causes plague, typhoid and other diseases.
2. *Diseases caused by contact with water:* Taking bath with polluted water affects eyes and causes skin diseases and diarrhoea.
3. *Diseases based on water:* Many types of micro organisms exist in polluted water, the use of which spread infectious diseases. Some

disease bearing micro organisms and diseases caused by them are described below:

(i) *Virus:* Infectious diseases like hepatitis, jaundice and polio diseases are caused by virus. About 2000 persons suffered from jaundice due to polluted water in 1978.

(ii) *Bacteria:* Bacteria cause typhoid, paratyphoid, plague, diarrhoea, gastroenteris, dysentery and other diseases.

(iii) *Protozoa:* Stomach and intestine related different diseases are caused by presence of protozoa in drinking water. Main diseases among them are thrombosis, giardiasis, dysentery, amoebic, hepatitis, apses etc.

(iv) *Worms:* Parasites enter intestines in the body through drinking water and give birth to various kind of worms including round worms, hip worms, thread worms, tape worms, escarps etc. these cause many diseases.

(v) *Others:* Apart from the above, wells disease in leptospiral diseases and systosamiasis in snail supported diseases cause adverse effect on human health.

Table 10.2

Population of Developing World affected by Water-borne Diseases

	Disease	*Affected Population*	*Deaths per Year*
1.	Plague	100 crore	20 lakhs
2.	Diarrhoea	150 crore children	40 lakh
3.	Typhoid	5 lakh	25,000
4.	Pilio Mitus	5 lakh	20,000
5.	Round Worm	100 crore	–
6.	Trachoma	90 lakh	-
7.	Lishmaniasit	1.2 crore	2 lakh
8.	Guinea Worm	1 crore	–
9.	Sleeping Sickness	20 thousand	–
10.	Filaria	9 crore	20 lakh
11.	Malaria	27 crore	50,000
12.	Revere Blindness	2 crore	–
13.	Yellow Fever	25 thousand	–
14.	Dengue Fever	6 crore	–
15.	Shistosomiyasis	20 crore	–

4. *Diseases related to water:* When water polluted from various sources is accumulated at a certain place, many mosquitoes, flies and insects are produced which cause malaria, philaria, dengue, yellow fever encephalitis and many other infectious diseases.
5. *Diseases caused by chemical substances mixed with water:* Many chemical substances like Fluorides, nitrates and mercury etc. exist in water in liquid form. They can prove harmful if their quantity increases due to human interference. Fluorides affect teeth and bones. Nitrates after joining with hemoglobin, reduce the oxygen carrying capacity of blood, and mercury causes constipation.

A World Health Organization (WHO) report of 1982 had indicated effect of water pollution on man as under:

- Every year 2,500 persons die in the world due to use of polluted water.
- Every year 2,500 persons and one crore children die in the world due to non-availability of sufficient water.
- In the world, 200 crore persons do not get pure and perfect drinking water.
- In developing countries, four out of every five children die because of water-borne diseases.
- Every year 50 crore persons suffer from water-borne diseases.
- 30 per cent population of the world suffers from dysentery.
- 40 per cent persons suffer from diseases caused by water-borne worms.

Thus man, the main creator of water-borne diseases, suffers most from it and its treatment is also in his hands. There is no other alternative of water on earth. Hence, man can secure his future only by keeping water fresh.

Effects on Animals

Polluted water affects creatures living in water as well as those using such water. They include fish, tortoises, water birds etc. which live in water and others which move on land, but are affected by use of polluted water. Sea creatures are affected more by thermal, oily, mercury and radioactive pollutants. Fish and other creatures die due to non-availability of sufficient oxygen. Due to pollution, 40 per cent sea creatures have reduced during the last 50 years. Iraq had burnt oil wells of Kuwait during the Gulf War, due to which biotic life had deteriorated to a large extent. Apart from this, creatures on the earth are also affected by polluted water, which occur when industrial, urban and agricultural effluents mix in drinking water sources. For example in Lucknow, many fish are dying due to sewage of urban areas mixing in river Gomti. Cows, buffaloes and other animals also die by drinking polluted water. In May 1999, 500 cows had died because of drinking polluted water. Agricultural effluents are also equally harmful. Many poisonous chemicals mix in water and cause diseases. Materials flowing from domestic effluent destroy micro organisms. Thermal pollution kills fish existing in lakes, tanks and seas.

Effects on Plants

Like human beings and other creatures, water pollution affects plants also. Nitrates and phosphate produced by agricultural effluent after mixing with water, result in developing many types of harmful algae which do not allow sunshine to pass. Many useless vegetation grow and harm water plants. Due to thermal pollution many algae and fungus develop which reduce the quantity of oxygen by causing decay. Vegetation is destroyed by this sewage by polluting water and it support the fast development of algae, fungus and bacteria. Excreta fungus mainly bacteria, spherotilas, netais, carkesium, polypinum, leptomitus etc. remain spread over water surface. Due to increasing lichens in polluted water, absence of transparency of sunshine is caused and the process of photosynthesis is hindered. This prevents growth of plants.

Thus, water pollution not only affects creatures but also the trees and plants which affect man. Vegetables are being grown by establishing sewerage farms near urban effluents. By consuming them, the food cycle of man is affected.

Effects on Agriculture

Crops are destroyed by irrigation with polluted water. Soils are also polluted in the process. Harmful elements found in polluted water enter into the soil and reduce its productivity. Apart from this, crops grown with polluted water are not beneficial for the health of human beings.

Ecological Degradation

Water is the basis of the ecological system. Trees and plants as well as other biotic organisms develop according to it. A water deposit is a complete ecological unit and all the living organisms found in it remain in a balance and a food chain is formed. This chain remains healthy till the water remains fresh in the natural form. Food chain becomes imbalanced as soon as water becomes polluted. Danger is created for the biotic community and many organisms, specially sea creatures become extinct. As Patrick has said, "Water pollution reduces aquatic biodiversity which causes destruction of life in river". Similarly, due to weeds like water hyacinth and water ferns etc. found in water, vegetation having roots in water and algae are destroyed. Photosynthesis is not possible due to water pollution and it also destroys vegetation in the water.

Grave effect of water pollution is caused on sea creatures. In the absence of availability of proper geographical conditions to planktons, many species of fish are on the brink of extinction. Similarly on the earth, every element of ecological system is being affected by water pollution. Its effect would be worldwide rather than being at the micro level.

Methods of Controlling Water Pollution

Control of water pollution is the demand of the day. Cooperation of the common man, social organizations, national government and non-governmental organizations, is required for controlling water pollution through different curative measures. The government should also take suitable steps for controlling water pollution by framing laws and enforcing them strictly. Hence, the following steps should be taken for controlling water pollution:

1. Industrial and urban wastes should not be disposed of in water sources.
2. Ban may be imposed on mixing of pollutants in drinking water sources for which a boundary wall etc. may be erected around drinking water sources.
3. Domestic waste and sewage may be treated before flowing them directly in drinking water sources. Use of such treated water may be made for cooking vegetables, which may be grown on sewerage farms.
4. Bathing and washing should be banned near drinking water sources because detergents used in washing clothes pollute water. It is difficult to treat water mixed with detergents.
5. Bathing of animals near water sources should also be prohibited.

6. Use of treated filthy water should be made for irrigation because as compared to filthy water, it is more productive.
7. There are many creatures which purify water on the earth. Hence, they should be protected.
8. Chemicals and pesticides should be used in agriculture in a balanced quantity.
9. Control may be exercised on unnecessary weeds growing in water, because they affect the process of photosynthesis and pollute water.
10. Sanitary latrines may be arranged in urban areas, so that defecation in open areas is restricted.
11. People may be made aware about dangers created by water pollution so that they feel their responsibility and take steps in this direction.
12. Chlorine, potassium permagnate and other bacteria preventing medicines may be inserted in drinking water sources from time to time so that they remain free from pollution
13. Dead bodies of humans as well as animals should not be flown in water.

The Government of India has constituted the Central Water Pollution Control Board under the Water Pollution Control and Prevention Act, 1974, which has given priority to the following facts:

1. Supervision may be kept on disposal of industrial effluents.
2. Survey may be done regarding pollution of water sources.
3. Cheap methods for treatment of polluted water may be developed.
4. Research may be done regarding environmental pollution.
5. Awareness may be generated regarding pollution.

11

River Water Pollution

Fresh water is only 2.7 per cent out of total water available in nature. The remaining 97.3 per cent is saline water. Out of the total fresh water (2.7%), 0.003 per cent water exists in rivers, whose total volume is 108 cubic km while the volume of water existing in lakes is 1,26,070 cubic km. Thus, the water deposit in lakes is more than the water deposit in rivers. But even though water of rivers is lesser than that of lakes, river water is more important because it is distributed over a larger part. River water continues flowing and does not remain in any river for more than two weeks whereas, water of lakes remains there for years.

River water is used to a maximum extent because of its common reach to the biotic community. Maximum economic activities concentrate around rivers. At present, 50 per cent population of the world resides in 250 river basins, where they use river water for different activities including agriculture and industry. In India, dense population is settled near the banks of rivers Ganga, Yamuna, Damodar, Hooghly, Cauvery, Godavari and Chambal. This has polluted these rivers to a large extent. Rhine river of Europe, literal meaning of which is pure (*Rhine=Pure*), is polluted to a great extent. In Ruhar river basin, at the junction of river Rhine and river Ruhar, death is caused by bathing and swimming there. Every hour 1,200 ton polluted salts are disposed off in river Rhine. Due to pollution, Rhine is called "Europe's Sewage Ditch'. Water of river Rhine sustains about 20 million persons who are suffering health hazards continuously due to pollution of water.

The main river Seine of France on whose banks the city of Paris is located, is being polluted incessantly. Heaps of dead fish and sewage foam are spreading near its banks. According to one study, in the total length from Paris on river Seine to La Harre (Port located in the English Channel), about one million cubic metre polluted sewage water is disposed. Similarly in Po river of Italy, about 300 million ton wastes are disposed off every year. Rivers of Scandinavian region are being polluted by acid rain caused due to winds coming from the side of Great Britain and Germany, which carry sulphuric acid in them.

In India, river pollution has crossed the mark of crisis. Three important river systems of the north like Indus, Ganga and Brahmaputra are suffering from pollution. Maximum populated areas of the world are settled in all the three basins. These river systems originating from snow peaks of Himalayas are the basis of prosperity of India in different forms. In southern India, river Godavari, Cauvery, Krishna and Mahanadi are highly polluted. Rivers of India are no more rivers but have been converted into filthy drains. The Comptroller and Auditor General (CAG) of India has entrusted the responsibility of examination of quality of water to the Environmental Research Laboratory (ERL), Lucknow. ERL has divided tested water in five categories, i.e., A, B, C, D and E. Description of these categories is given below:

- Category (A): Suitable for Drinking
- Category (B): Suitable for bathing, swimming and entertainment
- Category (C): Potable after traditional treatment
- Category (D): Suitable for fish and forest animals
- Category (E): Suitable for irrigation, industrial cooling and waste disposal

In India, river pollution has extended in every context. In the hilly portion of Kulu Valley of Himachal Pradesh, water level of river Vyas has degenerated from 'A' level to 'B' level, though Manali is a famous tourist place of Kulu Valley. Lakhs of tourists from foreign and local places come here. Due to increasing terrorist activities in Kashmir, the number of tourists has increased in Kulu valley. The water of river Beas is used for drinking purposes in Manali and Kulu but the attention of planners is not drawn towards the hotel and restaurant business, developing at a very fast rate where there is no proper arrangement for disposal of sewerage and waste.. Plastic and other non-perishable materials are continuously disposed off in this river.

Table 11.1
Quality of River Water

Name of Element	*Quantity*
Alkaline Benzene Sulphonates	0.5 milligram per litre
Arsenic	0.1 milligram per litre
Copper	1.00 milligram per litre
Chloride	2.50 milligram per litre
Cyanide	0.01 milligram per litre
Manganese	0.05 milligram per litre
Nitrate	45.00 milligram per litre
Pifnals	0.001 milligram per litre
Sulphates	250.0 milligram per litre
Total Soluble Remnants	500.0 milligram per litre
Cadmium	0.01 milligram per litre
Cromium	0.05 milligram per litre
Lead	0.05 milligram per litre
Selenium	0.01 milligram per litre

Pollution in River Ganga

Water of river Ganga, which is considered as nectar in India, has become poisonous today. What to talk of drinking, it cannot even be used for bathing. A dangerous virus named Bacteriophase is found in the Ganga. The quantity of mud is increasing continuously. The cities of Haridwar, Bijnaur, Farukhabad, Kanpur, Allahabad, Benaras, Ghazipur, Ballia, Chhapra, Patna, Barauni and Munger are settled on the banks of river Ganga, dispose of sewage and industrial waste in the Ganga, spreading dangerous pollution. Due to presence of such pollutant materials, ERL laboratory of Lucknow has placed water of river Ganga in 'D' category, according to which it water is not suitable for drinking and bathing. It can be used only for fisheries and forest creatures.

Our relation with river Ganga is also from the historical point of view and not just from the religious angle. Our civilization and development is connected with this river basin. Jawaharlal Nehru had remarked that "The Ganga is Life of India". In the beginning, the water of river Ganga was very pure, soft and healthy. Our ancestors had framed rules for preserving the purity of its water but revolutionary economic development and increasing population during the 20th century have broken these rules. *Brahm Puran* scriptures written between 325 to 400 AD clearly stated that "Keep Ganga Clean". Throwing dirty water, throwing flowers after worship, washing filthy clothes, throwing of hair, rowdyism, doing vulgar activities, throwing of dirty clothes etc. were prohibited. We could not think of pollution of river Ganga 1,500 years back. Today, as a result of it, water of river Ganga is going beyond the reach of the biotic community and we have played an important role in this. Mainstream of river Ganga flows through four states, i.e., Uttaranchal, Uttar Pradesh, Bihar and West Bengal, but its companion rivers bring water from Himachal Pradesh, Punjab, Haryana, Delhi, Rajasthan, Madhya Pradesh and Chhatisgarh.

In the Ganga, mainly silt, biotic and chemical pollutants are found. Silt is received from soil through soil erosion. About 150 crore ton silt is deposited in the Ganga every year. Soil erosion can be controlled by dense tree plantation in watershed areas. This can help get rid of the silt problem. Organic and chemical pollutants come from cities situated near the banks of river Ganga. The waste of 29 big cities having population of more than one lakh and 23 medium cities having population ranging between 50,000 to one lakh situated on the banks of river, is mixing in it. In most of these cities, there is no sewerage system. Chemical pollutants from waste coming out of the industrial units situated on banks of the river also mix with the water polluting it.

The Ganga flows through densely populated areas of India. Among the big industries located on the banks of Ganga river, 86 are in Uttar Pradesh, three are in Bihar and 43 are in West Bengal. In Uttar Pradesh, 59 out of the 86 industries are leather industries, which dispose off poisonous chemicals in heavy quantities. Poisonous industrial wastes including acid, alkaline, sulphate, nitrate etc. also directly mix in the Ganga without any treatment. Maximum domestic waste mixes in Ganga in West Bengal. Thus, on an average, among the pollutants mixing in Ganga, 80 per cent is domestic waste and 20 per cent is industrial waste. Domestic filth of the metropolitan city of Kolkata and the waste of nearby textile industries, paper industries, tanneries etc. is disposed off in the Bhagirathi-Hooghly river.

Ganga Action Plan

Ganga Action Plan was started in 1986 for control of water pollution in the Ganga. The main function of this plan was to make river Ganga free from disposal of waste of the cities settled on the banks of the river. The scheme is to make Ganga pollution free from Rishikesh to Kolkata. The Central Pollution Control Board had prepared a five year project for the action plan in 1984. The Central Ganga Authority was formed for its implementation in 1985 and Ganga Action Plan was then launched to make the Ganga pollution free.

First Phase of Ganga Action Plan

The first phase of the Ganga Action Plan (1986-1993) was inaugurated by late Rajiv Gandhi at the Rajendra Prasad Ghat of Benaras. A National Protection Agency was constituted for its implementation. During the first phase of Ganga Action Plan, 261 schemes involving an expenditure of 462 crores were undertaken in Uttar Pradesh, Bihar and West Bengal. Up to the year 1999, 254 schemes had been completed. Under this plan, a scheme was prepared to divert 134 million litre out of 873 million litre waste water disposed by 25 cities declared as 'A' class in 1985.

Table 11.2

Ganga Action Plan: Schemes Implemented in First Phase

Scheme	*Uttar Pradesh*	*Bihar*	*West Bengal*	*Total*
Check Sewage and Change it	40 (40)	17 (17)	30 (31)	87 (88)
Sewage Treatment Plants	12 (13)	3 (7)	14 (15)	29 (35)
Sanitation Schemes of Low Cost	14 (14)	7 (7)	22 (22)	43 (43)
Electricity Operated Cremation Grounds	3 (3)	8 (8)	17 (17)	28 (28)
Development of River Banks	8 (8)	3 (3)	24 (24)	35 (35)
Other Schemes	28 (28)	3 (3)	1 (1)	32 (32)
Total Schemes	*105* *(106)*	*41* *(45)*	*108* *(110)*	*254* *(261)*

The objectives of Ganga Action Plan are as under:

1. Convert drains carrying filthy water from different cities settled on the banks of river Ganga and divert sewage water to sewerage treatment plants. After treatment in refining plants, use water in fish farms and for irrigation and generate power from the remaining waste after treatment
2. Establish sewage treatment plants in cities settled on the banks of river Ganga.
3. Construction of community latrines in cities located on the banks of river Ganga so that most of the sewage can be controlled at the source itself. Construction of electrified cremation grounds and suitable disposal areas for urban wastes.
4. Prevent situation pollution by controlling soil erosion from banks of the river.
5. Create awareness regarding pollution of river Ganga and make people conscious about its purification.

Table 11.3

Chemical Position of Water of the Ganga

City	*Temperature (OC)*	*Dissolved Oxygen (DOmg.lt)*	*Demand of Organic Oxygen (BODmg.lt)*	*Chloride (mg litre)*	*Hardness (mg litre)*
Rishikesh	15.5	8.5	6.6	4.5	63
Haridwar	18.0	8.0	7.2	3.0	68
Narora	29.0	7.4	7.2	5.0	76
Kanauj	29.0	6.0	7.0	6.0	136
Kanpur	29.0	6.0	10.2	9.0	120
Allahabad	30.0	6.5	9.3	23.5	138
Varanasi	31.0	5.8	10.0	32.5	154

Special stations have been created at 27 places starting from Rishikesh in Uttaranchal to Utuberia in West Bengal for testing the quality of water of the river Ganga during the first phase of Ganga Action Plan. These stations are being operated by experts from Bharat Heavy Electricals (BHEL) Haridwar, National Environment Engineering Research Institute (NEERI) Nagpur, Patna University Research Institute etc. Great achievements were expected from Ganga Action Plan in the first phase from 1986 to 1993, but in spite of all such great works on a large scale executed during such a long period, river Ganga has not been purified. In the name of making Ganga pollution free, hundreds of crores of rupees were wasted in the polluted water of the river, also starting economic pollution. Ecologists also think that the condition of the ecology has deteriorated during the operation of Ganga Action Plan. Since the first phase of the Ganga Action Plan could not prove effective, the Government of India started the second phase of the Action Plan in the year 2001. The role of Uttar Pradesh Water Corporation, Central Pollution Board and Central Public Works Department and PWD still remain important in carrying out the plan.

Yamuna Pollution

The Yamuna river too has changed from its original form. In the ancient period, its water was called 'Water of Faith' due to its purity, but today its water is not considered fit for drinking or bathing. The latest research was conducted by Environment Research Laboratory

Table 11.4

Contribution of Different Cities in Pollution of Yamuna

City	*Flow of Water (crore lt. per day)*	*Demand of Organic Oxygen (BOD) (ton per day)*
Yamuna Nagar	2.80	4.15
Karnal	2.30	3.45
Panipat	2.50	3.75
Sonepat	1.90	8.10
Gurgaon	2.00	3.03
Faridabad	10.20	15.15
Delhi	110.0	132.00
Saharanpur	4.50	6.75
Ghaziabad	18.00	27.0
Noida	7.5	11.25
Mathura	3.1	4.95
Agra	9.0	13.50
Itawa	1.5	2.25

Source: Centre for Science and Environment's Reports on Homicide by Pesticides.

(ERL) Lucknow, as reported by Comptroller and Auditor General (CAG) of India. At present, Yamuna water is suitable only for fish culture and for comsumption by animals.

Yamuna is the main linking river of Ganga. It originates from the western slope of Bandar Poonchh of Yamunotri glacier. After starting from Yamunotri, it travels 1384 kms and joins river Ganga near Allahabad. During this flow route, Yamuna river is polluted by many polluting sources. Yamuna carries flow of chemical fertilizers and pesticides in large quantity from farm lands of Haryana. Not only this sewage water of Yamuna Nagar, Panipat, Sonepat, Karnal, Delhi, Gurgaon, Faridabad, Mathura, Agra and Itawa etc. cities also mixes in Yamuna.

Delhi Municipal Corporation contributes the most in polluting Yamuna river. Wastes and excreta water in large quantity from Okhla Industrial Estate is disposed off in Yamuna river. Apart from Delhi, disposal of sewage in large quantity is also done by Mathura and Agra. Like Delhi, industrial wastes along with sewage water pollute Yamuna here also. In Delhi, Yamuna water contains Benzene Hexa Chloride (BHC) at the rate of 218 nenogram per litre whereas, in Agra it contains BHC at the rate of 1733 nenogram per litre, which is six times more than that found in Delhi.

Due to eutrophication in Yamuna, the water has become greenish. Eutrophication is the creation of favourable conditions for development of plants. Increase in concentration of nutritive elements is a favourable activity for eutrophication of plants , due to which maximum growth takes place in algae and phytoplankton. As a result of it, the water becomes greenish.

Yamuna Action Plan

For making river Ganga free from pollution, its companion rivers were also placed on priority to make them pollution-free. Yamuna Action Plan was sanctioned in April 1993 to keep Yamuna clean. It was estimated to spend Rs 510 crores on this Action Plan. Different schemes covering 21 cities of Haryana, Delhi and Uttar Pradesh are being implemented to remove pollution from the water of river Yamuna.

Damodar River Pollution

Damodar river is being affected by the big industries and mineral environment in Jharkhand and West Bengal. Industrial areas of Dhanbad, Sindri, Bokaro, Asansol, Burnpur, Raniganj and Durgapur are situated on its banks. Dhanbad coal area has the biggest coal fields of India spread over 450 sq km. Coal ash in a large quantity from here is spreading a thick layer over the water of river Damodar.

Besides the above rivers of Northern India, rivers Godavari, Chambal, Cauvery, Krishna and Mahanadi in South India have also come in the grip of pollution. To keep all these rivers free from pollution, various schemes are being executed with the assistance from many countries.

12

Industrial Use of Water and Trend of Water Supply

Industrial water use is closely linked to the economy of a country. Water has traditionally been considered a common property and industry is included in this rubric. According to the latest scenario regarding use of water for different purposes by the 630 crore population of the world, domestic consumption is only 8 per cent. The maximum use of water, i.e., 70 per cent is made for agriculture while 22 per cent is used in industries. This estimate has been published in World Water Development Report (WWDR), 2003 of United Nations Development Programme. Simultaneously, it has been projected that in the coming times, industries would invade on distribution of water in both these sectors. At present, this tendency in industries is becoming visible in view of their increasing demand. In reality, use of water in industries in high income countries has already reached 59 per cent out of the total water portion of fresh water and it is double the use of water used in the agricultural sector. Hence, people are giving priority to industry as compared to agriculture as it is considered as the main indicator of economic development.

Increasing demand of water in industries at present is really a cause of anxiety, because developing countries are following the industrially important countries. Hence, it is probable that in the coming two decades, present use of water in industries (22%) would be double of the existing demand. According to (WWDR 2003), during the year 1995, annual consumption of water in industries was 752 km^3, which would become 1,170 km^3 per year by 2025. Maximum increase would be in demand by the industrial sector and in developing countries like India, industrial demand would be more because it is lesser today as compared to developed countries. (Table 12.1). Since, very long time, there has been a tendency of shifting industries from developed countries to developing countries.

Table 12.1

Use of Water

(in %)

Area	*World*	*Countries having Low and Middle Income*	*High Income Countries*
Agriculture	70	82	30
Industry	22	10	59
Domestic	8	8	11

Source: World Bank, 2001.

Use of water in industries is directly related to the economy of any country. Hence, if gross domestic production of India is increasing, then it is clear that use of water is also increasing in Indian industries. Industries not only use water but by polluting it, they also cause deterioration in quality of water. According to *WWDR– 2003*, 70 per cent industrial waste materials are disposed of directly without treatment in developed countries. Consequently, these industries are polluting the water supply that is fit for use. Thus, the matter regarding use of water in industries is interrelated with two issues – first, use of water and second, water pollution.

Table 12.2

Increasing Demand of Industrial Water Use in India

(in billion cubic metres)

Category	*1990*	*2010*	*2025*	*2025*
Irrigation	460 (88.6 %)	536 (77.3 %)	688 (73 %)	1008 (70.9 %)
Domestic	25 (4.8 %)	41.6 (6.1 %)	52 (5.5 %)	67 (4.7 %)
Industries + Energy	34 (6.6 %)	40.1 (6.0 %)	80 (8.5 %)	121 (8.5 %) 141 (10.1%)
Total (including others)	519	693	942	1422

Source: National Commission for Integrated Water Resources Development Plan, Ministry of Water Resources, 1999.

Per capita availability of water in India is reducing. In 1951, the per capita availability of water per annum was 5177 cubic metres, which reduced to 1869 cubic metres in 2001. It is expected that per capita availability of water per annum would reduce to 1,340 cubic metres by the year 2025. This would cause enormous shortage of water because 50 per cent capacity of many rivers has already been exploited. About 80 per cent of the gross domestic water supply depends on groundwater, but even then we are continuously polluting this water. Consequently, continuously decreasing water level and problem of pollution are becoming burning issues.

According to the Central Water Resources Ministry, maximum water supply in the country is made in agriculture. It is 80 per cent of the total water availability, whereas domestic consumption is only 5 per cent of total fresh water. Thus, industrial sector is at second place after agriculture.

Water Consumption in Industrial Sector in India

Various national and international agencies have presented estimates regarding use of water in industries in the Indian context. According to Ministry of Water Resources, about 40 billion cubic metre water is used in industrial areas of the country, which is about 6 per cent of total availability of water. According to Central Pollution Control Board (CPCB) of India, about 500 billion cubic metre water out of the total available fresh water is used in industries annually. Out of this, about 10 billion cubic metres water is used by processing industries and 30 billion cubic metres is used for refrigeration purposes. Thus, this part of water used for industries comes to about 6 per cent of total available water in the country. On the other side, according to World Bank, the demand of water for industrial use and power generation is increasing at the rate of 42 per cent per annum. By the year 2025 it would become 228 billion cubic metres as compared to only 67 billion cubic metres in 1999. Thus, water used in industries is about 13 per cent of total available fresh water at present. It is clear from different estimates that demand for water in industries has been continuously increasing.

Need for Rational Use

The effect of increasing demand for water by industries is caused on balanced availability of water resources in two ways- first, to maintain organized supply of its increasing demand and second, to maintain the quality of water in its natural form, so that balanced availability is maintained in all sectors. But due to increasing industrialization, such possibility seems very little because water used in industries is proving to be a double-edged sword harming both the sides.

Thus, it has become a cause for worry to maintain quantity as well as quality of water. For that, different industries should be categorized. Water may be reused by recycling and similar category of water may be utilized according to level of industry. Fresh water may not be directly used in every category of industry. Various studies have clarified that small and medium industries are polluting water to the maximum extent. Central Pollution Control Board has given an estimate of waste water disposed of by industries and similar type of research has also been conducted by the Centre for Science and Environment (CSE), New Delhi.

Table 12.3

Indian Scenario of Water Pollution by Industries

Industrial Sector	*Annual Waste Water Discharge*	*Annual Water Consumption (million cubic metres)*	*Proportion of Water Consumed in Industry*
Thermal Power Plants	27,000.9	35,157.4	87.87
Engineering	1,551.3	2019.9	5.05
Paper and Pulp	695.7	905.8	2.26
Textile	637.3	829.8	2.07
Steel	396.8	516.6	1.29
Sugar	149.7	194.9	0.49
Fertilizer	56.4	73.5	0.18
Others	241.3	314.2	0.78
Total	30,729.2	40,012.0	100.0

Source: Estimated by CSE Based on the Waste Water Discharged Data published by CPCB in Water Quality in India 1990-2001, *Down to Earth*, February 29, 2004, p. 66.

Demand for Water in the Industrial Sector of India

The use of water in industry depends on the nature of the industry. Thermal power plants, textile, paper and pulp, and iron and steel industries need maximum water. All the important thermal power plants of the country are under the control of the government. On an average, thermal power plants having a capacity of 1,000 kilowatt consume 80 cubic metre water. Paper and pulp industry in India also consumes more water as compared to that in other developed countries. Textile industry is also consuming additional quantity of water. In most of the textile mills of India, water is not used for washing of clothes in the form of a continuous flow but instead fresh water is used at every level.

Table 12.4

Demand of Water in Industrial Areas in India

Sector	*Average Water Consumption in Indian Industries*	*Globally Best*
Thermal Power Plants	Average 80 cubic metre per megawatt (mwh)	Less than 10 cubic metre (M^3) per mwh
Textile Industry	200-250 M^3/per tonne cotton cloth	Less than100 M^3/per tonne cotton cloth
Paper and Pulp	• Wood base mills: 150-200 M^3 ton/per tonne • Waste paper based mills: 10-80M^3/per tonne	• Wood base mills: 50-75 M^3 ton/per tonne • Waste paper based Mills: 10-25 M^3/per tonne
Total Iron and Steel Plants	10-8 0M^3 per tonne of finished product (Average around 25 M^3)	5-10 M^3 per tonne of finished product. Best Practice less than 0.1 M^3 waste water per tonne finished products.
Distilleries	75-200 M^3 per tonne Alcohol Produced	Figures NA
Fertilizer Industry	• Nitrogenous Fertilizer Plant 5.0-20.0 M^3 per tonne • Straight Phosphatic Plant 1.4-2.0 M^3/per tonne • Complex Fertilizer 0.2-5.4 M^3/per tonne	An effluent discharge of less than 1.5 M^3 tonne as P2 O5

Source: Down to Earth, CSE, New Delhi February 29, 2004, p.67.

The Iron and Steel industry in India the country is much ahead of other industries regarding water consumption. About 80-85 per cent fresh water is used in it. In United States of America, however, 95 per cent fresh water is used in manufacturing of iron and steel.

Water Scarcity

As a result of increasing demand of water in different industrial areas, a situation of water scarcity has been created, leading to closure of many plants in the year 2002. During that year, Harihar Polyfiber Limited in Karnatak and Indian Rayon Plant at Nagda in Madhya Pradesh had to be closed down due to scarcity of water, during the summer season when scarcity of water occurs in rivers. Community tension has also been created due to overexploitation of groundwater by industries. Effect of overexploitation and pollution of water is also being caused on water supply by agriculture and domestic sectors. In most of the industries there is no arrangement for treatment of water, hence fresh water is used at every stage rather than cycling the used water.

Cost of Water

Low cost of water used for industries is also responsible for its deterioration. At many places cheaper water is made available by giving subsidy on it. Three ingredients are important as regards cost of water:

First: Water tax should be paid to Pollution Control Board and its assessment should be based on quantity of waste water disposed off.

Second: Water supplying agencies like the Municipal Board, Urban Council or the private sector should charge purchase price of water.

Third: At many places, river water or groundwater etc. is used without paying the cost of water. Only in urban areas is the cost of water recovered.

European Economic Community and United States of America decide the cost of water on the basis of average cost or marginal cost. In India, charges of water supply to industries are decided on average cost basis, whereas the fact that pollution of surface water and groundwater is caused by industries is ignored. Thus, no pollution tax or effluent tax is paid by any industry. Indian industries are taking advantage of this situation and are continuously causing quantitative and qualitative deterioration of water sources.

Centre for Science and Environment (CSE), New Delhi, has studied the environmental effects of Indian industries through the 'Green Rating Project'. Recently CSE has completed rating of paper and pulp industry. According to them, plants situated in water scarcity parts of southern and western India consume lesser water as compared to plants situated in water sufficiency parts in northern and north-eastern areas of India. South Indian paper industry consumes about 160 cubic metre (M^3) water per tonne of paper manufactured whereas the same industry in north India consumes 165 M^3 water per tonne. Paper factories situated in northern India consume about 185 cubic metre (M^3) water per tonne production of paper, whereas in north-eastern parts the same industry consumes 205 M^3 water per tonne. It becomes clear from this analysis that there is no commonly agreed decided standard measure for consumption of water but it is consumed according to availability of water in the area.

Need for Government Policy

So far no specific policy has been decided in India as to what should be the form of water in industries, how much tax may be charged and what should be the responsibility of industries as regards pollution of water. At present, light has been thrown only on public health and availability of water. In industrial areas of many countries of the world, standards have been fixed regarding consumption of water. For example, in China, targets of consumption of water by important industries have been laid down. According to Water Conservation Agency Report of China, pressure was built for saving 6 billion cubic metre water in first water quota (quantity) by the year 2005. Similarly, in developed countries also, policies have been decided relating to saving of water in industries. India has not framed any rule according to which the limit for consumption of water by industries might have been laid down. Though Central Pollution Control Board has decided the level of consumption of water for some industries but they are only recommendations. They have not been implemented in legal form although some rules have, of course been framed as regards exploitation of groundwater. In our country, any person who is owner of a part of land, is also treated as owner of the groundwater reserves under the land. In such a situation, one uses groundwater as per his requirement rather than consciously, by exploiting the groundwater.

National Water Policy and Water Allocation in Industries

National Water Policy, 2002 has conducted studies at many places regarding use of water in industries. Industries have been described six times in the document containing 6,000 words. The following description regarding industries has been made:

1. Before disposing of industrial effluents in any natural water source, it should be treated at the appropriate level. In this context, the burden of pollution has not been discussed. Present level of industrial effluents is centralized.
2. In management of polluted water, the principle of payment will have to be followed, but more details of payment have not been elaborated.
3. Availability of water for activities like agriculture, industry, urban development and economic development will have to be kept in view.
4. Expertise of all types of uses of water should be adopted with optimistic inclination, awareness should be generated regarding water scarcity, and education regarding awareness of water conservation should be encouraged.
5. Resources should be conserved and efforts should be made to maintain their availability.

Solution of Water Problem in Industries

Demand for water is continuously increasing in industries, because Indian industries make little use of polluted water. Neither do they practice reuse of water nor recycling of water and process technology. After making a single operation of water, it is disposed of without utilizing it for a second subsequent use. Many a time, less polluted water is also not reused. Thus, unpolluted water also becomes polluted after mixing with polluted water. In India, specially, a large quantity of water is used for cooling in thermal power plants. Hence, for maintaining sufficient quantity and quality of water in industries, optimum management of water is essential. The following measures can prove useful for it:

1. *Improved Process Technology:* Improved and developed process technology can be helpful in reducing the demand for water in industries.
2. *Reuse of Process Water:* This process includes reuse of water at different levels. Where water is not thrown away after using it once, there will be lesser demand because it would be used in a chain.
3. *Re-circulate Process of Water:* The same water can be re-circulated after treatment and again used for a similar purpose. It is a cheaper process of saving water by incorporating latest techniques.

4. *Rain Water Harvesting:* Industrial units can use rain water in different forms. This water supply can be complementary to the traditional supply of water and it would save water. It can also be implemented through legal provision.
5. *Increasing Availability:* By balancing demand of water through its use in qualitative form, its availability can be increased. It may be made obligatory for all industries to dispose off water used by them only after its treatment, at a suitable place, and fresh water shall not be used at every level. Thus, water can also be conserved through its conscious use.

Such strategies may be determined for industrial areas in the country so that their demand is also fulfilled and availability of water is also maintained. For this, pressure will have to be built on industries for treatment of polluted water. Special Indian based standards for effluents should be adopted so that industries can save water. They should take steps to control pollution. A limit for taking water from water sources may be fixed so that reuse and recycling gain importance. Stress has to be laid on conservation of rain water. There should also be legal provisions regarding use of water at all levels so that there is a sustainable base for water and water supply in industries is maintained for the future as during the past.

Trend of Water Supply in Rural and Urban Areas in India

Accessibility to drinking water has been considered as a fundamental right of citizens in India but as far as the question of availability of water resources is concerned, uniformity does not exist everywhere. People get water as per their strength and influence. Thus, distribution of water resources and supply of water is controlled by economic status. In India too, as in other countries of the world, policy regarding providing fresh water to all and giving maximum priority to rural areas has been adopted. According to World Water Supply and Sanitation Evaluation Report 2000 A.D., target has been laid down that 50 per cent population facing problem of drinking water supply would get rid of it by 2015 and all persons would get safe drinking water by 2025 A.D. The same was repeated in the World Environment Conference held at Johannesburg in 2002 A.D.

In India, as per the above policy, stress has been laid on drinking water supply to rural areas, and among uses of water, the policy has given first priority to drinking water and has promised to provide fresh drinking water to all citizens of the country.

If all the water resources of India are estimated, it becomes clear that there exists an estimated annual flow of 1,86,9000 crore cubic metres water in 10,360 rivers and companion rivers having a minimum length of 1.6 km each, but due to geographical circumstances, only 32 per cent (6,90,000 crore cubic) water of these sources is available for use. Secondly, India has in-completing groundwater capacity of 4,39,00.9 crore cubic metres. Water supply for domestic consumption in rural areas is generally made from groundwater sources. In our country, the responsibility for providing drinking water supply in rural areas vests with the government but the central government, through its drinking water supply department, assists in the efforts of the state governments by providing financial assistance, policy guidelines and technical support. In this process, complete investment is by the government and private participation is almost negligible. The central government had constituted a National Drinking Water Mission in 1986, and in 1991, its name was changed to Rajiv Gandhi National Drinking Water Mission. This mission was entrusted the responsibility of supplying drinking water in rural areas. After setting up the mission, the country has made a good progress in the field of rural drinking water supply.

However, in spite of continuous efforts to provide drinking water, we have been able to solve the problem in only 86.9 per cent rural settlements in India. About 11.7 per cent settlements are still without safe drinking water supply. Through a recent report it has been made clear that in settlements in rural areas, 15 per cent persons are suffering from problems related to quality of water. Among these problems, the main problem is of fluoride, arsenic, iron element, and pollution due to nitrate and salinity.

In spite of so many efforts for providing drinking water in rural India, water supply system is disturbed on account of the fact that the common man has developed the presumption that water is a social asset and responsibility for providing it vests completely with the government. They forget that water is a rare economic resource whose management should be made on a community basis by organizations at local levels. The tendency of rural people for remaining dependent on others is completely responsible for it. Hence, participation of local people should be ensured in rural water supply arrangements and responsibility of maintenance of water sources should also be entrusted to them. At present, water supply arrangements are operated by Public Health Engineering Department or Water Boards. Hence, there should be participation of people in planning, selection of technique and operation of the system. Participation of local bodies, private sector and non-government organizations (NGOs) should also be there.

According to a present estimate, an investment of about 45,000 crore rupees is required for providing fresh drinking water services at the minimum level to all rural settlements. It will be difficult for the government alone to manage it due to various obstacles, hence other concerned beneficiaries and related persons will have to join in for the implementation, operation and maintenance of the project. The role of NGOs would be particularly important in it. Presently, this strategy is being treated as an important one in the country. Under this strategy, the participation of the rural community in the rural water supply programme is felt necessary and the following projects have been started under this programme:[1]

Community-based Perspective

The work of planning and execution of these systems was done by the government department and it was implemented for beneficiaries at the end. These systems were implemented, but they kept the local needs, priorities regarding technology, its feasibility and desire of people to bear the cost of services in mind to a limited extent only. The result was that while preparing the outline and design of water supply schemes, attention towards financial aspects like capital cost and maintenance cost of proposed systems was not paid. The local community should be included in the following ways for stability of the system:

1. Community initiative and capacity building.
2. Community participation in capital investment.
3. Community ownership and implementation.
4. Operation, maintenance and community management.
5. Reliable water source to fulfill community needs.

Regional studies have indicated that when consumers are participants in the operation and maintenance work of the system, they are prepared to happily contribute money also for it. When they are confident that they would be able to manage the amount needed for operation and maintenance, and reliable arrangement of water supply would be possible, they willingly join the process. In the traditional government programme, people neither know about technical aspects nor do they have any choice for selecting an alternative out of the available technologies. By not making local people participants in preparing the outline of the project, mostly sub-standard materials are used, workmanship is also sub-standard, and proper management of maintenance is also not possible.

World Water Supply and Sanitation Evaluation Report, 2000, has identified the following regional obstacles. They are as follows:

1. Financial obstacles
2. Institutional problems
3. Shortage of sufficient human resources
4. Lack of political commitment
5. Lack of community participation
6. Insufficiency of operation and maintenance
7. Lack of education regarding health and sanitation
8. Lack of quality of water
9. Lack of communication and sufficient information.

All these facts apply to India.

Sector Reform Project

After 73rd Constitution Amendment Act in India, responsibility for rural water supply was entrusted to Panchayati Raj Institutions (PRIs) within the areas of their rights. The Government of India is assisting PRIs for empowerment and capacity building, so that they are able to discharge their responsibility satisfactorily. A better solution would be to empower the Gram Sabha, which is a constitutional body and all adult voters are its members. Gram Sabha should be made responsible regarding participation in implementation and review of projects and drinking water system.

For achievement of desired targets, the Government of India made policy changes in 1999 regarding reforms in the rural drinking water supply sector. Broad changes have been made in this area. Stress has been laid on empowering the local community, planning, selection of alternative techniques, implementation of the techniques, and participation of people in operation and maintenance of supply of water.

Stress is also being laid on increasing awareness of people about their effective participation in programmes regarding water supply. The community should have interest in operation and maintenance of water supply schemes because they are its owners. About 10 per cent of capital investment and total cost of operation and maintenance should be borne by them. Under this new policy, the role of the State is that of an 'assistant' rather than a 'donor'. The amount under Sector Reform Project is directly released to the District Water and Sanitation Mission, which is a multi-disciplinary organization and has been constituted with the objective of control, supervision and directing the execution of the projects. In some districts where Panchayati Raj Institutions are sufficiently active and where they have properly stabilized, they are implementing projects instead of District Water and Sanitation Mission. After getting the experience of the first phase, more reforms would be made in it and would be implemented in the remaining districts of the country also accordingly.

Institutional Set-up

A national level organization which includes experts from different fields acts as the National Project Sanctioning Committee. The secretary of the Water Supply Department is its Chairman. This organization is responsible for complete coordination of the activities of the organization and preparing its directives. At the state level, the present institutional structure or specially constituted State Water and Sanitation Mission is responsible for preparing policy guidelines and directions, establishing coordination with the central government, other state governments and regional assistants, supervision and evaluation of project implementation work, audit of accounts etc. District Water and Sanitation Mission or Panchayati Raj Institutions themselves implement the projects. They are also responsible for taking decisions on projects, looking after their planning and management, receiving central funds and auditing them, selecting partners in private sector and NGOs, constituting village water and sanitation committees along with selection of Panchayat and government functionaries, convincing the community, capacity building, implementing of project activities, operation and maintenance of projects after their preparation, etc.

Operational Process

Under the test project, demand has been generated as a result of conducting awareness campaign and capacity building. Village Water and Sanitation Committees would be constituted in these villages and they would be responsible for preparing the actual plan of drinking water system, implementing it and thereafter for its operation and maintenance. It would be a sub-committee of the village Panchayat, which is the lowest body in the three-tier system of Panchayati Raj. The functioning of Village Water and Sanitation Committee is performed according to the rules and procedures laid down in the Panchayati Raj Act/Rules. This body is solely responsible for implementation of projects and is the main base for reform processes. Sufficient representation is given to women and members of weaker sections of society in this committee, and other beneficiaries, non-government organizations and subject matter specialists are also included in this committee.

Scheme Cycle

A 36-month cycle is prescribed by this project. It includes:

1. Phase of Organizational Forming (3 months).
2. Sanitation and Identification Phase (12 months).
3. Village Water and Sanitation Scheme/System Planning Phase (9 months).
4. Implementation and starting of work (12 months).

During the project period, other activities like testing of quality of water, institutional community arrangement, evaluation of different aspects of project/scheme and other related matters would also be organized.

Conclusion

As per the existing estimate, an investment of Rs 45,000 crores is required for providing the minimum level of satisfactory services for supply of fresh drinking water in all rural settlements. It would be difficult for the government alone to arrange for it because of different obstacles. Hence, it would be necessary to make other concerned beneficiaries and related persons as partners for implementation of the project, and its operation and maintenance. The role of non-government organizations would also be important in this respect. At present, this strategy is being given importance in the country and under it, realizing the importance of community participation in rural water supply, the following projects have been started.

After achieving the target to a considerable extent through 100 per cent government investment, now this scheme has been formulated according to which demand based and community oriented programmes would be executed very early. This change would depend on the time spent on success of test projects implemented under the Sector Reform Programme. Through these projects such models would be prepared which could be replicated. Arrangements are being made so that through sector reform project experiments, the standard of living of common people specially of rural areas, is improved to a large extent. Villagers would be thereby encouraged to take over responsibility of other community programmes of rural development also on their shoulders. It is hoped that a revolutionary change would come in all development-oriented activities on a long term and permanent basis.

After development of a model in a particular area through test projects, many projects similar to them would be prepared in the country. Not only would the government make constructive contribution in such projects, but common people, representatives of political parties, NGOs, public communication media and private sector would also play an important role. But test projects of sector reform are quite complicated. Their scope is also vast. Implementation of 63 test projects itself is a challenging task.

Swajaldhara Scheme

The central government had started this scheme on 25 December 2002, with the objective of ensuring cent-per-cent coverage of water supply in rural areas of the country. This is a demand based drinking water supply scheme which includes development of pipelines at suitable places as per requirement of the people, and renovation of traditional water conservation methods like *baoris*, tanks, ponds, lakes etc. and such other sources, so as to increase availability of water. The central government would implement this scheme in the whole of the country. In the beginning, this scheme was implemented in the States of Andhra Pradesh, Madhya Pradesh, Maharashtra, West Bengal, Uttar Pradesh, Haryana, Himachal Pradesh, Orissa and Rajasthan. In all, 882 projects of these states had been sanctioned.

Since, this project is need-based and to be implemented through community participation, the beneficiary group of the water source has to form a committee and pass a resolution through the village Panchayat with the consent of the people of the village. It is essential to register it under Societies Registration Act, 1860. It would be the responsibility of the Committee to establish the water source and operate and maintain it. In all, 10 per cent of the total investment of the project is to be borne by the people of the village or the panchayat, whereas 90 per cent is borne by the government. In case there are 50 per cent or more Scheduled Caste or Scheduled Tribe population in any village as per 2001 census, the people's contribution would be only 5 per cent.

The Government of India has adopted this strategy of rural drinking water supply by giving it an institutional shape so as to ensure community participation. The Swajaldhara scheme is implemented on the basis of three following important principles:

First: To ensure community participation of rural people through preparing an outline relating to drinking water facilities, financial control and capacity to take management decisions.

Second: Role of the government as service facilitator rather than as service provider.

Third: Participation by sharing cost in cash or partly in cash and partly in kind and taking over complete responsibility of operation and maintenance by beneficiaries.

Objectives of the Scheme

This scheme has the following objectives:

1. To establish a suitable drinking water source for sufficient water in all rural habitations as per demand of local residents of rural areas.
2. To arrange safe drinking water for all rural people as per their requirement.
3. To ensure installation of handpumps, and look after their repair and maintenance with the coordination of villagers.
4. To create health awareness about diseases caused by unsafe water by providing safe drinking water, thereby reducing water borne diseases.
5. To ensure drinking water supply in schools, villages and habitations suffering from scarcity of drinking water.
6. To give importance to renovation of traditional water sources.
7. To save time consumed by rural women in fetching water from far off places and convincing them for making better use of such saved time.

Internal Navigation in India

The share of internal navigation in transportation in India is only one per cent. There are about 14,500 navigable waterways in India which include rivers, canals, backwaters and narrow gulfs. At present, there is only 3,700 km length of flat base which is navigable by mechanically operated ferries, of which only 2,000 km is presently being used for navigation. Mechanical boats can be propelled only in 900 km out of 4,300 km long navigable ways. Approximately 180 lakh ton material is transported through these waterways.

Indian Internal Waterways Authority was constituted on 27 October 1986 for development, maintenance and regulation of national waterways of the country. It also provides advice to the Central and state governments regarding development of internal navigable waterways. The Authority operates projects regarding development of basic infrastructure for internal navigation on national waterways. Its head office is in Noida and it has regional offices at Patna, Kolkata, Guwahati, Kochi and branch offices at Allahabad, Ballia, Bhagalpur, Farrakka and Kollam.

National Waterways

For giving a important place to internal navigation in national transportation system and for development of internal waterways, the government has declared four important waterways as national water

Figure 12.1

National Waterway No.1

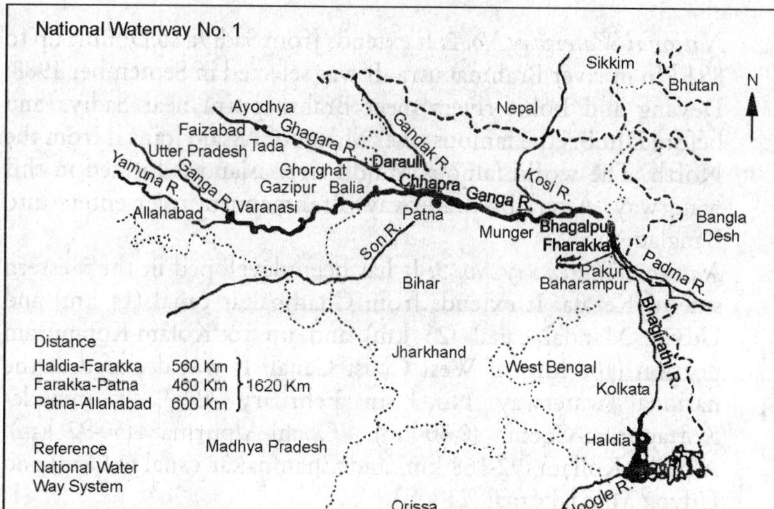

ways. Ten other waterways are still under consideration for such selection. They are the following:

1. *National Waterway No.1:* This water way has been developed on the Ganga-Bhagirathi and Hooghly system from Allahabad to Haldia which is 1,620 km long. It was selected in October 1986. It begins from the confluence of Ganga and Yamuna rivers in Allahabad and joins with Ghaghra and Kosi rivers in the North and Sone river in the South. Farakka barrage has been constructed over it.

Table 12.5

Important National Waterways of India

Waterway No.	*From and To*	*River System*	*Distance (km)*
N.W. 1	Allahabad to Haldia (Uttar Pradesh, Bihar and West Bengal)	Ganga-Bhagirathi	1,620
N.W. 2	Sadiya to Dhubri (Spread in Assam and connects North East area with Haldia)	Brahmaputra	591
N.W. 3	Kollam to Kotapuram (Kerala)	West Coast Canal	205
N.W. 4	Kakinada to Marrakanam	Godavari-Krishna	1,100

2. *National Waterway No. 2:* It extends from Sadiya to Dhubri up to 891 km in river Brahmaputra. It was selected in September 1988. Devang and Lohit rivers meet Brahmaputra near Sadiya and before Dhubri the famous river Manas of Assam joins it from the North. The world famous Mujoli river island is located in this waterway. After this waterway, Brahmaputra river enters into Bangladesh.
3. *National Waterway No. 3:* It has been developed in the western side of Kerala. It extends from Champakar canal (14 km) and Udyog Mandal canal (23 km) and up to Kolam-Kotapuram portion (168 km) of West Coast Canal. It was declared as the national waterway No. 3 in February 1993. It includes Kottapuram-Cochi (0-30 km), Cochi-Alpurma (30-92 km), Alpujha-Kollam (92-168 km) and Champakar canal (14 km) and Udyog Mandal canal (23 km).
4. *National Waterway No.4*: This waterway extends from Kakinada to Marakkanam in Krishna-Godavari rivers up to 1100 km.

Figure 12.2

National Waterway No. 2

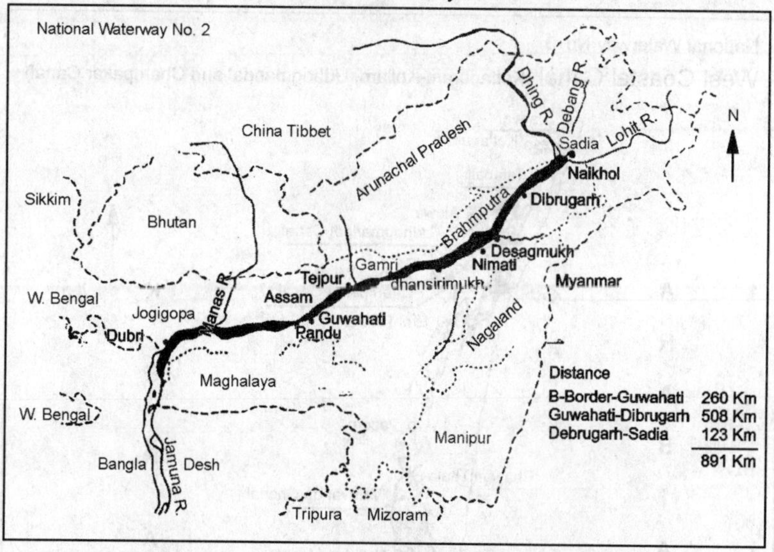

Apart from it, 10 other internal waterways have been identified for internal navigation. Special study has been made of Godavari, Sunderban, Krishna, Barak and Brahmani rivers, East Coast canal and Damodar Valley Authority canal etc. The studies report them as feasible.

Except the national waterways, Mandvi and Juari rivers of Goa, Mahanadi delta and the backwaters of Kerala are also being used for internal navigation. Buckingham river and Kambarjua rivers of Andhra Pradesh and Tamil Nadu are also important navigable canals.

Continuous efforts are being made by the government for development of national internal navigation. Public undertaking in the form of Central Internal Navigation Corporation was set up in 1967 for this purpose. It has its headquarters at Kolkata. This Corporation is mainly engaged in transporation of goods in Ganga-Bhagirathi-Hooghly, Sunderban and Brahmaputra rivers.

Weaknesses and Advantages of Internal Navigation

As per an estimate of Ministry of Transport, Government of India, the length of nine inland waterways of India is 19,000 km, out of which 14,500 km long are of rivers and 4,000 km are of canals. Out of

Figure 12.3

National Waterway No. 3

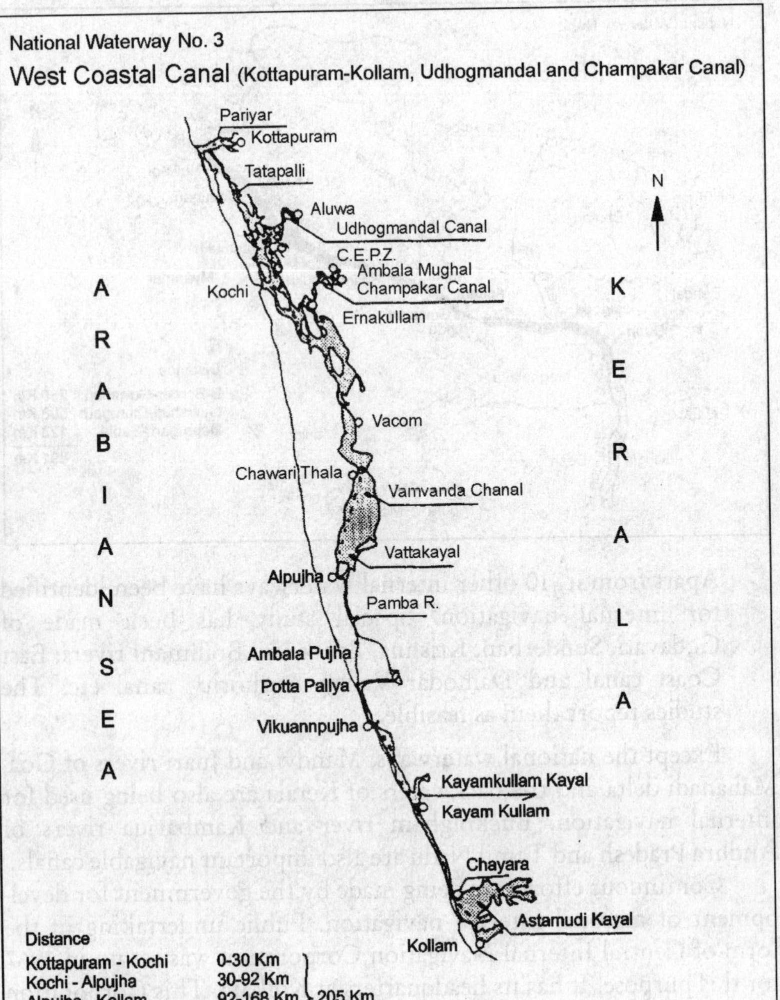

total navigable ways, 10,000 km waterways remain fit for navigation throughout the year whereas, the remaining 5,000 km are fit for navigation only for a few months. Steamers ply in 3,500 km, mechanically propelled boats ply in 2,500 km in rivers, and large local boats can ply in the remaining 5,700 km long way. Like other means of transport, inland waterways also have some advantages and weaknesses.

Weaknesses

1. In most of the rivers of India, floods are caused during rainy season when water current is so torrential that it becomes difficult to ply boats.
2. Most of the rivers remain dry during the summer and whatever water becomes available in rivers, it is used for supply of water to the vast canal system during the beginning of winter and summer seasons. Due to use of river water for irrigation, there remains no water for navigation during the summer season.
3. Rivers of the South India flow in rocky areas, hence they are not fit for plying boats because there are waterfalls in the way.
4. As compared to boats, railways are quicker, safer and cheaper means of transport.
5. Sometimes rivers also change their way and due to that only a small stream flows on one side and cannot be used for navigation. On banks of most of the rivers, thick layer of sand is deposited. Hence, loaded steamers or boats can reach the bank with great difficulty.
6. Almost all rivers fall from shallow and sandy deltas. Hence, ships cannot sail from sea shores to inland parts.

Advantages

1. Inland navigation has been considered as the cheapest means of transport among the methods of transport. Initial investment on river services as well as expenditure on their maintenance is much lesser as compared to road and rail transport.
2. For more load and long journey, transport through waterways is cheaper than rail and road transport.
3. It is considered as the most useful method from the point of view of energy saving. One horse power can carry 4,000 kilogram load in water whereas, it can carry 150 kilogram and 500 kilogram load by road and rail respectively.
4. Water transport is most useful for heavy loads.
5. In north-east India, road and rail transport is blocked during rains. It is also difficult to construct roads or railways in these areas. Hence, water transport is more suitable in such region.
6. There is lesser pollution in water transport.

Hydropower in India

Hydropower was first started in India at Shivsamundram on river Kaveri in Karnataka during the year 1902. Hydropower generation requires geographical conditions of sufficient water and slopy areas but hydro power occupies second place after thermal power in the world. In India, rivers play an important role in power generation. From the geographical angle, role of rivers in power generation can be understood from the following description:

1. In peninsular India, rivers falling in Bay of Bengal generate 21 per cent hydropower.
2. Rivers falling in the western region of peninsular India, produce 10.5 per cent hydropower.
3. Rivers of central India generate 10.5 per cent hydropower.
4. Ganga river system generates 11.5 per cent hydropower.
5. Brahmaputra river system generates 32 per cent hydropower.
6. Indus river system generates about 15 per cent hydropower.

About one-third hydropower generation in the country is from Brahmaputra river system. Since 1951, there has been special increase in hydropower generation. Hydropower capacity in 1951 was only 60 megawatt, which has increased to 29,500 megawatt in 2005.

About 27 per cent power in the country is generated through water. From the point of view of generation, hydropower generation was only 2.5 billion kilowatt in 1951, which has increased to 73.79 billion kilowatt in 2004-05. Many states of India like Andhra Pradesh, Assam, Bihar, Gujarat, Jammu-Kashmir, Kerala and Madhya Pradesh are important from the point of view of hydropower generation. Hydropower generation started in India at Darjeeling in 1897.

Table 12.6

Power Generation Capacity in India, 2004-05

Source of Power Generation	*Capacity in Megawatt*
Hydropower	29,500.23
Thermal Power including Gas and Diesel	77,968.53
Wind Power	1,869.66
Atomic Power	2,720.00
Total	*1,12,058.42*

Source: India 2005, p. 239.

Only a little part of the the potential capacity of power generation has been exploited in India. More emphasis has been laid on power generation in India only after 1951. In 1974, 42 per cent share was of hydropower. But later on, the share of hydropower went on decreasing due to greater emphasis on thermal power. Many multipurpose projects have been executed in India. Tungabhadra, Bhadra, Shiv Samundram (Karnataka), Iddiki (Kerala), Tata, Koyna (Maharashtra), Manchkud, Nagarjun Sagar, Shailam (Andhra Pradesh), Paikara, Mettur (Tamil Nadu), Bhakra Nangal (Punjab), Sharda, Rihand (Tihri-Uttar Pradesh), Uttaranchal, Damodar Valley Project (West Bengal) and Hirakud (Orissa) are among the important hydro electric power projects.

Note

1. Dinesh, C. (2000), Community Rural Water Supply Management, *Employment News*, New Delhi, pp. 1-3.

13

Flood Management

Flood is a situation when there is flow of water in a river more than its capacity and the water overflows the levees and spreads in nearby areas. Thus, flood is a part of normal conditions of flow of river water. The water spreads on the ground where the normal stream of water does not reach normally. Flood is caused due to different reasons in different areas, but basically the use of the word 'flood' is made in the folowing situations (Bharati 1997:122):

1. Excess flow of water in the river more than its capacity within its banks so that water flows on the nearby land.
2. Amalgamation of companion rivers on the mouth of a main river in ultimate flood position.
3. Excess rains than the capacity of the river.
4. Existence of icebergs or landslide in flow route of river, which results in flow of water over the banks.
5. Coming of high waves from higher planes.
6. Heavy rains at local level.
7. Storm or cyclone.

Flood is a part of the natural water cycle. It is a natural catastrophe which is directly related to rains, which affects water management. In case it rains excessively in any area, rivers become imbalanced and come in spate, which creates a situation of flood. Effect of such an environmental situation can be seen on the ecology of such an area. The normal meaning of flood denotes continuous waterlogging of vast ground level for a long duration. Hence, flood is such a natural incident which is the result of excessive rain. A total of 3.5 per cent area of the world is affected by floods, while it is inhabited by 16.5 per cent population. Out of the geographical area of 32.80 crore hectares, about 4 crore hectares (one-eighth of total area) is flood affected area. About 80 per cent of rainfall in India occurs in summer through south-western monsoon. Brahmaputra and Ganga rivers of India give birth to more floods that affect large parts of Assam, Bihar, Uttar Pradesh and West Bengal. Due to insufficient disposal of rainwater, water management projects are not always successful and they result in creation of a situation of flood.

Causes of Flood

Flood is the result of unfavourable combination of meteorological conditions and physical structure. Situation of flood is caused by rains in excess of disposal capacity of its catchment area. Mainly, nature is responsible for flood but flood is also caused due to human activities. Thus, the following reasons are responsible for floods:

Large Catchment Area

Large catchment areas conserve water in vast areas, due to which they receive water from different sources at diferent rates. When there is excess water, it resuts in flood situation. Mississippi and Missouri rivers of United States of America are of such nature, where devastating floods are caused. In March 1927, floods had affected 18 million acres area here. Among other rivers with large catchment areas are Ascambia and Tennesse rivers of Alabama province, Colorado river of Arizona province, White, Red, Arkansas and Mississippi rivers of Arakansas province, El and Clamuth rivers of California province, Savanna river of Georgia province, Wabash river of Indiana province, Niogi and Kansas rivers of Kansas province, Ohio river of Kentucky province, Potomac river of Maryland province, Missouri river of Missouri province, Dilawoeir river of New Jersey province, Miami river of Ohio province, Wilamet and Columbia rivers of Oregon province, Tennesse river of Tennesse province, Columbia river of Washington Province etc. Brahmaputra, Ganga, Kosi and Damodar are such rivers in India.

Tropical Cyclones

Cyclones play an important role in floods caused in coastal parts. In India, Andhra Pradesh and Orissa coasts in the east and Gujarat coast in the west are highly affected by problem of floods due to cyclones. Cyclones cause high tidal waves whose water affects coastal ground areas. In November 1982 and 1983, cyclones had affected 27 dams of

Saurashtra. Floods in Indian river basins are also caused by rainstorms which are generally associated with low pressure systems like well marked lows, depressions or tropical cyclones. During past 109 years (1891-2000), over 1,000 tropical cyclones and depressions originating in the Bay of Bengal and Arabian Sea moved across the Indian subcontinent. Passage of such storms in quick succession over a river basin invariably leads to severe floods. The soil generally gets fully saturated during the passage of the first storm. The rainfall following the subsequent storm leads to flood situation. Parts of the country, mainly coastal areas of Andhra Pradesh, Assam, Orissa, Tamil Nadu and West Bengal, experience such cyclones leading to extensive flooding. The flood due to super cyclone combined with heavy rainfall during October 1999 in the coastal belt of Orissa is an example.

Siltation

Due to deforestation of high ground areas, soil erosion occurs very fast, which results in collection of silt in lower areas. Due to silt collection, water in rivers comes up resulting in lower storage capacity. In Bhabar area of northern Nepal, annual siltation is about 15-30 cm. at the base of rivers. Base level of Kosi river of Bihar is at a very high plane. Problem of siltation is there in Ganga and Brahmaputra also due to their tributaries. Due to such siltation, natural levees of the river are also formed.

Cloud Bursts

Due to particular climatic conditions, some parts of the country experience sudden unprecedented heavy rain known as cloud bursts. During the year 2000, coastal districts of Andhra Pradesh experienced such a flood situation. Cyclonic circulations during monsoon conditions too sometimes lead to cloud bursts leading to floods. In July 1981, such a condition developed over East Rajasthan. Many places around Jaipur recorded a three-day rainfall more than their annual normal average. Annual rainfall at these places varied from 550 mm to 700 mm. But during this period, the daily rainfall was 250-590 mm, two daily rainfalls were of 430-800 and three daily rainfalls were of 450-900 mm, which led to the flood.

In arid and semi-arid regions, sudden and unexpected torrential rain also causes floods because such areas normally get very little rain and natural catchment is not in proper condition. Hence, these rivers are not able to absorb heavy rain water. For example, flood situation was caused in Jaipur city and nearby areas due to the unexpected 836.4

Box 13.1
Causes of Flood

World Scenario: Situation of flood is caused by various reasons, among which are the following:

Climatic extremes, specially heavy rain for long duration.

- Melting of snow and ice
- Breaking of dams
- Landslides

In the Indian scenario, the following are the reasons of floods:

1. *Meteorological Conditions*
- Cyclones
- Cloud Burst
2. *Physical Conditions*
- Large catchment areas
- Narrow outlets in flow route
- Shallowness in flow route due to siltation
- Absence of properly developed flow route
- Obstruction in flow due to landslide
- Development of meanders on river banks
3. *Anthropogenic Factors*
- Deforestation
- Deposition
- Faulty methods of agriculture
- Fast urbanization

millimetre rainfall from 17th July to 21st July 1981. Clouds generally burst in hilly areas where wind rises up fast during the monsoon season and winds rising high do not allow water to drop. Water collects on the surface of the air. Due to sudden rise and stoppage of air, a thick layer of water falls on the ground and its rate per hour is not less than five inches. On 16th July 2003, 100 persons had flown away in the Puliya nalla of Gadsa valley in district Kulu of Himachal Pradesh due to flood caused by cloud burst. Similarly, on 8th August 2003, 60 persons had died due to sudden flood caused by cloud burst in Kangdi nalla near Rohtang Pass.

Heavy Rainfall

Heavy rainfall for long duration also causes flood situation because increase of quantity of water in upper catchment area cause flood in lower places. If time of concentration for water is less, severity of flood increases. This situation depends on size and morphology of the catchment area. Such flood is often caused in rivers of India. Sudden and violent ecological storms also affect it. Rains of seasonal nature also give rise to floods, e.g., most of the rains during summer in tropical climate and during winter in mediterranean climate cause flood.

Deforestation

Deforestation on a large scale in the upper catchment areas of rivers increase the intensity of flood. These anthropogenic factors are seen in two perspectives:

Due to deforestation, most of the ground surface in upper catchment area of rivers becomes naked, which reduces infiltration of water underground and surface water flow is the maximum. Dense forest cover prevents flow of water. Forest canopy becomes an obstacle in allowing the drops of water to go the ground level. Shivalik, Lower Himalaya, Plateau of Chhota Nagpur and Western Ghats etc., have become a regular feature of flood. Similarly, Tista and Torsa (West Bengal), Gandak in Madhya Pradesh, Kosi and other rivers of Bihar, etc. are regular victim to floods. During last century, 86 per cent forest cover has reduced in India.

Earthquake

Natural incidents like earthquake also cause floods. After the earthquake in Assam in 1950, situation of flood was caused in river Brahmaputra, which affected large areas.

Inadequate Drainage System

Flood is caused due to inadequate drainage system. Generally, rivers get lesser water, due to which their flow is obstructed and flood is caused during the monsoon period. This flow route is obstructed in the following forms:

1. *Slow Development of Drainage Channel:* Proper development of drainage channels has not taken place in western and north-western India. Due to absence of proper natural drainage system in arid and semi-arid regions, rain water does not get sufficient outlets during the monsoon season.
2. *Reduction in Carrying Capacity of Rivers:* Due to deposition of silt in the flow route of rivers, water carrying capacity of rivers is reduced. This silt comes from catchment area after soil erosion. As a result of it, a situation of flood is created in adjacent plain

areas. Flood is caused in eastern Uttar Pradesh and north Bihar due to deposition of silt in Narayani and Kosi rivers due to soil erosion of hilly slopes of Himalayas.

3. *Obstruction in Natural Flow due to Landslide:* Because of landslide caused by earthquake or other natural reasons, natural flow of river water is obstructed. Sometimes, natural dams of rivers are damaged due to landslides, which causes a situation of violent floods. In 1978, landslide formed a dam amidst Bhagirathi in Himalayas, resulting in violent flood situation. In the year 2002, landslide caused flood situation in the upper portions of Yamuna river, and in 2005, in Sutlej river.
4. *Meanders on River Banks:* Formation of meanders and sand bars also cause floods. Normally, sand bars are formed on banks of rivers and they reduce capacity of water flow. Such a problem is found in delta areas of West Bengal and Orissa.

Faulty Agricultural Practices

Planning Commission of India has divided India into 15 agro-climatic zones and recommended cultivation of suitable crops as per climatic conditions. Soil erosion increases due to faulty agricultural practices. Faulty irrigation methods also develop flood situation. Irregular irrigation system causes situation of waterlogging due to which the ground level is not able to absorb excess water and it causes flood.

The main reason for incessant floods and continuous increase in its frequency is reckless deforestation in catchment areas. Non-availability of funds for maintenance of modern water management practices is also one of the reasons. Havoc of floods directly affects water management system because the problem of waterlogging would arise in flood affected areas, due to which the quality of salinity and alkanity in soil would develop.

Flash Floods

Flash floods are characterized by very fast rise and recession of flow of small volume and high discharge, which causes a huge damage because of their suddenness. These occur in hilly regions and sloping lands where heavy rainfall and thunder storms of cloud bursts are common. Severe flash floods in Arunachal Pradesh (see Box 13.2) and Himachal Pradesh (see Box 13.3) during the year 2000 are examples of floods resulting from such situations.

Large reservoirs downstream in areas prone to flash flood thus can absorb the resultant flood wave, protecting the progression of flood. Had the proposed storages upstream of Passighat been available, the damage to Passighat in this event could have been reduced.

Littoral Drift in River Estuaries

The flood problems of deltaic regions are attributed to various causes like flatter slope of drains and back flow due to tides. Littoral drift of sand in the form of sand dunes formation and consequent choking of outfalls of rivers into the sea, is one of the causes for flood in deltaic regions. Biccavole and Tublabhaga drains in Godavari Eastern delta and the Panchanadi, Lower Kowsika, Vasalatippa and Kunavaram drains in Godavari central delta are some of the problem areas. Straight cuts into the sea with a view to make the slope steeper in outfall reaches is sometimes considered as one of the effective measures to overcome the problem.

Glacial Outbursts in Himalayan Glaciers

Snow melt is a gradual process and usually does not result into major floods. Glacier melt is usually slower than snow melt and is not capable of causing severe floods. But sometimes glaciers hold large quantity of bounded water or impound lakes, which may be suddenly released with melting of ice jam resulting into Glacier Outburst Flood (GLOF). The rivers originating from Himalyas in the northern part of the country are also fed by snowmelt from glaciers. In 1929, the outburst of the Chong Khundam glacier (Karakora) caused a flood peak of over 22,000 m^3/sec. at Attock. Glacial outburst is one of the suspected reasons for the flash flood experienced in Sutlej river in July-August 2000.

Human Activities and Floods

Directly or indirectly human activities also sometimes result in floods. Sudden breaches in dams/reservoirs cause devastating floods. Many dams in the world have failed due to various reasons. Some of the major dam failures in the world are (i) Puentes Dam (Spain) (69 m high, 291 m length, masonry gravity): constructed in 1791 failed in 1802 due to internal erosion, leading to abandonment of the project itself. (ii) Gallinas Dam (USA) (29 m high, 210 m long, masonry gravity): constructed in 1910 failed in 1957 due to submersion (iii) Banqiao Dam (earth fill 24.5 m high, 2.02 km long) in China, constructed in 1956, failed in 1975 due to insufficient capacity of spillway and submersion. The Machhu-II Dam (Earthfill/concrete gravity, 24.7 m high, 3.9 km long) in Gujarat, constructed in 1972, failed in 1979 due to over pressure of silt etc., causing severe damage to Morbi town downstream. However, the record of India in regard to dam safety has been very good. Detailed guidelines and manuals for dam safety have been prepared and dam safety services have been established.

After the construction of dams, the chances for floods downstream are is reduced to a great extent. This creates a sense of false security whereby people tend to occupy more and more of the flood plains, thereby reducing the actual carrying capacity of the river. Due of this reason, even small releases from the reservoirs sometimes create flood situations affecting the people living in the flood plains, which otherwise could have been avoided if the area was not encroached. Reservoirs like Damodar, Ukai etc., where large scale encroachments in the downstream area have been made, are examples of this. Further information on the above is given in Box 13.4. In flood plains protected by embankments, a false sense of security prevails, and the fact that the embankments could be over topped, say once in 25 years or so, is overlooked. The rapid economic development in such areas increases the damages in unusual events.

Flood Hazards

The problems of flood is found in many parts of the world. The effect of its hazard is different in different countries, the main reason being its management. Calamity of flood is a grave problem of any natural type, resulting in many disorders. On one side, new deposits are collected due to flood whereas on the other side, fast erosion also takes place. Flood damages natural ground morphology and man made structures as well as causes loss of life. According to UNEP Report (1992), 6,592 persons lost their life due to floods from 1960 to 1990.

From the point of view of flood hazards, Mississippi and Missouri rivers of United States of America are of devastating nature. Terrible floods caused by these rivers create a scene of ruin. The catchment area of these rivers spreads over 40 per cent of the geographical area (except Alaska and Hawaii Islands) of United States of America. Flow route of this river is very comprehensive. Except meanders, the direct distance of Mississippi is only 960 km, whereas the actual distance (as per actual route) is 2,720 km. The average slope of its flow route is 3 centimetre per km. In its speeded meandered way, catastrophic floods are caused due to torrential rain etc. on account of atmospheric turbulence (Hurricane and Tornado). In March 1927, the effect of its flood was to the optimum limit. About 180 lakh acre area was affected and 313 persons died. Its companion rivers Missouri, Ohio, Tennessee and Arkansas also contribute to floods caused by this river.

Table 13.1

Important Maximum Historic Floods in the World

Country	*River (Gauge Station)*	*Date*	*Flood Discharge m^3/s*	*Drainage Area km^2*
Australia	Budekin (Clare)	17.02.1968	27,043	
Bangladesh	Brahmaputra (Jamuna)	01.08.1955	93,500	530,000
	Ganga (Hardinge Bridge)	Not available	70,790	1,073,070
Brazil	Amazon (Obidos)	22.05.1944	227,000	4,688,000
Burma	Irrawaddy	1877	63,700	4,688,000
Burma	Irrawaddy	1877	63,700	360,000
China	Yangtzekiang	Not available	84,950	Not available
Egypt	Nile (Aswan)	Not available	13,500	30,00,000
Ghana	Volta (Senchi)	1963	14,272	394095
India	Godavari (Dowleswaram)	17.09.1959	88.350	314.684
	Brahmaputra (Pandu)	Not available	73.620	934.990
	Narmada	Sept. 1970	70.790	98.420
Khmer (Cambodia)	Mekong (Kratie)	1939	75,700	646,000
Madagascar	Mangoky (Banian)	09.01.1956	14,800	Not available
USA	(Mississippi (Columbus)	27.02.1937	70,792	2,387,950
Ohio (Cairo)	04.02.1937	55,218	528,300	
USSR	Yenisey (lgarka)	03.06.1937	132,000	2,440,000
	Voga (Volgograd)	29.05.1926	51,900	1,350,000
	Amur (Kombomolsk)	1876	50,000	1,700,000
	Lena (Kusur)	11.06.1944	194,000	2,430,000
Zambia	Zambezi (Kariba)	05.03.1958	16,990	633,040

Source: 'Flood control in the world' by Shri Framjio & Garg (1978).

Flood situation in America is mainly caused by cyclonic wrath. Due to the Hurricane during 19th to 27th June 1972 in United States of America, there was 150-230 millimetre rainfall in southern provinces, and 350 millimetre rainfall in Virginia, Maryland and Delaware provinces. It caused heavy damages in Pennsylvania.

Hwangho river of China is called 'Sorrow of China' because of its flood hazard. Devastating floods are caused in Yangtse river also. In June 1999, with the first rain of the monsoon, the calamity of heavy flood devoured about 725 persons. According to media, 4,100 persons died at that time and 5.5 million people were affected. As per Ministry of Urban Affairs of China, 11.31 million hectare crop area was damaged due to floods in 23 provinces of China.

In the continent of Europe, flood hazard in 2002 caused a great damage in mid Europe. Water level of river Danube in Hungary crossed all danger marks. The one hundred year old Parliament building of located in the capital of Budapest had to bear the risk of flood. Before water receded, it left a thick layer of mud and filth and loss of thousands of millions of dollars. The water level of Danube river had reached a record rise of 8.48 metres in Budapest. Huge devastation was caused during this year by Waltava river in Czek and Prague Republic, Drasden, Elbe river in Germany and Danube river in Hungary. Not only was there loss of lives and goods but valuable historical and cultural heritage was also destroyed due to floods. Loss of 1,000 crore dollars was estimated alone in Germany. Drasden of East Germany looked like a big lake due to floods. A few tall houses looked like small islands at some places. Elbe river water had drowned Raisa Torgau town of Drasden. Water level of Muld river in Roitushj near Liepzig had also crossed all danger marks.

Floods in India

The Rashtriya Barh Ayog (RBA) made an assessment of the extent of the country exposed to the possibility of floods. The maximum area affected due to flood in a state in any one of the years is taken as the area liable to flood in that state. The total of such maxima of the various states is considered to be the area liable to floods in the country. On the above basis and using the figures available for the period from 1953 to 1978, Rashtriya Barh Ayog (RBA) assessed the flood-prone area in the country as 40 m. ha. The states are to work out area prone to floods in a more systematic and scientific manner by marking out the area inundated on maps and working out the flood-prone area on sub-basin and basin-wise.

In India, about one-eighth part of the area (about four crore hectares) is affected by flood. The following are the flood-affected areas of India:

Eastern Region

This area extends from east of Ghaghra river to Dibrugarh and beyond. Eastern Uttar Pradesh, North Bihar and West Bengal, Manipur, Arunachal Pradesh and Assam are included in this region. Yamuna, Ganga, Damodar, Brahmaputra, Dihang and Lohit are important rivers of this region.

Brahmaputra is the main river of the eastern region where devastating floods are caused every year. Important rivers of Brahmaputra area are Brahmaputra, Barak, Lohit and Dihang. Heavy rainfall up to 600 cm. per annum occurs in this region. Hilly portions are unconsolidated here, hence soil erosion is fast. Deforestation on a large scale has been caused due to mining activities in Meghalaya. Deforestation has also occurred due to 'Jhum' (shifting) cultivation system prevalent in these areas. Rivers of this region flow through narrow valleys, due to which problem of discharge of water exists. Sediment is being deposited in Brahmaputra river. Two rivers Torsa and Jaldhaka of Brahmaputra drainage area flow through areas of West Bengal bringing sediment in large quantity every year. Because of their changing route, they also cause floods. Thus water dispersal, obstacles in discharge of water, landslides and change in flow route of rivers are the main problems of the Brahmaputra area.

Brahmaputra river originates in friable and is susceptible to erosion and thereby causes exceptionally high silt charge in the river. In addition, the region is subject to severe and frequent earthquakes which causes numerous landslides in the hills and upsets the regime of the rivers. The predominant problems in this region are flooding caused by the spilling of rivers over their banks, drainage congestion and tendency of some of the rivers to change their courses. In recent years, the erosion along the banks of the Brahmaputra has assumed serious proportions.

Considering the individual states in the region, the main problems in Assam are inundation caused by spilling of the Brahmaputra and the Barak and their tributaries and erosion along the Brahmaputra river. In northern parts of West Bengal, the rivers

Figure 13.1
Flood-prone Areas of India

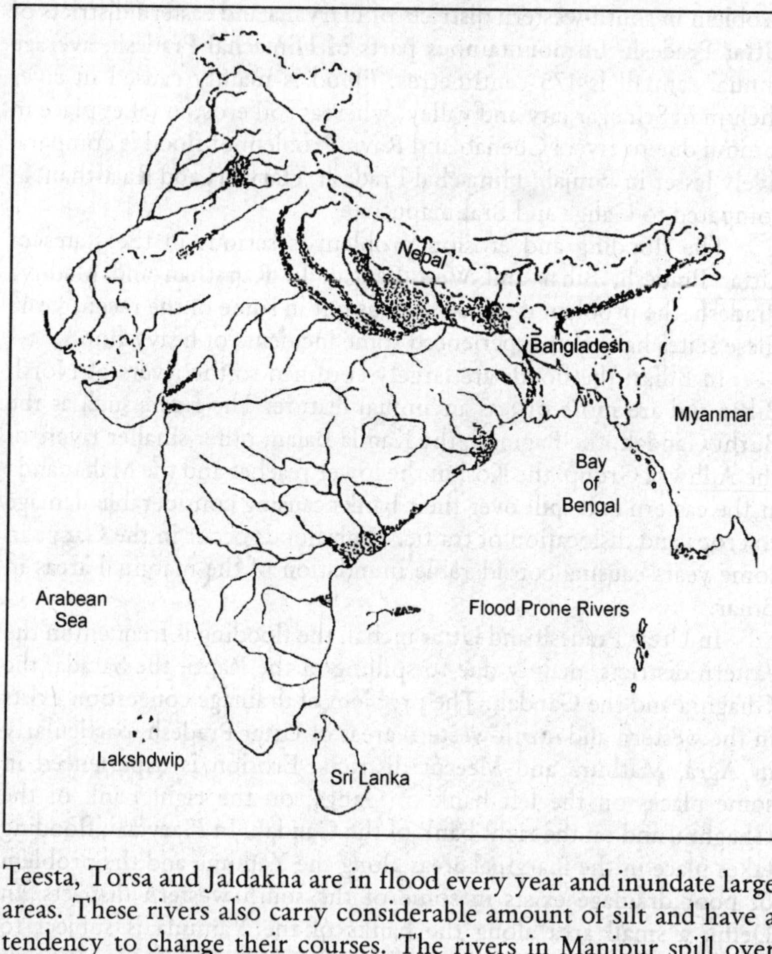

Teesta, Torsa and Jaldakha are in flood every year and inundate large areas. These rivers also carry considerable amount of silt and have a tendency to change their courses. The rivers in Manipur spill over their banks frequently. The lakes in the territory get filled up during the monsoon and spread to large marginal areas. In Tripura, there is the problem of spilling and erosion by rivers.

Northern Region

Rivers Jhelum, Ravi, Beas, Chenab, Sutlej, Yamuna and Indus of Jammu & Kashmir, Himachal Pradesh, Punjab, Haryana, Uttaranchal and Western Uttar Pradesh, are included in this region.

In Haryana, flood is caused in areas adjoining river banks. Flood also occurs in Sahibi and Nazafgarh nallas. Water disposal is the main problem in south-western districts of Haryana and eastern districts of Uttar Pradesh. In mountainous parts of Himachal Pradesh, average annual rainfall is 175 centimetres. Flood is mainly caused in river Jhelum in Srinagar city and valley, whereas soil erosion takes place in Jammu due to rivers Chenab and Ravi. Problem of flood is comparatively lesser in Punjab, Himachal Pradesh, Haryana and Rajasthan, as compared to Ganga and Brahmaputra.

The flooding and erosion problem is serious in the states of Uttar Pradesh, Bihar and West Bengal. In Rajasthan and Madhya Pradesh, the problem is not so serious but in some of the recent years these states have also experienced some incidents of heavy floods.

In Bihar, the floods are largely confined to the rivers of North Bihar and are more or less an annual feature. The rivers such as the Burhi Gandak, the Bagmati, the Kamla Balan, other smaller rivers of the Adhwra Group, the Kosi in the lower reaches and the Mahananda at the eastern end spill over their banks causing considerable damage to crops and dislocation of traffic. High floods occur in the Ganga in some years causing considerable inundation of the marginal areas in Bihar.

In Uttar Pradesh and Uttaranchal, the flooding is frequent in the eastern districts, mainly due to spilling of the Rapti, the Sarada, the Ghaghra and the Gandak. The problem of drainage congestion exists in the western and north-western areas of Uttar Pradesh, particularly in Agra, Mathura and Meerut districts. Erosion is experienced in some places on the left bank of Ganga, on the right bank of the Ghaghra, and on the right bank of the Gandak. In Haryana, flooding takes place in the marginal areas along the Yamuna and the problem of poor drainage exists in some of the south western districts. In Delhi, a small area along the banks of the Yamuna is subject to flooding by river spills. In addition local drainage congestion is experienced in some of the developing colonies during heavy rains.

Southern and Central Region

In this part of the peninsula, Narmada, Tapti, Godavari, Krishna, Kaveri and Pennar are main rivers which cause losses due to flood in delta areas. These rivers have mostly well defined stable courses. They have adequate capacity within the natural banks to carry the flood discharge except in their lower reaches and in the delta area. The lower reaches of the important rivers on the East Coast have been embanked, thus largely eliminating the flood problem.

This region covers all the southern states, namely, Andhra Pradesh, Chhattisgarh, Karnataka, Tamil Nadu and Kerala, and the central states of Orissa, Maharashtra, Gujarat and parts of Madhya Pradesh. The region does not have very serious problems except for some of the rivers of Orissa, namely, the Brahmini, the Baitarni, and the Subarnarekha. The Delta areas of the Mahanadi, Godavari and the Krishna rivers on the east coast periodically face flood and drainage problems in the wake of cyclonic storms.

The Tapti and the Narmada arc occasionally in high floods affecting areas in the lower reaches of Gujarat.

The flood problem in Andhra Pradesh is confined to spilling by the smaller rivers and the submergence of marginal areas along the Kolleru lake. The most important coastal problem is presented by the Kolleru lake and the numerous rivers that rise in the ghats and flow into the lake. Rivers like Budameru and the Thammileru not only overflow their banks along their course to Kolleru lake but also cause a rise in the level resulting in inundation of adjoining lands. The problem is being tackled systematically since 1969 on the basis of the recommendations of an expert committee of engineers known as the Mitra Committee. The execution of drainage improvement programme is being coordinated by Krishna Godavari Pennar Delta Drainage Board.

Orissa Region

Mahanadi, Brahmani, Baitarni and Subarnarekha are the main rivers of this region. Flood is a grave problem at the origin of these rivers. Mahanadi is called "Sorrow of Orissa", but the construction of Hirakud dam over it has given some relief from floods.

In Orissa, damage due to floods is caused by the Mahanadi, the Brahmani and the Baitarni, which have a common delta where the flood waters intermingle and when in spate simultaneously cause considerable havoc. The deltaic districts are densely populated and receive heavy precipitation in the eastern ghat region, and this creates problems in one or more of these rivers. The problem is accentuated when the flood synchronizes with high tides. The silt deposited constantly by these rivers in the delta area raises the flood level and the rivers often overflow their banks or break through new channels

Table 13.2

Flood-affected Areas of India

Name of the State	Flood-Affected Area (in lakh hectatres)
1. Andhra Pradesh	13.9
2. Assam	31.9
3. Bihar	42.6
4. Gujarat	3.9
5. Haryana	23.5
6. Himachal Pradesh	2.3
7. Jammu & Kashmir	0.8
8. Karnataka	o.2
9. Kerala	8.7
10. Madhya Pradesh	2.6
11. Maharashtra	2.3
12. Manipur	0.8
13. Meghalaya	0.2
14. Orissa	14.0
15. Punjab	37.0
16. Rajasthan	32.6
17. Tamil Nadu	4.5
18. Tripura	3.3
19. Uttar Pradesh	73.36
20. West Bengal	26.5
21. Delhi	0.5
22. Pondicherry	0.1
Total	*335.16*
Flood-Prone area during the year 1953-78 according to figures of states	34 million hectares
Secure Area since 1978	10 million hectares
Total	*44 million hectares*
Flood affected area where security arrangements failed, hence could be included in notified flood affected area	4 million hectares
Total Flood-Prone Area of the Country	*40 million hectares*

Source: Bhagirath, April-June 2002, p. 72.

causing heavy damage. The lower reaches of the Subarnarekha are affected by floods and drainage congestion.

Problem of flood is caused in India due to quantity of rainfall and meteorological variations during monsoon season. During this period, 90 per cent rainfall occurs in India. The form and intensity of the problem of floods depends on the nature of different catchment areas. The most complex problem is faced in Brahmaputra and Ganga basins. Every year, during floods, 77.5 lakh hectare land is waterlogged and thousands of cattle and hundreds of people lose their lives in the country.

Soil erosion is being caused at a very fast speed by rivers at the points of their origin on high mountainous places as well as in lower areas due to deforestation. As a result, of which burden of sediment is increasing day by day, thereby reducing water carrying capacity of the rivers. Flood is caused every year due to this reason.

Flood losses are caused every year in the north-eastern states. In Meghalaya, Cherapunji gets highest rain of the world on Funnel Shaped hills where there are limestone rocks. Due to mining of limestone, reckless deforestation has taken place, due to which most of the rain water flows away causing flood situation. Cherapunji faces drought during summer. Devastating floods are caused in Uttar Pradesh due to rivers Ganga, Yamuna, Ramganga, Ghaghra, Gomti, etc. rivers. Due to flood in Yamuna river, nearby areas of Delhi, Mathura, Agra and other towns are affected. River Ganga affects areas in Allahabad, Varanasi, Gazipur, Balia, Patna etc. Gomti river causes flood losses every year in Jaunpur and Lucknow. In Lucknow, floods occurred during 1923, 1960, 1971 and 1980, while in Jaunpur floods occurred in the years 1974, 1830, 1871, 1894, 1903, 1915, 1922, 1936, 1949, 1955, 1960, 1971 and 1980.

Damodar river, called as "Sorrow of Bengal", is considered as a wild monster from the point of view of destruction. There is maximum effect of cyclonic rainfall here. During 26 to 29 September, 1978 there was rainfall of 600 millimetres. Its tributaries have been controlled by construction of dams over them like Panchet Dam, Maithan Dam, Konar Dam and Tilaiya Dam. In Bihar, 193 persons died in floods during 1999. Kosi (Sorrow of Bihar), Bagmati, Kamla, Balaan, Puppun and Gandak are the main rivers which have completely destroyed the kharif crop in this state in an area of 3.75

lakh hectares. During this period, flood also caused losses in Assam, Gujarat and Himachal Pradesh. About 3 lakh persons were affected by floods in Assam. In the famous 'Kajiranga National Park' of Assam, many wild bears and rhinoceroses died in the flood. Landslide happened in Himachal Pradesh due to heavy rains. Water level of Beas, Ravi and Sutlej crossed danger marks whereas, 34 persons died in Gujarat due to floods.

Flood Control and Management

Many systematic steps are being taken for control and management of floods. The quantity of water and rate of its flow is controlled in water flow system. The following steps are effective in flood control:

1. *Reduction in run off:* Surface run off increases due to unexpected heavy rainfall. This has been increasing with depletion of forest. Due to forest cover, rain drops do not fall on the ground intensely and ground surface absorbs them and flow of water does not gain speed suddenly. By increasing the rate of infiltration in the catchment area, run off can be reduced. It can be done by afforestation on a large scale. Dense vegetative cover forms a canopy between ground and rain water by creating interception. Their roots, leaf litter and humus hold water in them.
2. *Control on deforestation:* Deforestation may be banned in catchment areas of rivers and afforestation should be resorted to. Forest cover would reduce soil erosion and reduce sedimentation of rivers due to its deposition on account of soil erosion.
3. *Reduce flood level:* Level of flood in water flow route can be reduced through the following means:
   (i) *Canal system* may be developed in nearby areas as per capacity of the stream.
   (ii) *Improvement in stream* can be made to reduce level of flood. This improvement can be made by deepening the base of water flow and increasing the width. Many problems are also faced in these activities like increase in soil erosion capacity, effect on water creatures and living organisms, flood water immediately reaching lower streams etc.
   (iii) *Flood diversion:* Effect of flood can be reduced by diverting flood water towards marshy lands, depressions and lakes etc. Many types of engineering methods can also control

floods. Construction of flood control storage reservoirs is the main method by which surplus water can be stored. Such types of water reservoirs were constructed in the Miami river of United States of America between 1913 and 1921.Tennesse river of United States of America too was suffering from recurrent floods, waterlogging and various diseases caused on its account up to 1933. Fast soil erosion had made lands barren and unfertile. However, since 1933, Tennessee Valley Authority (TVA) controlled the above terrors by constructing many dams and water storage reservoirs. On the pattern of TVA, efforts were made for flood control in Damodar Valley Corporation (DVC) in India. It constructed four dams (Karonar, Panchet, Tilaiya and Maithan) on Damodar and its companion rivers Barakar and Konar, as well as water reservoirs having water storage capacity of 1,603 million cubic metres at the time of floods.

(iv) *Meanders and diversions in rivers* cause obstruction in flow of water. Hence, more diverted routes should be made straight. Meanders are a natural phenomenon. United States of America made meanders of Mississippi river straight near Greenville during 1933 and 1936, thereby reducing the meandered route of 530 km to 185 km.

(4) *Embankment and construction of dams:* By construction of embankments on sides of rivers, excess water can be controlled at the time of flood. Such types of engineering works have been executed in north eastern India, Gangetic plains and Bangladesh. Embankments to the extent of 12,265 km were constructed on a large scale under flood management activities between 1954 and 1979. Embankments were constructed in 246 km by the sides of Kosi river banks, 249 km by the sides of Bagmati river banks, 208 km by the sides of Mahananda river banks and 317 km by the sides of Budhi Gandak river banks.

The Central Flood Control Board had demarcated flood affected areas in 1957. After identification of flood affected areas, flood water flow routes should be interconnected for land utilization and after study of flood cycles, mapping of large affected areas should be done.

5. *Forecast of floods:* Devastation due to flood can be prevented by flood forecasting. Central Water Commission (CWC) has established an all-India system of warning and forecasting of floods.

Table 13.3
Length of Embankments in India

State	*Length of Embankments (km) (1945 to 1987)*	*Secure Area up to 1987 (in million hectares)*
Assam	4448	1.56
Bihar	2756	1.87
Andhra Pradesh	507	1.00
Haryana	578	1.70
Gujarat	408	0.43
Kerala	92	0.03
Himachal Pradesh	58	0.01
Jammu & Kashmir	56	0.06
Madhya Pradesh	13	0
Karnataka	—	—
Goa, Daman, Diu	9	—
Maharashtra	26	—
Manipur	279	0.08
Meghalaya	106	0.09
Orissa	1007	0.46
Punjab	1047	2.66
Rajasthan	141	0.04
Sikkim	—	—
Tamil Nadu	35	0.08
Tripura	114	0.03
Uttar Pradesh	1711	1.42
West Bengal	974	1.75
Delhi	83	0.08
Pondicherry	61	0.01
India	14,511	13.36

CWC makes announcements regarding forecast of estimated flood situation through 157 centres located in 62 river basins and sub-river basins. Out of them, 109 centres are functional in Ganga, Brahmaputra and Meghna river systems, 15 centres are working for rivers flowing in the west. Similarly, eight centres are established in river Krishna, three centres in river Mahanadi, 13 centres in river Godavari and nine centres are established for rivers flowing in the east. Out of these centres, the following are more important:

(i) *Guwahati:* For Brahmaputra and its companion rivers and Berar river.
(ii) *Jalpaigudi:* Teesta river.
(iii) *Varanasi-Azamabad:* Ganga and its companion rivers of Uttar Pradesh.
(iv) *Buxar-Patna:* Ganga and its companion rivers in Bihar.
(v) *Bhuvaneshwar:* Suvarnarekha, Burhabalang, Baitarni and Brahmani rivers.
(vi) *Bharaunch:* For flood of Narmada river.
(vii) *Delhi:* Yamuna and Sahibi river up to Delhi.
(viii) *Hyderabad:* Godavari and companion rivers.

Flood Plain Zoning

For zoning of flood-prone areas, identification of flood routes is necessary regarding land utilization. Detailed mapping of flood affected areas is also required which is possible only after a long study of flood cycles. It is also necessary to adopt non-structural methods like exchange of economic activities because construction of embankments alone will not give relief from floods. From this point of view, a model draft law was sent to the states for legislating it. So far, only Manipur and Rajasthan have enacted such a law.

Flood Management Programmes in India

Central Flood Control Board was established in India in 1954 and on its basis, Provincial Flood Control Boards were also established at the state level. During the First Five-Year Plan itself, importance was given to construction of embankments and big dams. It was written in the draft plan that big dams could prove useful in storing excess water brought during floods.

The Ganga Flood Control Commission was constituted in 1972 with the main objective of preparing detailed projects of different river systems of Ganga river basin to prevent floods. The Commission has taken steps for getting early sanction of detailed flood management projects of Son, Punpun, Budhi Gandak, Bagmati, Mayurakshi and Jalangi rivers.

It is possible save 320 lakh hectares area out of 400 lakh hectares of flood affected area in India, from floods. Up to March 2000, protection has been assured to 157 lakh hectares from flood hazard by construction of channels for embankments, urban security works and settlement of habitations on a higher level. A total of Rs 7,239 crores have been spent over it. 37,904 km new embankments and 37,904 km disposal channels have been also constructed. Apart from it, protection has been provided to 2,337 towns along with additional villages and 4,713 villages have been raised above the flood level.

Though flood management is a fundamental subject of the states, even then central government provides special central assistance in technical and developmental form. Central Water Commission provides flood forecasting of interstate rivers through 166 forecasting centres and 25 intermediate forecasting centres (*India 2005*, p. 740).

14

Drought and Dry Farming

Drought is the cumulative effect of scarcity of water, which affects agriculture, natural circumference and related activities in the form of a natural calamity. Its effect goes on increasing continuously creating conditions of famine. In scientific language, drought is an abnormal state of season spread over a part of land, where there might be absence of rain in spite of its probability. Due to increase in dryness for a long duration, the state of drought is created. In this state, the quantity of rain, normal average annual rainfall, its distribution and demand for water have intimate relationship. British Rainfall Organization (BRO, 1936) has classified drought into three categories – first, *absolute drought,* when normal rainfall is below 0.01 inch daily continuously for 15 days, second *partial drought,* when average rainfall is more than 0.01 inch or more at least for 29 days continuously, and third, *dry spell* when rainfall is less than 0.04 inch continuously for 15 days daily.

According to Indian Meteorological Department (IMD), drought is a condition when actual rainfall is 75 per cent less than average normal rainfall in that geographical area. Indian Meteorological Department has divided drought in two forms: *(1) severe drought:* when absence of rain is 26 to 50 per cent of normal rainfall; and *(2) normal drought:* when absence of rain varies between 50-25 per cent. According to C.G. Betej, a situation of drought is created when annual rainfall is 75 per cent or less than the normal annual rainfall and monthly rainfall is 60 per cent or less than the normal monthly rainfall. J.C. Hoyat, on the other hand, has written that drought condition is created at a time when annual and monthly rainfall is less than 85 per cent of the normal rainfall. During drought condition, there is also an adverse effect on probability of groundwater because it is less than the expected rainfall. Normally, drought is visible in nature in following forms:

1. Meteorological Drought
2. Hydrological Drought
   (i) Surface Water Drought
   (ii) Groundwater Drought
3. Agricultural Drought
4. Socio-economic Drought
5. Soil Induced Drought
6. Ecological Drought

Meteorological Drought

Metrological drought situation is created when actual rainfall is clearly less than expected, according to the climatic conditions. In other words, rainfall is lesser than average. Average rainfall of India is 1,170 millimetres. It is normally sufficient rainfall for any geographical region but drought conditions are created due to great ups and downs and regional disparities in it. Sometimes rainfall starts before time and sometimes it comes late. Various government agencies have presented a formula after long analysis for delimiting drought-prone areas. It is based on coefficient of variation in rainfall.

$$\text{Coefficient of Variation (CV)} \frac{\text{Standard Deviation}}{\text{Average}} = \times 100$$

The value obtained by the above formula is called coefficient of variation. After measuring the rainfall for many years, its average is found out, which indicates lesser or excess quantity of rainfall.

Coefficient of variation in India remains between 15 to 30 per cent. Rajasthan and Gujarat are among the highly variable areas of rainfall where variation remains between 50 to 80 per cent. Internal peninsular India and rain shadow region behind the Western Ghats are the areas having variation between 30 to 50 per cent. Indian Meteorological Department (IMD) has clarified drought conditions on the basis of conditions of rain and severity of drought:

1. *Excess than normal:* +20 per cent or more than the average of last 70 to 100 years.
2. *Normal:* +19 per cent to 19 per cent of average of last 70 to 100 years.
3. *Deficient:* 20 per cent to 59 per cent of average of last 70 to 100 years.
4. *Scanty:* 60 per cent or less than the average of last 70 to 100 years.

Long-term drought areas are those areas where probability of drought year is more than 40 per cent. Drought year is considered when there is 75 per cent less rainfall.

Hydrological Drought

Hydrological drought is considered when surface water sources like lakes, rivers, and water reservoirs become dry. Its severity also causes shortage of groundwater. It occurs when meterological drought continues for a long duration. It can be divided in two classes:

1. Surface Water Drought
2. Groundwater Drought

Surface Water Drought

When meteorological drought continues for a long time, scarcity of water is caused and surface water sources start drying up. But there are many other reasons for drying up of surface water sources besides meteorological drought, mainly excessive deforestation in catchment areas of hilly areas and hydrological shifting. Rivers also dry up before monsoon. Normal rainfall is also affected due to drying up of surface water, which is what happens in Cherapunji. Due to absence of forest cover, all of the rainwater flows away and after sometime drought is faced. Thus, Cherapunji, which gets the highest rainfall in the world, becomes drought affected in September each year, because all the forests have been cleared during the last 30 years. The mountains of Cherapunji arc completely barren today. Except a few Baltu trees, there seems to be no greenery there. In Cherapunji, the hills have limestone in abundance. This is found just below the ground, hence these hills completely soak rain water up to its depth. The ecological system of Cherapunji remains affected by grave drought during March-April. Infiltration paths have been destroyed here due to the hydrological cycle and flood situation is created here immediately after rains.

Groundwater Drought

Scarcity of water is being caused due to continuous depletion of water level. Due to over exploitation of groundwater, groundwater drought occurs in spite of normal rainfall, because exploitation of groundwater is more than the water received from rainfall, making it scarce. Except Indus, Sutlej and Ganges basin in India, groundwater is found in a limited the quantity in the rest of the country. Normally, most of the exploitation of groundwater is done from 400 to 500 feet deep aquifers. Due to rural electrification at a great speed and construction of a large number of wells, exploitation of groundwater has increased. After the green revolution in agriculture, requirement of water has also increased due to cultivation of commercial crops.

The main reason for fast depletion of groundwater level is the increasing demand for water in agriculture, domestic and industrial sectors. This demand is met by groundwater. Due to deforestation also, intensity of monsoons has been disturbed and quantity of rainfall has been irregular and insufficient. Destruction of traditional water sources has caused reduction in the rate of seepage of groundwater, causing most of the rainwater to flow away.

Agricultural Drought

Agricultural drought happens when moisture of soil and scarcity of rain is insufficient for the growth of crops. This kind of drought is also called drought of soil moisture or water drought. Its severity depends on the nature of plants as well as soil. After the green revolution, importance was given to intensive agriculture, which led to cultivation of crops demanding more water.

Hybrid seeds have played an important role in agricultural drought. They require more water, whereas they contain lesser quantity of organic material. Organic material increases water-retaining capacity of the soil. Apart from it, change in crop rotation also causes agricultural drought. Traditional crop rotation remains in balanced condition according to agricultural climatic conditions but since crops requiring more water are included in such rotation, shortage occurs in soil moisture resulting in agricultural drought. Such effect is seen in Punjab, Haryana and western Uttar Pradesh.

Socio-economic drought

Economic and social impact on account of hydrological drought and agricultural drought is called socio-economic drought. This drought is caused as a result of natural incidents and agricultural activities. It is the result of non-availability of sufficient water. In this situation, except agriculture cattle and other creatures also face scarcity of water.

Soil Induced Drought

When scarcity of water and decrease in soil moisture is so much that crops decay, such a situation is called soil-induced drought.

Ecological Drought

Due to greater deterioration in environment, when productivity decreases in ecology, it is called ecological drought.

Causes of Drought

Drought is a natural calamity, which is directly interconnected to the meteorological conditions of any geographical area. However, man is also responsible for it to some extent. By excessive deforestation on account of fast industrialization, man has affected favourable conditions of monsoon. Normally, there are the following reasons for drought:

1. Delay in monsoon and its early departure.
2. Scanty rainfall due to absence of low pressure in Thar desert.
3. Long dry spell in monsoon rains.
4. El-Nino and Southern Oscillation (SNSO).
5. Re-establishing of southern branch of monsoon due to jet stream.
6. Effect of cold stream of Somali on Arabian Sea water.
7. Fast deforestation.
8. Effect of latest disturbances (El Nino etc.) on monsoon.
9. Change in traditional crop pattern.
10. Depletion of groundwater.
11. Scarcity of water in surface water sources and deterioration in its quality.
12. Absence of management of rain water in a planned way.

Impacts of Drought

Drought affects the whole biotic world because plants as well as living organisms both are dependent on water. Long duration drought affects any geographical region from ecological, economic, demographic and political points of view. Normally, the following impacts are perceived in any natural region:

1. Many species of creatures as well as plants are destroyed because they cannot bear the condition of scarcity of water.
2. Living organisms shift to other places for protecting themselves. This results in excess population in new places and the ecological balance is disturbed.
3. Groundwater level is depleted, affecting the whole agricultural system.
4. In drought affected areas, animals die of hunger due to shortage of fodder.
5. Agricultural production decreases, which affects cattle breeding as well industrial production. Thereby economic development is prevented.
6. Unemployment spreads.
7. Adverse effect is caused on health.

From the geographical point of view, impacts of drought can be divided in the following categories:

1. Physical Impacts
2. Impacts on Agriculture
3. Impacts on Animals
4. Socio-economic Impacts

Physical Impacts

One physical impact of drought is that soil moisture deplets in drought-prone areas. Groundwater level goes deeper and many large plants are destroyed due to lack of moisture. The whole area looks like barren land. Another adverse physical impact is caused on living creatures due to drying up of water in water reservoirs. Wild animals especially are most affected. Agriculture suffers also because the wells become dry. Surface water of rivers also dries up in summer.

Impacts on Agriculture

Indian agriculture is dependent on rains. Its success depends on time and condition of monsoon, because in drought-prone areas soil moisture is stored only through rainwater, and quantity of ground-water resource is directly influenced by monsoon conditions. Thus, impact of drought on agriculture becomes visible directly in the form of shortage of food grains. In India, 73 per cent area was affected by drought in the year 1918-19 resulting in 32.3 per cent loss in production of food grains. Similarly, loss of production of food grains was 18.8 per cent in 1965-66, 7.7 per cent in 1972-73 and 17.0 per cent in 1979-80, as compared to its average production. Moisture is possible only due to rains in the area and in its absence, the following impacts are visible:

- Loss in agricultural productivity
- Lack of pasture land for cattle
- Scarcity of water for animals
- Lack of decided crops
- Diseases in crops
- Loss of crops by creatures
- Loss of productivity of forests
- Loss of fisheries

Impacts on Animals

In a situation of drought, there is lesser greenery due to scarcity of water. Surface water sources dry up causing scarcity of water as well as fodder for cattle. In western Rajasthan, thousands of animals died and lakhs of animals migrated elsewhere in the year 1999-2000 due to scarcity of water and fodder. Humans also suffered and lakhs of people migrated to other places in search of food, water and employment. According to National Crime Record Bureau, 700 farmers were compelled to commit suicide during the years 1999 to 2001.

Due to incessant droughts in Rajasthan, 22 rare species of vegetation in Shekhawati region have vanished and natural habitation places of Kurja birds have been affected by scarcity of water. Thousands of Kurja (migratory) birds come to Rajasthan each year but their number is reducing due to water shortage. Apart from them many birds like Vultures and Fakhta are also finishing. Due to scarcity of water, either they are dying or they are migrating. Siberian cranes and other birds have reduced in the Kevladev Bird Sanctuary of Bharatpur in Rajasthan.

Socio-Economic Impacts

The following social and economic impacts become visible due to drought:

1. Lack of employment opportunities in agriculture sector.
2. Shortage of fodder.
3. Shortage of number of animals.
4. Problem of malnutrition.
5. Loss of agricultural production.
6. Increase in prices of food grains and other articles.
7. Loss of per hectare productivity of crops.
8. Lack of purchasing power.
9. Drinking water crisis.
10. Depletion of groundwater level.
11. Compulsion for sale of cattle.
12. Spreading of various types of diseases.
13. Compulsion to sell permanent assets (Land, House, Ornaments etc.).
14. Migration in search of employment elsewhere.
15. Deaths due to hunger, disease and malnutrition.
16. Increase in social crimes, eg. looting of food grains and other assets.

Figure 14.1

Impacts of Drought

Level	*Agriculture*	*Economic*	*Social*
World Level	World Food Supply	Price of Food grains Foreign Economy	Famine (Social Friction)
National Level	Food Supply	Price of Food	Effect (Balance of Trade)
Regional Level	Regional Production	Economic Productivity	Pressure (Migration etc.)
Local Level	Yield	Income from Fields	Burden/Pressure (Health)

Drought-affected Areas

The Sahel region of the tropical zone of Africa, which is the hot, arid and desert portion of Sahara in the north of North Africa and the area spreading from west to east amidst Savana region in South, remain in the grip of parmanent drought. This region includes parts of Maritonia, Senegal, Mali, Burkina Faso, Niger, Nigeria, Chad, Uganda and Ethiopia countries. The Sahel region is really tropical grass area where the climate is termed as famine climate. Surface water sources and groundwater resources of this area are completely based on rainfall. Drought began here in 1968 and was most severe during 1971-72. It affected people till 1975. Due to its cumulative effects, the stage of extinction of humans had been reached in 1974. Groundwater reservoirs dried up and wells became dry. Animals purchased at high costs had to be disposed off at throwaway prices. Many farmers became refugees. Thousands of persons in this region died on account of starvation, thirst and diseases in spite of all possible cooperation from the government. About 5 lakh cattle died because of drought conditions. Only in Ethiopia, 50,000 persons became victims of death due to famine.

Australia also often faces drought. Condition of desert exists in 75 per cent area of this country. During 1895 to 1902 there had been acute drought, number of sheep which was 1,060 lakh in 1891, reduced to only 540 lakh in 1902. Thus, the number of cattle was reduced by 50 per cent. Severe droughts during 1895-1902, 1911-16 and 1919-20 caused large areas under cultivation to decrease, which improved later due to good rainfall. There are also such areas on earth where drought does not occur normally but sometimes a severe one may occur. The drought condition in England during 1975-76 was one of such type. Rainfall was 40 per cent less than average during at year, creating severe water crisis. Agricultural production was also affected. Production of wheat, barley, oat (jai) and potato came down in the province of Wales. The British Government had to appoint a Famine Minister that year to look to the famine affected people.

Many areas of United States of America are in the grip of drought at present due to lack of sufficient rains for the last four years. Conditions have become so serious that rationing of water has been introduced in many provinces. Dry branches are burning into ashes in forests. Colorado, New Mexico, Arizona, South Dakota, Atlanta, Texas, Georgia and Denver are some of the affected places. According to agricultural expert Jeff Trine of Colorado University, it seems that the cows are disappearing. Jeff says that there were 2,800 cows in Hurfani county of Colorado, but their number is continuously going down. Decreasing number of cows and other animals describe the severe impact of drought in any place.

Meteorologists say that this period is the most dangerous after drought conditions during the decades of 1930 to 1950. 32,00,000 acre forest area has been converted into ashes in Colorado, New Mexico and Arizona. In more than a dozen cities of Georgia and Atlanta, rationing system of water has been introduced. People get water supply only for three days in a week here. Water level of Lake Lanier and other water reservoirs has gone down by 12 feet and Rio Grande river of Texas received only 2 per cent water. Water level of Las Vegas and Meade Lake has reduced to below 60 feet. Farmers and breeders of these areas are being given loans to provide money for livelihood and arranging fodder for cattle on the insistence of the government.

In all 16 per cent area of India is drought-prone. North western desert and semi-desert area, Maharashtra, Karnataka and shadow rain area of Andhra Pradesh are included in the drought-prone areas of India where annual rains are very irregular. Apart from these, Tirunivelle in the south of river Vega, Coimbatore, Saurashtra, Kutch, Mirzapur, Palamu, Purulia, Kalahandi districts also are drought affected. According to Indian Irrigation Commission, there are 77 such districts where rainfall is 75 per cent less than normal rainfall. These districts include 34 per cent of the total cultivated area. Besides them, there are five other districts where probability of drought always exists. Following areas of India are drought affected:

1. Rajasthan, Gujarat, Punjab, Haryana, Western Uttar Pradesh, Madhya Pradesh and parts of Chhattisgarh. Among these, western Rajasthan and the Kutch area are maximum drought affected areas.
2. Central Maharashtra, internal Karnataka, Rayalaseema, South Telangana and some parts of Tamil Nadu.
3. Some parts of north western Bihar and some parts of eastern Uttar Pradesh (Mirzapur etc.).
4. Some parts of Orissa adjacent to West Bengal.

Maximum frequency of drought exists in the Thar desert of the north western portion of India, which basically falls in Rajasthan. The main reasons of drought in Rajasthan are temperature inversion due to dust (sand) particles in Thar desert and existence of Aravalli hills. The desert is also extending with time. Increase in the rate of

Figure 14.3
Drought-affected Areas in India

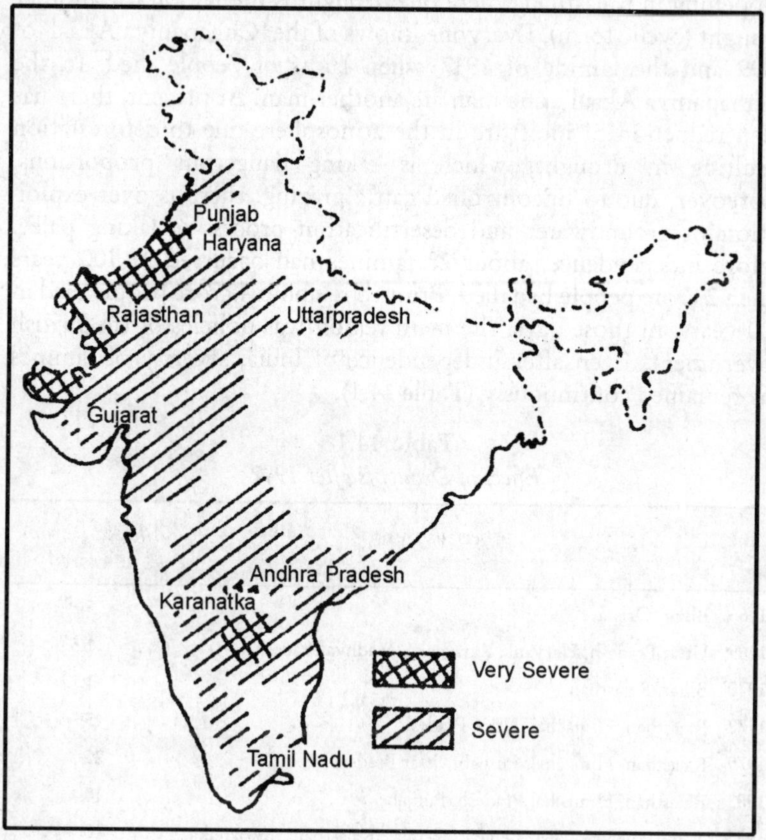

deforestation, increasing organic pressure, human and cattle population has also been responsible for drought in this region.

The basic reason for frequency of drought in Rajasthan is its natural condition. Rajasthan has a vast desert, no mountains to prevent monsoon winds, no perennial rivers, no large water reservoirs, insufficient canal system, poor quality of the soil etc. Due to these reasons, the impact of drought in Rajasthan has been more severe as compared to that of floods in the north eastern states. There has been so much deterioration of social and economic environment in the state that problems of poverty, malnutrition, unemployment and starvation have been created. From the geographical angle, if drought continues for long or over exploitation of groundwater reservoirs takes place, then drought takes a cumulative form. The same is happening in Rajasthan where one drought is the reason for another drought (cyclic form). Everyone knows of the 'Chhapaniya Akaal' of 1899 and the famine of 1917 when lakhs of people died. In the 'Chhapaniya Akaal', one man ate another man. At present, there has been reduction of moisture in the atmosphere due to deforestation resulting in drought, which is taking dangerous proportions. Moreover, due to uncontrolled cattle grazing, there is over exploitation of groundwater and desertification process is taking palce. Before independence, about 28 famines had occurred in 100 years when 2 crore people had died. Prior to it, about 4 crore people died in 100 years. In those days, the main reason was policies of the British government. Even after independence of India, severity of famines has remained continuously (Table 14.1).

Table 14.1

Effect of Droughts after 1947

Year	*Affected Area*	*Affected Persons (in crores)*
1966	Bihar, Orissa	5
1969	Uttar Pradesh, Haryana, Karnataka, Madhya Pradesh	1.5
1970	Bihar, Rajasthan	1.72
1972	Rajasthan, Himachal, Uttar Pradesh	5
1979	Rajasthan, Himachal, Punjab, Uttar Pradesh	20
1982	Rajasthan, Himachal Pradesh, Punjab	10
1983	Tamil Nadu, West Bengal, Kerala, Rajasthan, Karnataka, Bihar, Orissa	10
1987	Complete Northern, Western and North Eastern India	30
1992	Rajasthan, Gujarat, Orissa, Andhra Pradesh, Madhya Pradesh	NA
2000	Rajasthan, Gujarat, Orissa, Andhra Pradesh, Madhya Pradesh	10
2001	Rajasthan, Gujarat, Orissa, Madhya Pradesh, Andhra Pradesh, Maharashtra, Chattisgarh, Himachal Pradesh	17

In India, during 1999-2000, about 12 crore people were affected by drought in 12 states. Famous press correspondent Rohit Brijnath had described the scene by saying that "With the dashing of hot air, vultures sorrowfully fly up slowly and again come down in the same speed in a deadly mood. The ground below has become dry, naked trees and their withered branches look like a patient. Streets of small towns are lonely, as if a curfew has been imposed there. Wrath of sun God has shadowed all the four sides. Animals are lying down on places where they used to walk. Bones are peeping through their skin, which has become like a film. Bad smell has spread around. Temperament of nature has spoiled. Almighty also sometimes plays a cruel joke. Only six months ago, cyclone and severe flood water had devoured many lives and now its absence is taking lives of people. About 10 crore population of the 12 states, specially Rajasthan, Gujarat, Madhya Pradesh and Andhra Pradesh has come in the grip of drought slowly by slowly and in such state, lamp of hope also seems creeping far away." Reality of figures is that 23,406 villages in Rajasthan and 17,431 villages in Andhra Pradesh are in the grip of drought. Production of grains in Gujarat has reduced by 30 per cent. There is huge shortage of fodder for 345.6 lakh cattle in Rajasthan.

Table 14.2

Frequency of Droughts

Meteorological sub-divisions	*Recurrence of the period of highly deficient rainfall*
Assam	Very rare, once in 15 years
West Bengal, Madhya Pradesh, Coastal Andhra Pradesh, Maharashtra, Konkan, Kerala, Bihar, Orissa.	Once in 5 years
South interior Karnataka, Eastern Uttar Pradesh, Vidarbha.	Once in 4 years
Gujarat, Eastern Rajasthan, Western Uttar Pradesh, Tamil Nadu, Kashmir, Rayalaseema, Telengana	Once in 3 years
Western Rajasthan	Once in 2.5 years

Source: The India Meteorological Department studied the periodicity for 1871-1980 and indicated the frequency.

The drought that occurred in 2001 was the result of cumulative effect of cruel monsoon, which again came in the subsequent year of 1999-2000. According to Mihir Bhatt, Director of Disaster Mitigation Institution, Ahmedabad "Severity of drought is far beyond that point of imagination where such civilization has come up in which every one fights with each other". This drought affected about 1,700 lakh people of about 1,00,000 villages of India. Even today, about 18 per cent of the total area of India is badly drought affected. According to the Central Ground Water Board, scarcity of water has remained most severe during the current decade. Groundwater level has gone down by 2-4 metres as compared to last year. It was told in the meeting of Management Group on 17th January 2001, that the drought of the year 2001 was the most severe drought of the whole century.

It had become clear by July 2000 that monsoon had completely failed, whereas the government machinery committed the same mistake of the previous year (1999-2000) and could not understand it. Due to delay in relief, full benefit could not be taken, which resulted in great loss of resources. Government should have started relief works on a very large scale by January. Secretary of the Planning Commission, N.C. Saxena had written that "During the current times three companion systems of life-land, water and forest, had degenerated, hence administration should take concrete steps in the direction of integrating them".

Management of Drought

Drought management can be done in two forms: First, Management in the context of basic reasons. Cumulative reasons for drought should be found out and it is necessary to manage it accordingly so that a permanent solution of drought is done. Second, by providing relief to drought-affected areas.

Meteorological hardness, ground unevenness, hydrological state and agricultural conditions are among the basic reasons for drought. In such context, management of drought can be made in the following manner:

Management of meteorological drought

Meteorological It is the result of meteorological conditions of India. India has a tropical climate, which is affected by jet stream, El Nino and lack of depression of air (creation of depressions) etc., that cannot be easily managed. Forecasting of changing climate is the only way of safety from drought. Timely forecasting of changing climatic conditions can reduce severity of drought. Forecasting can be made whether rainfall would be excessive or meagre or medium. Science has made enough advancement in forecasting drought. The coming of satellite on computer has made forecasting all the more easier.

According to Indian satellite imageries, vegetative cover in Rajasthan has reduced by two-thirds and it has reduced in the whole of Gujarat.

Management of hydrological drought

Management of hydrological drought can be done in the following manner:

1. Afforestation should be done by organic activities; tree plantation at new places should be done under social forestry, agro forestry and other programme.
2. Deforestation should be controlled.
3. In single-crop, different tree plantation should be done in which Xerophytes tree cover is more so that loss of moisture from soil is less.
4. Recharging of groundwater should be made in uneven areas specially aquifers in hilly areas. Excess water should be stored in water reservoirs.
   (i) Water harvesting structures should be constructed for storage of surface water at suitable places so that excesses of rain can be stored in them. In Rajasthan, there are many structures of such type as *baoris, tankas, ponds, tanks, lakes* and *jhalras.*

Management of agricultural drought

There is great difference between traditional crop rotation and crop rotation adopted after the green revolution. Crop rotation should be followed only according to agricultural climatic conditions. Management of agricultural drought can be done in the following manner:

1. Dry farming should be scientifically developed in drought-prone areas; coarse grains etc., should be cultivated there. In this method, deep ploughing is done before rains so that moisture during rainy season can be absorbed in it. Irrigation methods are also adopted as per dry environment.
2. Short duration crops should be sown.
3. Such crops may be given priority, which can bear drought. Cotton varieties PRS-72 and PRS-74 mature in 120 days. Arun variety of castor, Pusa and Baisakhi varieties of Moong, CHS-1 and CHS-3 varieites of jowar, HB-1, HB-2, HB-3 and HB-4 varieties of Bajra are very useful. Kalyan Sona variety of wheat matures in 120 to 130 days.

4. Rain water should be systematically used in dry farming. Small dams, farm ponds, tanks, anicuts and earthen check dams should be erected for efficient use of rain water.
5. Contour bunding from the top for soil protection, terracing of fields and tree plantation by the side of crops is beneficial.

Management of drought areas is possible through watershed development programme by physical and organic activities. Storage of water can be made at suitable places at the local level through various physical activities.

In the second type of drought management programmes, relief works are undertaken during and after drought. After independence, the following programmes have been undertaken for providing relief to drought-prone areas:

1. *Community Development Programme (CDP):* This programme was started in 1951-52 for providing employment in rural areas.
2. *Intensive Agriculture Development Programme (IADP):* It was executed during 1956-61 for generating employment opportunities in agriculture.
3. *Rural Development Programme (RDP) and Rural Manpower Programme (RMP):* These programmes were implemented during 1961-65. They were the first programmes from the point of view of famine.
4. *Drought-Prone Area Programme (DPAP):* Its main objective was to provide relief to drought-prone areas and to conserve soil and water. This programme was started in 1974.
5. *Desert Development Programme (DDP):* It was started in 1977-78 in the desert areas.
6. *Food for Work Programme (FFW):* This programme was executed at the time of green revolution in 1977.
7. *Minimum Needs Programme (MNP):* this was started in 1975-76.
8. *National Rural Employment Programme (NREP):* This programme was supplementary to DPAP for providing employment, since 1980.
9. *Rural Landless Employment Guarantee Programme (RLEGP):* This scheme was started in 1983.
10. *Integrated Wasteland Development Project (IWDP):* This was started in 1989. It is fully financed by the Government of India, which is executed in Non-DPAP and Non-DDP areas.
11. *Jawahar Rozgar Yojna (JRY):* It was started in 1989 by amalgamating NREP and RLEGP.

12. *Employment Assurance Scheme (EAS):* It was started in 1983.
13. *Swarn Jayanti Gram Swarozgar Yojna (SGSY):* This was started for rural poor in1999.
14. *Sampooran Gram Rozgar Yojna (SGRY):* It was started in 2001 by merging JRY and EAS.
15. *Food for Work Programme 2001:* Under this programme, the central government allotted 0.50 million tonnes foodgrains for drought affected areas in January 2001.
16. *National Rural Employment Guarantee Programme (RLEGP):* This programme ensures 100 days employment guarantee to one member of the family. It has been executed in 200 districts of India and is likely to be extended all over India during the coming five years. In case a worker is not provided work for 15 days, he/she would be paid guarantee amount of Rs 30 per day even without work.

The Central and state governments have been providing relief through the above programmes in drought-prone areas but the required benefit is sometimes not given because of diverting funds to wrong heads. From the point of view of drought management, the role of DDP and DPAP programmes has been important.

The Hanumantha Rao Committee gave valuable suggestions in 1994-95 for amending the DDP and DPAP programmes. Considering the failure of these programmes, he suggested to execute both of them on watershed area development basis. Relief may be provided under DPAP by construction of water harvesting structures. Watershed area unit of 500 hectares was suggested in which development works were to be executed for four years. Subsequently, this period of execution was increased to five years. Watershed is a geographical unit in which water disposal is at one point.

After independence, India has suffered from many natural calamities including drought, which is the main one. As compared to other calamities, efforts made for drought mitigation are praiseworthy. There is strong interrelation between drought and the capacity of the south-western monsoon. In the month of May the Indian Meteorological Department announces forecast regarding precipitation from its 35 meteorological sub-divisional centers. Everywhere in the country, an eye is kept on rains. Figures of capacity of monsoon in previous years are used for forecasting of rains for the current year. As compared to other calamities, the effect of drought is not sudden but cumulative. Hence, by adopting possible steps, its severity can be reduced.

Dry Farming

Dry land is that arid and semi-arid geographical area where there is scarcity of seasonal water. Arid areas have short duration rains in which variations are found at the local level. In agricultural geography, study of dry farming is done in the context of rainfed agriculture because only rainfed land is available in the form of dry land. 67 per cent part of Indian agriculture is covered under rainfed area. Under this area 44 per cent foodgrains are produced, which maintain 40 per cent population. Two-third cattle wealth of the country is also in this area. Small and marginal farmers are mostly found here, who usually practice traditional techniques. These areas are located in arid and semi-arid climatic zones and they extend to south western Uttar Pradesh, Maharashtra, Rajasthan, Gujarat, Karnataka etc., where average rainfall is betwwen 500 and 750 mm. Main crops of these areas are jowar-bajra, barley, gram, wheat, oil seeds and cotton. Dry farming can be called a system of new scientific technique in the context of latest agricultural practices and changing human circumstances. Dry farming has been defined by different geographers in the following forms:

According to Geographical Terminology Dictionary (India), "Agriculture in semi-arid regions, in which crops are grown without irrigation by storing available limited moisture. Due to scarcity of rainfall, continuous efforts are made to maintain soil moisture and develop it. Deep ploughing is done for it, and effort is made to prevent evaporation. Crops requiring lesser moisture and those of short duration are sown in it. Such type of cultivation is specially done in Mediterranean regions and on Columbia plateau in United States of America." According to Oxford Educational Dictionary, "Soils are prepared in dry farming in such a way that whatever rain falls in semi-arid regions, they absorb in them so that after storing, it is received by roots of plants." After that in 1983, Kutz defined dry farming in his book *Development of Applied Geography* as "it is defined as techniques using no artificial irrigation under conditions of inadequate rainfall." Thus, Kutz has called it "Pure off Culture" in complete absence of rains, and Jatzold (1979) has termed it as "Water Concentrating Culture".

E.J. Russel (1950) in his book *Soil Conditions and Plant Growth* has told that "dry farming is agricutural system in scarce water geographical circumstances."

According to Webster, "production of crops without irrigation in areas having scanty rainfall is called dry farming, which is an agricultural system mainly developed by conserving moisture in arid areas."

Dry farming system is not a new technique but an art of agriculture since ancient times. In ancient times also, all lands were favourable for agriculture, hence nomadic tribes and backward castes always practiced agriculture in arid portions. During present times when population has been fast increasing, demand for foodgrains is also increasing continuously, hence it has become essential to develop dry farming on a scientific level. In the era of scientific development, new seeds of such types may be sown which can develop even in different and unfavourable environment. From such a point of view, new agriculture is in areas having scanty rainfall. Hence, the main object of dry farming is intensive use of lands where there is lack of rainfall.

Dry farming system is a special type of system in itself, which indicates that special type of environment and plants may be prepared for it, which should be able to retain their existence in a condition of scarcity of water. Thus, it is clear that:

1. Such type of crops may be grown in arid areas which require lesser water.
2. Areas of completely rainfed agriculture system may be delimited. If agricultural development were done on the basis of the above facts, it would itself be a scientific technical view point.

Hence, in view of the second angle, rainfed or marginally rainfed areas have to be identified. For that it would be necessary to find out the quantity of average rainfall of any area, quality of soils and existing moisture in them. The above work can be easily done by a geographer on the basis of tried physical conditions. Hence, a geographer along with an agriculture scientist can give it a more scientific shape after making some effort. Rainfall is the only source of water in arid areas and it depends completely on probability of monsoon.

Dry Farming Methods

For giving a scientific basis to dry farming system, agricultural development is made in view of geographical conditions of the area. Its main bases are:

1. Prevent loss of water
2. Develop methods to maintain moisture of soils

For systematic management of the above facts, the following dry farming practices are adopted in rainfed and arid areas:

1. *Watershed Based Development:* Watershed areas are considered suitable for dry farming because there is moisture to a sufficient extent in such areas. In watershed areas, moisture is conserved through management of all resources through physical and organic controls. Water conservation is done through check dams and gully control, whereas crop production, dry horticulture, agro forestry and fodder production is encouraged for land use planning. The programme was started in the Seventh Five-Year Plan. Its success is dependent on community participation hence, it is also called people's participatory programme. National Watershed Development Programme (NWDPRA) is doing an important work for rainfed areas. Hence, after successful implementation of activities under watershed development programme, a concrete shape can be given to dry farming in rainfed and arid agriculture. Soil conservation and biomass development also take place along with moisture conservation, which are necessary ingredients of dry farming. In Watershed Development Programme, different physical structures are developed to balance water distribution, which conserve moisture in spite of absence of rain for long duration.

2. *Mechanization of Small Farms:* It is difficult to sow crops at proper time in arid areas. Use of improved implements in different regions is also important. Of late, many types of implements have been developed and their use is made for sowing crops, applying fertilizers etc. For harvesting different type of crops also, implements are being developed. Such type of implements should also be cheaper so that they can be adopted by every type of farmer.

3. *Crop Diversification:* Pulses and oil seeds are important crops in agricultural management of arid lands. For increasing income and production, crop diversification is also necessary. It can be developed in the form of agri-horticulture, agri-silviculture and silvipasture. This arrangement is adopted according to capacity of different lands.

Rainfed areas are also suitable for cultivation of medicinal plants, aromatic plants and dye plants. Hence, it is necessary that dry farming should adopt latest techniques of crop diversification.

4. *Water Conservation:* In arid areas, rain is the only source of water. Hence, fields should be prepared during monsoon in such a way that water obtained through rain is conserved to a definite extent so that there is no loss of moisture. In rainfed and arid areas, crops get moisture after monsoon during winter, hence moisture conservation is necessary according to the nature of crops. Hence, some conservation measures should be effectively adopted so that soils are able to absorb maximum water in them and do not permit evaporation. Deep ploughing of fields is required for it and slope of land should be such as to absorb stored rain water. Contour agriculture, contour bunding and vegetative live fencing etc. should be adopted.

5. *Agricultural Field System:* Other arrangements based on agricultural systems are developed in arid areas, where rural people have important contributions. Such type of agricultural practices include tenancy system, land ownership and land holding, on the basis of which the following practices have proved useful:

(i) Fallow land
(ii) Crop rotation

If a farmer himself adopts these practices, then land utilization takes place within limits and in sequence, whereas over exploitation is done under tenancy system.

Crop rotation is essential for productivity and moisture conservation of land. In crop rotation, crops with deep roots should be grown after crops with shallow roots, so that quantity of nutritive elements remains balanced. Crops requiring more nutritive elements should be grown after crops which require lesser nutritive elements. Selection of crops should also be made by keeping in mind the productive capacity of soil. Crops like bajra, moong, barley, gram etc. requiring less nutritive elements should be grown in less productive lands. Less water consuming crops should be grown after more water consuming crops, so that communication of air due to movement of water in land and organic activities of soil are not hindered. Crops included in crop rotation should not be from the same vegetative family. Crops without leaves should be grown after leafy crops since root knots of leafy crops contain rizobeium like nitrogen element, which is used by roots to some extent. The remaining portion is left in the soil itself. Hence, crop rotation should include such crops which utilize all the facilities like irrigation facility, labour, agricultural implements, fertilizers, seeds etc., available with the farmer, and they are utilized to an optimum level.

6. *Simple Water Conserving System:* Simple water conserving system is also a watershed based activity which is adopted in a particular topographical condition. Concentration of running water is done by this method. Rain water is conserved through different activities under watershed development programme, which mainly includes bank stabilization and reforestation activities. Stored moisture is found in lesser quantity in slopes as compared to plain area lands, which contain more moisture. Hence, terraced fields are prepared on sloppy lands by constructing a stone wall so that fields may become nutritive. Water is conserved by constructing steps from place to place. Water percolates deeper and is stored in terraced fields so that favourable conditions prevail in dry farming.

Structures by the name of 'Khadeens' are being developed in arid desert areas for moisture conservation. Embankment is constructed on one side of ponds where water collects and crops are grown in it after rainy season. Even rabi crops are being grown in them during winters now. Hence, it is innovation in dry farming based on indigenous knowledge. In hilly areas, groundwater is stored in watershed basins. Water is collected in fields by doing 'medhbandi'. Thus, sometimes fields are saturated even to the extent of 100 to 150 cm. depth. This is also called flow water agriculture in which cultivation is done by collecting moisture through flow water.

7. *Emphasis on Livestock Farming:* Livestock farming is the main source of subsistence of farmers in dry farming. It is found in arid and semi arid areas. In these areas, due to shortage of fodder there is lack of animal products also. Silvipasture should be developed on uncultivated lands where crop production is also minimum. These type of activities should be developed in such lands so that there is increase in animal products. During the Ninth Plan, National Agriculture Technical Project (NATP) made some progressive efforts in this direction by identifying rain fed areas for doing such activities. In this context, Indian Agriculture Research Institute, New Delhi, is also taking praiseworthy steps.

For development of dry farming practices, some important strategies should be adopted by which productivity of dry farming can be increased. They include: (1) cultivation after season, (2) cultivation of fields against the slope, (3) adoption of short duration plants,

(4) adoption of suitable moisture conservation in problematic soils, (5) suitable selection of crops to be grown in suitable season, (6) timely control of weeds, (7) selection of improved varieties of plants, (8) use of fertilizers as per recommendations of the department etc., as important strategies. The above methods assist in making dry farming more scientific.

Problems of Dry Farming

Dry farming in itself is cultivation fighting against problems. It has to face different types of problems related with meteorology, soil and crop management. Normally, dry farming has to face the following problems:

1. *Problems related to climate:* Climatic elements are very important in dry farming. Rain is the only source of soil moisture, hence it plays an important role in crop production. Unfortunately, annual rainfall is insufficient and duration of rain, intensity of distribution, nature of its begining and end, all are uncertain. Most of the rainfall is received from south-west monsoon. Normally, short duration drought or long duration drought is faced even during monsoon. Duration of rain is not certain. Sometimes it starts before time and sometimes it comes late but finishes very soon, due to which sufficient moisture is not stored in such areas. This causes adverse effect on dry farming. Effect of nature of rain falls on management of crops. Sometimes it rains very intensely and sometimes it rains very slowly. Low productivity of crops and their failure as well as variations in production are the problems of dry farming.

In arid regions, there is great variation in temperature and speed of winds. Temperature during the day in May-June rises as high as 40°-45°C whereas, night temperature falls to 20°-25°C. On the other side, due to high temperature and moving winds, evaporation takes place at a faster pace and it results in loss of moisture.

2. *Problems related to soils:* Different types of soils are found in arid and semi-arid regions. Due to these differences, problems are faced in proper management of soils. Crop production capacity of soils of rainfed areas is comparatively lesser than that of irrigated areas. Quantity of carbonic materials is continuously reducing in these soils due to high temperature. There is lesser water storage capacity in the soil. Arid lands are spread over in unlevelled condition, due to which quantity of rain water is not uniformly distributed. Due to high temperature in these areas, rate of evaporation is very high and the nature of soils becomes saline and alkaline

because of this. After rains, the soil becomes dry because of evaporation due to high temperature. Cracks develop due to dryness in land and its upper layer becomes hard. Hence, loss of soil moisture becomes faster, which becomes the main problem of dry farming.

3. *Problems related to crop management:* Farmers are presently growing traditional crops of low productivity and long duration in arid regions. Timely supervision of crops is also lacking. Use of fertilizers is negligible. Productivity is low due to absence of control over weeds. The cultivators neither adopt suitable crop rotation system nor make efforts in the direction of adopting methods of moisture conservation. Due to absence of adoption of moisture conservation methods in rainfed areas, farmers cannot ever follow inter-crop production.

Solution of Problems of Dry Farming

Crop production is lesser in arid fields as compared to irrigated fields. Awareness should be generated among farmers for using methods of moisture conservation and scientific cropping methods to solve problems of dry farming. These methods are as follows:

1. *Methods of Moisture Conservation*
   (i) *Reduce drainage and increase water retaining capacity of soil:* Land having low slope should be levelled so that quantity of rain water is uniformly distributed and the soil is able to absorb more water. Soil erosion will also be reduced due to slowing down of flow of water. 'Medhbandi' should be done in fields. Deep cultivating should be done immediately after harvesting rabi crop for moisture conservation and organic manure should be used for moisture conservation.
   (ii) *Reduce loss of moisture due to evaporation:* Evaporation takes place at a faster rate in arid regions due to high temperature, which causes loss of soil moisture. Plants start drying up because of absence of rain and moisture in plants. Hence, stored quantity of moisture should be prevented from evaporation and loss of moisture through unused areas should also be prevented. Refuse of crops, leaves, dry grasses, cow dung manure, compost manure, dust of wood and polythene sheets should be spread over the surface land to reduce the quantity of evaporation.
   (iii) *To reduce rate of evaporation:* The adverse aspect of dry farming is evaporation, which can be reduced by use of

evaporation preventing materials. These are such chemicals which are sprayed over the surface of trees and plants to reduce the rate of evaporation. Apart from it, plants of desert ecology nature should be planted.

(iv) *Increase water storage:* Cultivation is not done on all the lands in rainfed and arid regions due to meagre rainfall and absence of irrigation facilities. Methods of storage of water prove more useful here. Cultivation can be done in one part of the field and water can be stored in the remaining part, which can be flown in the cultivated area. Crops can be grown in the fields by constructing medh and drains.

2. *Agronomic Methods:* Though the main basis of dry farming is moisture conservation, but scientific cropping methods are also important as they help in utilizing the maximum moisture stored in soil. By selecting suitable types of crops for increased productivity, sowing is done at the proper time in the proper way. By controlling weeds, nutritive elements for the crops are maintained. In dry farming, all the required fertilizers in the proper quantity should be applied in rabi crops at the time of cultivation itself. Cultivation of crops should be done at the proper time. Seed rate should be 10 to 15 per cent more but number of plants should be 10 to 15 per cent lesser than those in irrigated fields. Due to lesser soil moisture, short duration crops should be grown in arid regions. For example, RC-19 variety of rice takes 60 days to mature and RHB-30 variety of bajra matures in 75-80 days, which utilize lesser soil moisture as compared to normal crops.

For success of dry farming, the following scientific cropping methods should be adopted:

(i) *Selection of crops:* Such type of crops should be grown in dry farming, which are favourable to bear the absence of moisture at the time of drought eg. jowar, bajra, Phaselies Mingo (Uradbean), moth bean, kideney bean, green gram (moong), Cowpea (chanvla), arhar, taramira, kusum and other crops, which can bear drought. Hence, these crops should be selected for cultivation.

(ii) *Selection of varieties:* After selecting drought bearing crops, such varieties of crops should be selected which mature in a short duration and yield more production.

(iii) *Ploughing of field:* In dry farming, Rabi crops in dry farming subsist on stored moisture. Hence, such land should be ploughed very deep and much before rains so as to absorb more rain water in the soil during rains. During summer, necessary moisture would be retained after 2-3 cultivations.

(vi) *Sowing:* Successful sowing in dry farming depends on seed rate etc. Growth of crops according to time depends on the method of sowing. There would be adverse effect on crop growth if sowing is not done at the proper time. For example jowar and bajra should be sown immediately after rains. About 10 to 15 per cent more seed rate should be sown because problem of lesser sprouting of seeds exists in arid regions.

(v) *Number of plants and supervision:* Number of plants should remain 10 to 15 per cent lesser than those in irrigated areas. If drought conditions are severe, such number should remain 40 per cent lesser. Seed treatment is necessary before sowing. During sowing, depth of sowing should be kept in mind since roots of plants develop properly by application of fertilizers in dry farming and deeper roots can get moisture from depth of the soil. Organic manure should be used once in three years and use of fertilizers should be done according to crops grown.

(vi) *Control of weeds:* Loss of moisture is expedited in the absence of control of weeds. Crops get destroyed before maturity, hence proper crop rotation should be adopted by controlling weeds. Timely weeding is also essential.

15

Water Conservation

Water crisis is continuously increasing in the world. About 50 per cent population of the world is experiencing water crisis directly or indirectly in the first decade of the 21st century. A total 80 countries having 40 per cent of the population of the world in them are facing serious water crisis. Rivers are the main source of water and 50 per cent population of the world lives in 250 river basins. Among them, many rivers flow fully only in one country, some rivers are inter-country, while some countries share water according to agreement.

Humans alone are not responsible for the origin of water crisis. Nature also increases it in different ways. For example, distribution of rainfall is not equitable in the world. On one side, the equatorial region gets 250 cm. rainfall in the year, while on the other, the desert of Sahara has scarcity of water almost throughout the year. Similarly, in India, there is 1170 mm. rainfall on an average every year, but 90 per cent of it is either absorbed in the earth, or evaporates and goes to seas. Hence, for meeting the water crisis, conservation of water will have to be done at different levels.

Seragolden, who has been associated with the activities of the World Bank and was the chief of the World Water Commission constituted at Hague, said on 11th August, 1999 that "the main reason for water crisis is population explosion". The present population of the world is 615 crores and till its stabilization, 300 crore more people would be added. Hence, for meeting the needs related to water for such a great population, systematic management of water sources will have to be made.

At the regional level as well as at the world level, all countries need to conserve large quantities of fresh water, for which suitable methods should be adopted according to different geographical locations and conditions. With changing time, due to the increasing variety of economic activities, the form of use of water has changed. Though water cannot be re-established in its original form but it can be conserved according to the changing context with change of time. The following methods would prove useful for it:

Protection of Water from Pollution

If the total fresh water available on the earth remains pollution free, it is sufficient to meet the drinking water needs of the existing population of the world, unfortunately a large portion of fresh water does not remain fit for use of the living world due to increasing economic activities, urbanization etc. Oceanic water in the form of ecological system of seas is an important environmental system, but during the last century pollution has spread in large proportions. Surface water is mainly found in rivers and lakes and underground water is found under land at different depths, but which have become polluted. Large cities located on banks of rivers are directly disposing off different wastes without treatment in rivers. Similarly, tourism has spread pollution at war speed on famous lakes and sea coasts. Man is greatly dependent on groundwater for his water related necessities, but some special industrial units have also polluted this amount of water stored in the security cover of the ground.

All the available water store on the earth should be kept pollution-free because pressure of demand for water is increasing on a large part of the earth. Normally, it is presumed that sea water being saline, is not fit to be used by human beings, but indirectly it is useful for maintaining living organisms with whom man is related, e.g., fish provide nutrition to a large portion of the world. Hence, it is necessary to have a cooperative policy at the international level for control of oily and radioactive pollution of the oceans. On 18th November, 2002, an oil tanker of Bahama met with an accident and drowned in the sea near the *'Coast of Death'* sea area 233 km from north-west coast of Spain. 77,000 metric tones of oil poured out from this tanker and spread over the nearby seas and polluted the sea water. During the Gulf War of 2003 also, the water of Red Sea, Gulf Persian and Mediterranean Sea adjacent to the Arabian Sea became polluted.

The origin of water crisis initially started due to its pollution and qualitative deterioration, which became acute due to increasing demand for it. For preventing water pollution, important water sources should not be made places for disposal of wastes. Industrial units should dispose off water only after its treatment. Bathing and other such activities should be prohibited near drinking water sources. Weeds produced in water should be controlled. Water should become re-usable after physical, chemical, mechanical and organic processing methods. Rivers and lakes should not be used as canals for transport of goods as is being done in Rhine river of Europe, Volga river of Russia, and the Ganges, Yamuna and Kaveri (Cauvery) rivers in India. Thus, the most important aspect of water conservation is control on water pollution.

Redistribution of Water

Water found on the surface of the earth is not equally distributed. Existing form of distribution also becomes a reason for the water crisis. In the African continent, though there is maximum hydro electric production because of excess availability of water in Mediteranean regions, but the Sahara desert situated in the north of Africa and the greatest desert of the world, suffers from water crisis for the whole year. Droughts faced by the Sahel region in the north of Sahara desert are world famous. Similarly, there is maximum rainfall of the world (1,187 cm. in Mawsynram) in north-east India, whereas there is only 50 cm. rainfall in the west. As a result of it, more than 60 per cent portion of water of Brahmaputra and its companion rivers in the north east flows to the saline seas uselessely, whereas rivers of western Rajasthan remain dry for most of the time in a year. Hence, by arranging supply of water from areas having lesser demand to the areas having greater demand, water crisis can be minimized. By construction of surface water reservoirs and storage of excess water in them, supply can be made to scarcity affected areas. This work can be accomplished by development of water reservoirs and canal network.

Excess rainfall water which flows away from rivers without being used, can be stored by construction of water reservoirs, from where it can be supplied for agriculture, industries, urban areas etc. Facilities of fisheries and transport also exist in stored water. These reservoirs are also constructed for protection from floods, and apart from flood protection, such water can also be used for different purposes.

Redistribution of water is also possible through canal system. Canal system transfers water from excess rain water areas to scarce rain water areas and conserves water for different uses. Indira Gandhi Canal is such a type of canal system which has brought water of Himalayas to western Rajasthan and changed the arid ecology. K.L. Rao, the then Irrigation Minister, had drafted this plan by the name of 'Ganga-Kaveri Grid'. Its length was decided as 2640 km in the beginning, which had to link river Ganges of north India with river Kaveri known as 'Ganges of the South'. It was also proposed to include central rivers Narmada, Tapti, Godavari, Krishna and Pensnar falling in the way. Former President of India A.P.J. Abdul Kalam took a lead in 2003 for its implementation and its work is being started by giving a modern shape to it. It will be able to transfer water from excess rainfall water areas to scarce rainfall areas every year.

Thus, by developing artificial canal system through artificial rivers, excess water can be conserved. For this, canal system should be developed from national level to micro level. Along with big rivers, canal system should also be developed by the side of small rivers to conserve large amounts of water. Though development of canal system will have to face geographical, economic, social, cultural (rehabilitation) and political problems, but its future results would be favourable from the geographical point of view.

Rational Use of Groundwater

Groundwater meets 25 per cent of total supply of water in the world. remaining 75 per cent supply is met by surface water sources of rivers, lakes etc. Demand for groundwater goes on increasing in proportion to its available quantity due to which quantity of groundwater goes on decreasing. After exploitation of groundwater, its re-infiltration takes a very long time to complete. Hence, groundwater exploitation should be only in proportion to its recharging capacity. In India, the maximum groundwater is utilized for agriculture. Instead of growing crops requiring water according to agricultural climatic conditions, crops of commercial importance requiring more water are grown, due to which overexploitation of groundwater is done. In Rajasthan, out of 237 blocks, 203 were in the safety zone from the point of view of groundwater in 1984, but due to excessive over exploitation of groundwater during the last two decades, the number of safety blocks reduced to only 49 in the year 2001. Remaining blocks have gone to the 'Dark Zone', where groundwater is not available in the required quantity for different uses in future.

By making only conscientious exploitation of groundwater, it can be proportionately conserved. For that, crops requiring lesser water should be given priority over crops requiring more water. Lesser water should be used by industries. For recharging groundwater, necessary drainage area should also be made available.

Population Control

Water crisis has assumed dangerous proportions due to fast increase in population and deterioration in quantitative and qualitative aspects of water resources. Demand for water is continuously increasing. Along with population increase, demand for fresh water has increased due to expansion of industries and increasing urbanization. Demand for water increased 35 times from the year 1700 to the decade of 2000. Demand for water has increased more in developing countries. For the year 2001, demand for quantity of water in the whole world was estimated to be 435 cubic kms. 60 per cent of this is required in agriculture, 30 per cent in industries and 10 per cent for cooking, bathing and drinking. Hence, demand for water can also be controlled through population control and at the same time, qualitative deterioration of water can also be prevented by population.

Population of the world would become around 800 crores by the year 2025. In this context, the United Nations Organization has warned that in case population control is not exercised, the whole world will have to face grave water crisis. United Nations Organization declared the year 2003 as the 'Fresh Water Year'.

Renovation of Traditional Water Sources

In India, traditional water storage places have been able to meet the demand of drinking water in many regions but they have been renovated from time to time. Water stored in traditional water sources has been used for both purposes, agriculture as well as for drinking. Important traditional water conservation methods for irrigation include Kuhul in hilly areas, Jing (Ladakh), Kool, Water Kundis called Khoop in Arunachal Pradesh, Zabo method of Nagaland, Aabi tanks of Haryana, Dong Pokhar of Assam, Bandhare of Maharashtra, Kere of Karnataka, Iree (Tanks) in Tamil Nadu, Jackwell of Andaman Nicobar and Nadi, Tanka, Kund, Khadeen, Kui, Bedi, Baori, Jhalra, Toba etc. of Rajasthan.

Beginning of traditional water conservation methods was made in Java (Jordan) in the year 3,000 B.C. by construction of a vast water reservoir. In India, a system of water conservation and drainage was found in Dhauliveera habitations (3000-1500 B.C.), which was constructed during the Harappan civilization. These traditional water conservation methods had been developed in India looking to the nature of rainfall in different regions, but the increasing population necessitated extension of agriculture, leading to deterioration of these traditional sources. Conservation of any traditional water source, instead of being limited to the water reservoir portion, extends to the whole drainage area system where rain water is stored after flowing. Due to increasing area of agriculture, drainage area of water has been destroyed resulting in reduction of water flow therein. This has caused crisis for the existence of traditional water sources. Water crisis can be prevented by renovating traditional water conservation sources. Unless complete conservation of rain water is done in any region, the dream of water conservation cannot be fulfilled. Hence, supervision has to be done regarding renovation of dying traditional water conservation places. Now, the whole world has agreed with this fact that water available in its vast form also shall have to be conserved to keep it within our reach. From this point of view, ownership of traditional water sources should be at the village level as has also been agreed by the World Bank.

Use of Modern Irrigation Methods

At the world level, 69 per cent water is used for the agricultural sector. Necessary requirement of water in agriculture is met by surface water sources and groundwater. Surface water used for agriculture is obtained through canals and tanks and underground water from wells and tubewells. Unfortunately, much water is wasted through traditional methods of irrigation. Canal irrigation does not distribute water in a balanced way and it creates problems of waterlogging.

A large portion of water can be conserved by adoption of modern methods of irrigation. Irrigation consumes double the quantity of water in comparison to all other uses. Sprinkler and drip irrigation methods save 50 per cent water. In drop or drip irrigation method, pipes with holes are spread over the surface of land so that the crop directly receives water. There is no loss due to evaporation in this system and almost 95 per cent water is utilized. Thus, maximum water is utilized by this method. Improved modern irrigation methods are thus useful for conservation of water.

Increasing Forest Cover

According to hydrological movements, water is received through rainfall every year in different quantities on the surface of the earth. This water flows on the surface and reaches the seas. Some part of rainwater is stored in stable water reservoirs (lakes and tanks), whereas some quantity of water infiltrates into the land and takes the form of groundwater. Due to increasing deforestation during the last century, most of the rainwater flowed away to the saline seas without infiltrating into the ground. Water crisis also developed during the last decade in Cherapunji, which gets highest rainfall in the world, because forest cover has been destroyed there due to mining of limestone. As a result of it, rain water flows away very fast to the rivers. A similar thing is happening in the Dehradun area of Uttaranchal.

The old tradition of tree plantation on the banks of rivers and tanks will have to be revived. Forest cover will have to be developed on uncultivable waste lands and hilly slopes on a large scale. Since trees bear drought conditions for a long duration as compared to crops, hence trees are helpful in reducing the demand for water along with recharging water sources.

Change in Crop Pattern

Excess water is not required if crops are grown according to agro-climatic conditions but in the present race of development, changing crop pattern with higher profits has replaced them. These commercial crops require more water than the traditional crops. In north-eastern part of Rajasthan, crops were not grown as per availability of water and intensive cultivation was adopted during the last three decades. Groundwater was over exploited because of non-availability of surface water and plantation of commercial crops requiring more water. It created serious water crisis. Hence, keeping in mind the experience of Rajasthan, crop rotation should be adopted according to agro-climatic conditions. Agro forestry and horticulture should be given priority in areas having scarcity of water.

Flood Management

A large portion of fresh water in the world becomes devastating due to floods in India, out of a total land area of 32.8 crore hectares, in India, 4 crore hectare land is flood affected, out of which 3.2 crore hectare land can be protected from floods. By construction of embankments and canals a large part of land can be conserved besides minimizing flood losses. Intensive afforestation can also provide security from floods. If will be helpful in absorption of water in the soil. Drainage areas of Ganges, Yamuna, Mahanadi, Damodar, Kosi and other rivers have been taken up in flood management and security to some limit has been provided to the 1.44 crore hectare land.

Use of Geothermal Water

Water is also received from hot waterfalls regularly at many places on earth. Scarcity of water can be met to a certain extent by using such geothermal water.

Conserving Water in Industries

About 23 per cent of total fresh water available on the earth is used by industries the world over. Some special industries consume water to a large extent, whereas some industries pollute the major portion of water. Dyeing industry and leather industry are such type of industries which pollute the water. To manufacture one ton of steel, 300 tons of water is required. Consumption of water in industries is done in quantitative as well as qualitative forms. Proportionate consumption of water in industries in developed countries is more (50%), out of which 75 per cent demand is met from surface water sources and 25 per cent from groundwater sources. Water has to be both protected from pollution in industries, and re-used after processing. Recycling of water should be developed because normally industrial units dispose off water on the ground surface after using it only once. Because of not re-using it after processing, it also pollutes other water sources. Demand for water can also be reduced in industries by recycling used water.

Reuse of Urban Waste

Demand for water has increased in cities due to increasing urbanization. There is no provision for waste water treatment in many big towns and cities of various countries of the world. Instead of being reused, it pollutes other water sources. Such condition is visible in cities of Delhi, Agra and Mathura on the banks of river Yamuna, whereas in many countries, urban water is used after treatment in nearby fields for growing vegetables and fruits. After use of water in urban areas, disposed waste water can be treated and conserved for use in agriculture in the peripheral areas of cities. Such policies should be incorporated while planning urban development.

Water Conservation by Municipal Bodies

Municipal bodies should manage both individual demand and supply of water as well as conserve water. Municipal laws should provide for collection of rain water from roof tops and implement it.

Individual awareness is very important in water conservation. Every individual should develop the attitude of water conservation and prevent every drop of water from being wasted. As far as possible, one should conserve rain water along with maintaining quality of the water. Depth of traditional water sources should be maintained by controlling its cleanliness. Since the maximum part of water is exploited in agriculture, hence water should be conserved in agriculture in different forms. Cultivating fields in off season helps in maintaining soil moisture. If cultivation is done up to 30 cm. depth, moisture can go up to 90 cm. depth, and capillary action prevents moisture coming out from the soil. Besides this, soil moisture can also be maintained by before-time sowing, proper use of fertilizers, pesticides and weedicides. Green manure and crop rotation should also be adopted.

Environment balancing is the main basis of water conservation. Changes coming in the world environment result in shortage in the quantity of fresh water. Due to rise in temperature of the world, fresh water in the form of snow is melting and drifting towards saline oceans. This change is visible from Antarctica to Gangotri Glacier in the Himalayas. Change in climate has also caused change in nature and quantity of rainfall, due to which existing unevenness in distribution of water has also increased. Thus, the mentality of World Environment Balancing has to be developed at the individual angle and conservation has to be worked on each stage.

The objective behind the constitution of 'World Water Commission' at the Hague in 1999 was to consider questions related to water conservation and creating an environment so that strategy of water management may be worked out to search out solutions to the impending water crisis. Five groups have been constituted here. First, for change in information system; second related to energy; third for considering issues related to organic technology; fourth to discuss institutional changes and the fifth panel is for consolidating serially future forecasts based on study of statistics. Apart from them, there is also a panel for traditional areas to consider water for food, water for human consumption and water for environment. One group also works on dams and national universality in international waters. The years 2025 and 2050 have been targated by the commission for fulfilling its objective. The reason for such a long duration is that the results of solutions in this context are visible after a long time.

16

Traditional Methods of Water Conservation in India

Tradition of water conservation has been in practice since ancient times. Man has made improvements in it according to his needs. Origin of water conservation is considered to be from the time when man started living jointly 6,000 years ago. Evidence for this is found in ancient literary inscriptions etc. During excavations at a place named Dhaulaveera in Gujarat, arrangement for improved water conservation system was found dating to the Harrappan period (3000-1500 B.C). Inscriptions-records of Kolhapur in Maharashtra, indicate improved irrigation system at that time. 'Nagarjunkonda Inscription' and 'Elephanta Cave Records' describe about irrigation system in Kalinga (Orissa) in the 2nd century B.C. Remains of expert water conservation system of first century have been found at Shringverpura near Allahabad. In medieval times, the work of water conservation was at its height. Bhopal tank of the 11th century spread over 65,000 hectares, was the biggest artificial reservoir of its time. The famous writer Kalhan has mentioned in his book 'Rajtarangini' that irrigation system was fully developed in 12th century in Kashmir.

Evidences of improved system of irrigation in India have been found in the Puranas, Mahabharata, Ramayana as well as in Budhist and Jain scriptures. Descriptions of canals, tanks, dams and wells are also found in them. Information regarding water conservation as available in *Arthashastra*, the famous work of Kautilya, secretary and teacher of Chandra Gupta Maurya (321-327 B.C.), is supported by inscriptions and archaeological remains. Description of water conservation in *Arthashastra* is found in the form of departmental works. It is said these that "man should construct arrangements for making use of water from natural and other water sources for consumption and irrigation and if any person constructs a new water source instead of using traditional water source or harms them, then he would be liable to punishment". Thus about 2500 years ago also, arrangements regarding natural activities related to water were done on the basis of direct experience. Even outside India, ancient evidences about water conservation have been found.

India is a country of geographical diversities where clear differences are found in relief, climate and vegetation. There are high mountains ranges in the north, plateaus in the south and coastal plains in the north and the east. There is also an unevenness in the availability of water sources and rainfall. On one side, there are places in the north-east having highest rainfall of the world, while on the other, in the arid region of the west, scarcity of rainfall prevails throughout the year. Systems of water conservation are developed according to availability of water sources. There is intensive drainage area in the Himalayas, hence water storage system is through flow of rivers. In Rajasthan, only rain water is stored. Water sources have intimate relationship with physical relief, climatic soil, vegetation and living creatures and organisms. Hence, water conservation systems in India have been developed and conserved according to geographical conditions in different geographical regions. Water conservation systems in India can be described according to its geographical regions. In the first series, four natural regions are:

1. Himalayan Mountainous Region
2. North Indian Plateau
3. Peninsular Plateau
4. Coastal Plains and Islands

The above four natural regions of the first series are again divided into 33 regions in the second series. Out of them, five are in the northern mountainous region, five in the northern plain areas, 21 in the peninsular region and two are in the coastal region. Geographical description of natural regions has been done only from the view of water conservation systems. Hence it is described with reference to the four natural regions of the first series:

Himalayan Mountainous Region

The states of Jammu and Kashmir, Himachal Pradesh, Uttar Pradesh, Uttaranchal, Sikkim, Arunachal Pradesh, Assam, Nagaland,

Meghalaya, Manipur and Tripura are included in the Himalayan Mountainous Region, where according to different geographical conditions and availability of water, various systems of water conservation are in existence.

'Guhal' system is found in Jammu and Kashmir as well as in Himachal Pradesh. It has been developed like a canal system, consisting of mountainous streams up to 15 km in length. There are also earthen drains connected with *'Guhal'* called *'Mogha'*. Through the *'Guhal'* system, melted glacier water flows and is the stored in

Figure 16.1

Mountain (C.S.E. Guhal)

tanks. The water flows at a very fast speed at the time of snowfall but during the cold season, lesser quantity of water comes in them. An investment of about Rs 3,000 to 5,000 is required for construction of each *'Guhal'*. A similar route has been developed in Ladakh where it is called *'Jing'* and distribution of water is done by an official named *'Churpun'*. In Lahaul and Speeti areas also, similar water routes were developed through which melted snow reached the villages in the form of water. There are very long routes in uneven regions. To prevent leakage of water on the way, stones are placed in the base as well as on sides. This system is called *'Kul'* and this system is dependent on mutual cooperation and partnership. However, due to its simplification during present times, its existence has been endangered. At many places, Himachal government is also carrying out repairs and renovations, but their form is being modernized. In Uttar Pradesh, old tanks and other small wells are called *'Noula'* or *'Houzi'*, which are traditional systems of water conservation. On both their sides, trees are planted to prevent evaporation of water. Religious sentiments are also attached to construction of 'houzi'. Such a system is also known as *'Hara'* system in Uttar Pradesh.

In Darjeeling of eastern Himalayas, irrigation is done from the falls called *'Jhoron'*. Water is taken over to terraced fields through bamboo pipes. *'Jhoron'* system has remained in use because of Lepch, Bhotiya and Gurung people. In Sikkim, water of falls and tanks (Khols) is used for drinking. Bamboo pipes are used to take water into tanks. People construct water ponds (Kundis) called *'Khoop'* in the compound of their houses for drinking water. In Arunachal Pradesh, irrigation is done through bamboo drains. In the Subansiri district here, Aptani tribals have constructed dams on river *Kele* by the traditional system. Water is stored from falls here where fish farming is also practiced. Aptani tribals construct bamboo drains by the side of villages so that animal and human excreta also mix in the irrigation water, which would become more fertile. *'Jabo'* system of Nagaland is famous in the north-east. It is also called *'Ruja'* system in the local language. It is a mixed system of agro forestry and animal husbandry, which is an effective system of water conservation. People of Kirkuma village are expert in water conservation. They collect water by the side of roads and take it over to tank. These water conservation places are also repaired from time to time. In Meghalaya, irrigation is

---

* In the Himalayan region, water of melted snow is taken to fields through a water route (a special canal) known as Guhal. In other hilly areas, it is also called mountainous Kuhal.

done through bamboo drains after taking water from falls. This system is prevalent in Garo, Var and Khasi areas. Bamboos of different radius are used for preparing water routes. Bamboos are first peeled and made thinner and their knots are removed.

Table 16.1

Traditional Water Conservation Systems of India

Climatic Zone	*System of Agriculture*	*System of Storing Drinking Water*
Mountain and Mountainous Region	(i) Water routes carrying water up to fields-like Kuhal or Guhal in Western Himalayas. (ii) Water storage system in water route for use in dry season like Jing in Ladakh.	(i) Water of falls is taken (ii) Rain water of roof tops is collected (iii) In north-east, water from falls and sources is taken through bamboo pipes to far off places
Arid and Semi-Arid Regions	(i) Tanks of rain water system irrigating areas in down level. (ii) Systems collecting water from rivers and sources, filled in sequence from one tank to another. (iii) Encircle flow of rain water on the ground to make soil moist, and unfertile land also gets fertile soil. Khadeen and Johad are its examples.	(i) Wells and baoris getting water from groundwater aquifers (ii) Construct tanks and lakes, so that wells and baoris on their banks get seepage water. (iii) Collect rainwater off roof tops. Tankas are its examples. (iv) Collect water in field ponds within the land by constructing artificial catchment eg. ponds in Rajasthan. (v) Special methods of separating fresh rain water from saline water of the ground, e.g., Virda of Kutch. (vi) Special system of central east, Surangam system, in which water falling on hills reaches wells from inside the hill portion.
Plain Regions	(i) In flood areas, water is diverted through channels by water lifting system. (ii) By strengthening field boundaries, water is prevented from flowing out of the field as done by Haveli system in Madhya Pradesh	(i) Wells
Coastal Regions	(i) System preventing sea water coming from above.	(i) Wells

Source: Boondon Ki Sanskriti, Centre for Science and Environment, New Delhi.

After that it is made smooth by pressure of local axe. Thereafter, small streams are used to take water from the main drain to distribute in different places. There are four to five steps from entry of water into the bamboo pipes up to reaching the plants. A.K.Singh, the main scientist, Water Technology Centre, Indian Agriculture Research Institute, New Delhi, has written in his book *Bamboo Drip Irrigation* that "reaching of water from one water route to the other, is the main specialty of this system. Flow of water is diverted at every stage after great care so that water reaches drop by drop near the root of plants." It is estimated that two workers have to work for 15 days to prepare this system for one hectare of land. Otherwise, this amount of labour depends on distance of land from the source of water, and number of plants to be irrigated. Once this system is installed, most of its material remains durable for a long time.

This system also prevails in the 'Var' area of Meghalaya but it is comparatively more prevalent in Khasi and Jaintiya hills. It also widely prevails in the Muktapuri area adjoining Bangladesh. There is

Figure 16.2

Jabo System (C.S.E.)

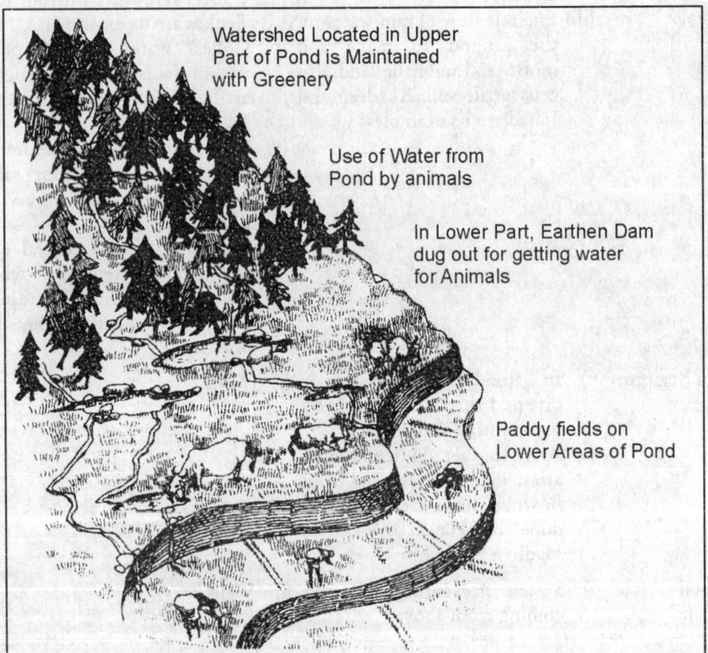

steep slope here and fields are rocky. It is not possible to construct drains on the ground to take water to the fields. Clans (*kabeelas*) have ownership of fields and cultivation can be done only by paying rent to senior persons of the clan. Seniors of 'kabeelas' have a right to give fields to any one and of any type. After payment of rent, land is given on lease for cultivation and duration of lease continues till the plants stand over it. Duration of betel leaves is longer because even after plucking of leaves, roots stand there. But if plants die due to any reason, the fields revert to the 'kabeela' and rent has to be paid again for taking it on lease.

Water for irrigating betel plants is also brought through bamboo made water flow system. Betel plants are planted in the month of March before rains and irrigation water is required only during winter. In Mizoram, water is collected from roof tops because drinking water crisis is faced during summer when waterfalls (Tuikhur) become dry. In Manipur, water is collected in tanks for irrigating terraced fields. Thus in the Himalayan region, different traditional systems of irrigation are in practice which should be continued according to the need of the day.

North Indian Plain

In India, the Gangetic plain is a vast area of small rivers. Rivers have played a major role in the civilization of this area. Traditional systems in Punjab are mainly canals, wells and ponds. The wells are called 'Jhalra', which are of greater width and lesser depth. The water of ponds is pumped out by electric motors for irrigation. In Haryana, apart from wells and canals, water is stored in '*Aabi*'. 'Aabi' is a type of tank in which rain water is stored. Surajkund of Delhi is an example of traditional water reservoir. It gets water from the northern portion of Aravalli hills. Water is also stored in the Hauz-e-Shammi reservoir (Shammi lake) near Mehrauli. Apart from it, Yamuna river is the main source of water in Delhi though it has now been polluted. River Yamuna flows for 48 km, its ancient form however has completely changed. In Uttar Pradesh water is stored in 'Johads'. These 'Johads' were constructed since the Mahabharat era. They were cleaned from time to time but they have now become polluted. Kachhla is the main 'Johad' among them.

'Ahar-Paen' System of Bihar is also a very good example of traditional water conservation. 'Ahar' is rectangular in shape and covered by water on three sides and water is taken to fields through 'paen'. Use of 'Ahar-paen' system first began in the 'Jatak' era. Its use was jointly made by people through collaboration but due to continuous floods, its ancient system has changed.

In the Brahmaputra plains water conservation is done through '*Jampoi*' system. This system is prevalent in district Jalpaiguri of West Bengal. Small drains called '*Dong*' are constructed for irrigation in this system. However, due to deforestation and excessive floods, their form has changed. Apart from it, water conservation was done in West Bengal from small rivers in Kaana, Kunti and Damodar, from where water was drawn out by 'Dhenkli'.

In Assam, the Ahom empire had made good efforts for water conservation. 'Pokhars' were constructed during their times. Tribals had constructed 'Pokhars' called 'Dong' on higher planes of hilly areas. Lahomiyas used to take water from these tanks by 'Chanvar-Chachaar' system.

Peninsular Plateau

In Plateau areas, '*Haveli*' system of Madhya Pradesh and '*Kere*' system of Karnataka are the main systems. In Jabalpur, rain water is stored through construction of high dams and it remains in the fields till crops grow there. Boundary (medh) called '*bandhare*' is constructed on four sides of the fields. It is repaired from time to time. In Madhya Pradesh, the Mughal period water management system is stilled followed. *Abdul Rahim* was entrusted with the scheme of ground-water management. In Burhanpur, water was stored in 'Bhandar', which were like anicuts of present times. Water used to move through underground drains called 'Jalikaranj' towards the city. At a distance of every 20 metres in the 'Jalkaranj', holes were made for letting in air. This whole system was based on gravitation. Unfortunately, this system also fell prey to pollution. In Jhabua, water was diverted within the hills by special types of drains called '*Paat*'.

In Maharashtra, '*Phad*' system is a suitable traditional system of water conservation. Its history is 300-400 years old. It was first developed in the districts of Dhule and Nasik. 'Bandhara' used to supply water to more than one village. Its maintenance was done by the owner of that land.

Phad System

Traditional system of water conservation in Karnataka is called as '*Kere*'. 'Kere' were tank form of water reservoirs in which water was brought through construction of anicuts over rivers. Overflow of one 'Kere' filled subsequent 'Kere' till the end. Aftercare of 'Kere' was done by the farmers of that area. This was the main source of irrigation in the districts of Shimoga and Chikmangloor in Karnataka. 'Kere' has different names according to location and nature, e.g., *Arakere* (half tanks), *Devikere* (for temples) and *Bovelkere* (tanks of prohibited area). Dams for conservation of water were called '*Katte*', which were also used for bathing and other domestic purposes. 'Katte' were tanks without drains.

Similar type of other such structure was called '*Kol*' which was circular or rectangular. Rain water was stored in it. Use of such water was made for domestic purposes or for religious purposes.

Coastal Plains and Islands

In west coast parts, groundwater is at a higher level in Kathiawar area of Gujarat and for water conservation, Baoris called '*Wav*' are constructed. Water is drawn for use from '*Wav*' through '*Chasak*'. During the 15th century, many tanks were constructed in Ahmedabad and most of them were constructed by Sultan Kutub-ud-din. Mansar Lake is famous here since water remains in it for the whole year. It has an area of 6 hectares. Water comes to it from a 'Kund'. There are a number of such tanks in Kheda, Baroda and Karavan. In the vast plains of Maharashtra, Vaitarni, Ulhas, Tassa, Savitri and Vishishti rivers originating from the Western Ghat flow there. Apart from them, rain water is also stored in lakes and tanks. Rain water is stored in fields through the traditional method by raising 'Medh' around the fields. In Konkan, there are many water reservoirs . They were constructed by Hindu rulers and philanthropic persons. Borala, Pathardi, Nageshwar and Kaashar reservoirs are important among them.

In Thane and Colaba of Maharashtra, water from sea tides was conserved in dams for irrigating fields. This system was known as 'Shilotri', which was first started by Maratha rulers. In Kerala, more than 60 per cent land is stony, which makes construction of tanks difficult. Hence, Kerala is thirsty in spite of being in water. There is absence of arrangements to conserve rain water of the monsoon period in Kerala.

In East Coast Plains, systems related to Palar river are important. Palar river originates from Nandi hills, flows for 350 km through Tamil Nadu and enters the Bay of Bengal. It has been used for recharge of groundwater in district Chittoor and its water is used for drinking purposes in Tamil Nadu. 'Kasam' system is also running near Palar river. In the 'Kasam' system, a large tank is constructed below the level of the river so that the water infiltrates and it is stored in the tank below. People had made many arrangements of water conservation in this river basin. Revu, Don, Odu, Gokunta, Kunta and Cherutvu are important among them. '*Revu*' was a system of collecting water from falls and it was constructed with stones, mud and grass. '*Don*' were small ponds which were formed by natural pits in rocks. Since its surroundings were also rocky, there was no seepage of water which could continue for a longer duration. In the '*Odu*' system, walls in the form of anicuts were constructed on streams of rivers and most of the water was used for irrigation. '*Gokunta*' were ponds constructed for use of cattle and they were normally constructed in forests near pasture lands. '*Kunta*' were ponds for storage of rain water for domestic use and for cattle. They were helpful in recharging groundwater as well as for control of floods. Tanks constructed in Palar basin for irrigation were known as '*Cherutvu*' which were constructed over streams (Vagu) by preventing water through anicuts. The present condition of all these traditional water conservation systems is deplorable since no maintenance was done by the government.

Traditional water conservation technique in Tamil Nadu is in form of tanks called '*Eeri*'. 35 per cent part of Tamil Nadu is irrigated from them. Due to control of floods through 'Eeri', soil erosion has been checked and rain water which used to flow waste, was stored. This also raised the groundwater level. The system of maintenance of 'Eeri' existed in the 16th century. It is evident from historical figures of Madurantakam in district Chengleput that 20 per cent part of total production of every village used to be spent on maintenance and irrigation of 'Eeri'. After the British regime, the downfall of these traditional water conservation systems started. Since the decade of eighties, campaign for modernization of 'Eeri' and responsibility for maintenance of 'Eeri' has come in the hands of the government.

The Islands

Andaman, Nicobar and Lakshadweep are included in the Islands. Since the effect of seas is found in islands, the Greater part of Andaman and Nicobar has equatorial nature of climatic conditions. Tribes of this area are expert in conservation of water through traditional methods. Shonpen and Jarava tribals conserve water with bamboo by tearing and ripping it. After tearing bamboo, it is fitted like a shield in lower places of the ground. Scattering rain water is stored in these shallow pits called '*jackwells*', which are connected to each other. Torn bamboos are also spread under trees to collect water from leaves. Jackwells are so connected that overflow of the earlier jackwell automatically goes to the subsequent jackwell and ultimately stored in a big jackwell. Besides the bambo technique, water is also conserved by keeping a pot or some utensil under a tree. Ongi tribals collect water from roof tops and conserve it. In Coral Heaven Lakshadweep of India, wells, baoris and tanks have been developed for conserving drinking water. There are about 800 wells in Kavratti. Tanks and baoris are mainly used for domestic purposes. However, even after such arrangements in Lakshadweep, water crisis exists there and it can be only solved by storing rain water.

The above traditional water conservation systems have stood the test of time. These systems have developed their effective nature due to different social, economic and political circumstances. They have been developed in special forms according to the local environment in different regions. Among them, the traditional water conservation methods of Rajasthan possess special characteristics because along with historical factors, different geographical factors had important effects on them.

Table 16.2

Recommendations for Improvement of Traditional Water Sources for Harvesting Rain Water in Different Agro-Climatic Zones

Agro-Climatic Zone	*Existing Method*	*Recommendation*
1. North-Western Mountaineous Region	• Collect water from roof tops. • Conserve water of falls by diverting it. • Tanks of village. • Conserve Water from slopes of hills.	• Improvement in structure of roofs by using corrugated sheets which can collect more water. • Right construction of structure to prevent pollution of water.
2. Lower areas of Himalayas	• Conserve water flowing from slopes of hills. • Tanks	• Make suitable arrangement for collection of rain water from roof tops.
3. The North-Eastern Region on wet and heavy rain areas	• Collect water from roofs. • Divert perennial falls and other water sources towards tanks.	• Make arrangement for collection of rain water from roof tops.
4. Plain areas of Assam-Bengal area having moisture (Lower Ganga)	• Tanks Anicuts/Dams. • Drains, Contour Bunds.	• Improvement in construction of tanks, cleaning of silt. • Prevent pollution.

contd...

contd...

5. Semi-Arid Alluvial area of Sutlej-Ganga	• Tanks, Drains, Contour Bunds	• Improvement in construction of tanks. • Prevent water from Pollution.
6. Semi-Arid and Arid areas of North-West (The Western Rajasthan)	• Nadi, Tanka, Khadin, Anicut, Drains, Contour Bunds	• Improvement in design of Tanka and Nadi. • Inland dams. • Slopy pond.
7. Semi-Arid areas of Central Vindhyas	• Tank, Dam, Contour bund, drains, inland dam	• Improvement in design of Nadi and Tanka. • Slopy Tank. • Construct bandhars.
8. Plateau area of excessive rain in Chhota Nagpur (The South eastern plateau)	• Tank, Aneicut, Dam, Drains, Contour Bunds	• Improvement in construction of tank to prevent water wasted by seepage and evaporation. • Cleaning silt of tanks .
9. Malwa Plateau having black soil and confirmed rain, and NarmadaValley area	• Tank, Storage Dams • Inland dams	• Improve construction of tank to prevent water wasted by seepage and evaporation. • Cleaning silt of tanks.
10. South-Central portion of uncertain rainfall in South Plateau of variable rain	• Tank, Dams, Seepage Dams, Bandhara, Drains, Inland Dams, Contour Bunds	• Tanks with levelled bed. • Improvement in system connected with implementation. • Improvement of existing systems in use.
11. Plateau Area of Chhatisgarh	• Tanks, Dams, Drains, Seepage Dams, Bandhara, Inland Dams, Contour Bunds	• Tanks with levelled bed. • Improvement in system connected with implementation.
12. South-Eastern Portion of Red Soil	• Tanks, Dams, Soakage Tanks, Inland Dams	• Tanks with levelled bed. • Improvement in system connected with implementation. • Better management of groundwater.
13. Areas of South having irregular rains and mixed rainfall	• Tanks, Reservoirs, Soakage tanks, Drains, Inland Dams, Better management of groundwater	• Tanks with levelled bed. • Improvement in system connected with implementation.
14. Two Types of Rainfall Areas of the South	• Reservoirs, Tanks, Soakage Tanks, Drains, dams, Contour bunds	• Improvement in system connected with implementation.

Traditional Methods of Rainwater Harvesting in Rajasthan

Once upon a time, waves of sea used to rise in Rajasthan, whereas waves of famine rise there now. At present, there is a vast sea of sand in place of a sea with waves. This change of nature into a desert from a sea took a long time to happen. But people of Rajasthan have not been able to forget their original form. They have reserved their traditional ways of water conservation in the depth of their mind. It is said that Lord Krishna had given a boon to this desert land that there would never be a drought here. This context is of the time of the Mahabharat war but the society of Rajasthan desert did not sit calmly after getting such a boon. The community organized itself in various forms in the context of water conservation.

Rajasthan is a region where there are no perennial rivers. All the rivers are rainfed. There are less problems related to water but more problems due to irregular rains and insufficient water in rivers. Nature and culture are inter-related here.

There are many stories in old scriptures regarding the origin of almost all natural sources of water like waterfalls etc. in Rajasthan. Origin of Baan Ganga is considered to be the place where the Pandavas had remained some time or the other. Arjuna had pierced an arrow in the earth and water had come out. The place where Bhima had pressed down his foot and a spring of water had sprung up there, it is known as Bhimlat. Water is available in such a small quantity in this region, that wherever any natural water source was found, people started worshiping it and many sources of water became places of pilgrimage.

Local people have constructed many artificial sources of water. There are many traditional sources of water in Rajasthan eg. Nadi, Tank, Johad, Bandha, Sagar, Samund and Sarovar. Whenever people talk of 'Nadi' in a village, they have a clear idea about it (how water collects in a Nadi, how its catchment area is prepared etc.). People of the village even know what type of clay is used in construction of a dam and what special type of excavation is done for it.

Wells are another important source of water here. Many types of wells are found in Rajasthan. Normally, there used to be a single person as owner of a well. Large wells known as '*Kohar*' were owned by the community. Apart from them, there are also 'Jhalra' or 'Baori'. 'Baoris' are considered important from the religious point of view and they used to be constructed for earning blessings.

In the desert of Rajasthan, there are places known as 'Paar'. 'Paar' is place where flowing water goes underground by infiltration into the earth after the water is collected there. People of the village know that if a well is dug at such a place, sweet water can be obtained. Moreover, people can estimate about 'Paar' due to growth of plants and trees there. The wells there are called 'Beri'. 'Beris' are used to make water available for both people as well as for animals. 'Beris' are also dug in valleys of dry lakes and rivers.

In the village Jaisindhi near the Pakistan border, there had been no rain since many years. There were more than 30,000 sheep in this village but since there was a good system of 'Paar' here, the sheep did not need to move anywhere else for grass or for water.

Traditionally, Rajasthan is divided into two parts. One, where there is *'Paalar' water*, and second, where there is *'Vaakar'* water. Rain water is 'Paalar' water, which is the purest form of water and can remain collected in a 'Tanka' for 3 to 5 years. 'Vaakar' is groundwater. Many elements are mixed in it. Crops grown in 'Paalar' water are entirely different from those grown by 'vaakar' water. Time of sowing and method of irrigation are also different for both of them.

People of Rajasthan have complete knowledge about matters related to their soil and they have given local names to different types of soil. For example, labourers digging wells give different names to different layers of earth and they also know after which layer of the earth water would be found or not. Labourers for digging of wells are different from other kinds of labourers.

In some wells of the state, there is excessive water. Such wells are called 'Saagar'. These wells are about 60 metre deep and they never become dry. Water is pure and perennial in them. There are about 60 such wells in village Borunda. Engines of 300 horse power capacity are fitted on them and they irrigate more than 100 acres of land. Separate names have been given to other wells eg. Seer-Ka-Kuan (Joint Well), Sajhe-Ka-Kuan or Jharare-Ka-Kuan. Groundwater which has infiltrated into the ground is taken out in Sajhe-Ka-Kuan. In Seer-Ka-Kuan, aquifer opens in the well itself. Different types of crops are grown from different types of water.

Traditional system for measuring the distance of ground is through '*Pavanda*'* (treading by feet). Sometimes it is measured with hands. For telling depth of wells, people used to say "it is 60 men". Water in rivers was measured according to time eg. water for 3 months or for 6 months etc.

Water was connected with beautiful and artistic things from the very beginning in Rajasthan. Potters used to prepare drinking water pots with a great devotion and labour. Persian wheel (Charas in local language) was prepared from very superior leather. There were 360 types of different joints in it. Pot for drawing water out of a well is called' Khatari'.

Traditional methods of water conservation in Rajasthan are of high standard since the very beginning. Religious and cultural resolutions of the state have important roles in their development. Nature and culture have remained co-existent with each other. The tradition of water conservation here has been associated with the social structure. Because water is regarded from the religious point of view, natural water sources are worshipped. Local people of all areas have constructed artificial sources of water conservation. They have developed public faith for management of every drop of water and it is due to such confidence that difficult situations have been made easier in life by conserving natural water.

Even today in western Rajasthan, people take bath while keeping the chowki (stool) in a Paraat (big iron pot) so that the remaining water may be used for other domestic purposes. The following methods of water conservation are important in the state:

1. *Tanks:* Tank (pond) has been a main source for collecting rain water. Many types of artistic features are there in tanks constructed during ancient times. They were developed in the form of beautiful and worth-seeing places. Many types of wall paintings are engraved on doors, windows and verandas over tank beds. In the valleys of some tanks, wells called '*Beri*' were constructed. Proper care and maintenance of tanks was done by the community. The Ranisar and Padamsar tanks of Jodhpur are the result of their proper care and maintenance, otherwise they would also have fallen prey to urbanization like other tanks. It is essential to give immediate attention to these tanks in the state because many wells and Baoris (Vapis) are able to get water due to them. Tanks constructed with religious feelings are well maintained. Forest tanks of Ranthambore, Sukhsagar Tank and Kalasagar tanks are very famous. All these tanks have natural catchment area. Padmini tank constructed by Rani Padmini is also

* Pavanda is 5 feet 6 inches and a man is approximately of similar height. But Puras is called a man means in which Puras, Pavanda and man (himself) Puras is measured by a suitable man (according to height) are considered having same height. Male is suitable person.

quite famous. It has natural catchment area on three sides and a pakka wall on the fourth side.

2. *Lakes:* Maximum conservation of water in the traditional form in Rajasthan is in the form of lakes. World famous lakes exist here. Rulers, merchants (banjaras) and common people all have jointly contributed in it. Lalsagar (1800), Kailana (1872), Takhatsagar (1932), and Ummedasagar (1931) lakes of Jodhpur, collect 70 crore cubic metre water, which is sufficient for fulfilling the needs of 8 lakh persons for 8 months. Their vastness can be calculated from it. Balsamand lake has been constructed to conserve water from hills of Mandore in Jodhpur. Similarly, water in sufficient quantity is conserved in the world famous lakes Jaisamand, Udai Sagar, Fateh Sagar, Raj Samand and Pichhola of Udaipur. Catchment of Pichhola is natural and is connected with Sisarma river. Water from around Naag hills of Ajmer is collected in Anasagar lake. World famous Pushkar lake is associated with religious sentiments. Water of these lakes is also used for irrigation. Moreover, water of lakes also infiltrates into Baoris which are used for drinking purposes.

3. *Nadi:* Nadi is a type of small tank (Pokhar) in which rain water collects. Its catchment area is not of any special type. Description of construction of a Pakka Nadi in Rajasthan is found in the year 1520, when Rao Jodha Ji had constructed a Nadi near Jodhpur. In Western Rajasthan, there can be found one Nadi in every village. While constructing Nadi, quantity of rain water and its catchment area is kept in mind while selecting a place for constructing water reservoir. Conserved water lasts according to its capacity. Location for

Figure 16.3
Nadi of Village

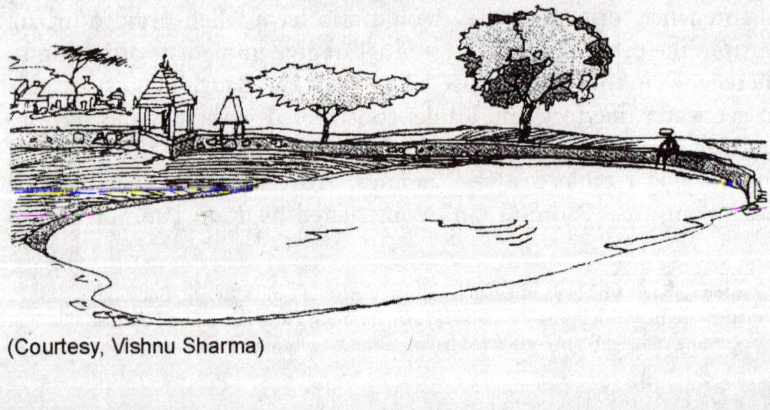

(Courtesy, Vishnu Sharma)

construction of Nadi decides its catchment and its disposal of water. Nadis are 3 to 12 metre deep in sandy areas. Their catchment area (Agor) is also large. Due to lesser leakage, its water lasts for 7 to 10 months.

Nadi in alluvial soil is bigger. It is normally 4 to 5 metre deep and water lasts for 8 to 12 months in them. According to a survey by Central Arid Zone Research Institute (CAZRI), Jodhpur, 37.06 per cent of total requirement of water in district Nagaur, Barmer and Jaisalmer is fulfilled by water from Nadis. According to the geographical situation and the people residing there, there can be found one or more Nadis in every village. Nadi is actually a natural pit on the surface of the ground where rain water comes and keeps on collecting. After some time, its capacity is reduced due to silting in it, hence it has to be desilted from time to time. To increase water storage capacity of some small Nadis, Pakka wall is constructed on one or two sides. Normally, a Nadi was named after the name of the person who constructed it or deepened it. Most of the Nadis are losing their original form due to silting in them or due to pollution. There would be lesser water according to size but more evaporation from it. Infiltration is also affected by its depth.

4. *Bavdi or Baoris:* Like a well or a Sarovar, there has been a very ancient tradition of construction of Vapi (Bavdis) in Rajasthan. Bavdis were constructed in Harappa civilization here. Description of Bavdi is also found in inscriptions of the first century. Even Vishwakarma's science of architecture has given knowledge about it. In the 74th chapter of Aparajit Prachha, four types of Bavdis have been mentioned. In ancient times, most of the Bavdis were constructed with the help of temples, one example is the Chand Bavdi in Abhaneri (Bandikui) constructed along with Harshat Mata temple.

Bavdis and Sarovar have remained important sources of drinking water and irrigation respectively since ancient times. When there were no public hand pumps and pipes inside the houses, housewives used to take drinking water from Sarovar or Bavdi every morning and evening. Even now, wherever there is no water supply scheme in any village, this scene can be observed at a public well. Housewives carry two to three artistic earthen pots on their head for fetching water. While fetching water they also exchange homely talks with one another and sometimes also sing melodious songs. These folk songs contain beautiful descriptions about emotions, feelings and contemporary social conditions as well as difficulties faced in fetching water.

Feeling of selfless service has remained an old tradition of our culture. With such spirit and enthusiasm, many rulers and capitalists got wells, Bavdis, tanks and sarovars and other drinking water sources constructed. We also find descriptions of selfless service in the folk songs of Rajasthan.

Wells and Bavdis are also connected with social activities. For example, the tradition of worshipping of well at the time of child birth and chanting related songs is prevalent even to this day.

The main objective of construction of Bavdis in Rajasthan has been collection of rain water. In the beginning, there were also such Bavdis in which residential arrangements existed. Poet Kalidas has described in Raghuvansh drama about a sports sarovar. In Meghdoot drama, he has written about the construction of a Bavdi in his residence by a Yaksha. Unfortunately, the condition of old Bavdis is not satisfactory in Rajasthan at present. If timely renovation of these Bavdis is done, these Bavdis can lessen to some the dangerous water crisis.

5. *Jhalra:* Jhalras have no water source. They receive water from soakage of tanks or lakes situated at a higher level. They do not have their own catchment area. Traditionally, the water of Jhalras was not used for drinking purposes but for completing religious customs, community bath and such other functions. The shape of most of Jhalras was rectangular having steps on three sides. Jhalras were constructed in Mahamandir, Jodhpur in the year 1660. However, most Jhalras are out of use today. The architecture of Jhalras was of a strange type but they occupied special importance from the point of view of water conservation.

There is immediate need to conserve Jhalaras today. Besides administrative and legal steps, public cooperation is also required for this purpose. Tree plantation in their catchment areas is also essential

6. *Toba:* Toba is a water conservation structure that is similar to Nadi in shape. It is deeper as compared to Nadi. In the Thar desert, Toba is an important traditional water conservation method. Land with dense composition is considered more suitable for construction of a Toba because it has lesser seepage. Its slope has to be downwards. The water of toba is used by human beings as well as animals. Because of moisture near toba, natural grass grows plentifully, which is used for grazing by animals. With the arrival of monsoon, people start shifting near Tobas by constructing their hamlets near it. Normally, water remains in a Toba for 7-8 months. After it is exhausted, people

shift to another Toba having water and use the water on mutual consent basis. Aftercare of Tobas is also done by mutual cooperation. Along with its deepening, sufficient greenery is also developed in its catchment area. This method of making water and grass available for cattle has traditionally developed jointly in collaboration with each other.

Tobas are constructed in every village by special castes and classes. In case of larger number, people of every caste construct their huts on respective Tobas. Hence, neither animals of one caste harm the other castes by entering into the boundary of other Tobas nor there is any dispute. Due to shifting of villagers near Tobas, pressure on local water sources is reduced.

Because of construction of Tobas, water remains available in the desert region for the whole year. Whenever there is an adverse effect of season and water gets exhausted, people use each other's water by mutual consent. Normally, About 20 families can use water from one Toba according to its capacity of water conservation. Toba protection is done by a special method. Pre-decided rules have to be followed and the catchment area has to be increased from time to time by excavation. Instead of widening, it is deepened to prevent evaporation and also to have more storage capacity.

7. *Kundi or Tanka:* Kundi is an important old traditional system of collecting rain water in rural areas in the desert of Rajasthan. It is also called Kund. Collected water is mainly used for drinking purposes. It is a type of micro underground pond covered from above. Mostly it is constructed by clay or at some places by cement. Kund is constructed in the whole of the desert because water is saline and not fit for consumption. Hence, rain water is collected in these Kunds, and the saying 'Flowing water is pure' is proved true in the desert of Rajasthan through construction of Kunds.

Kundis can be constructed at all places. They are constructed in forts over the hills, in valleys of hills, on roofs of houses, in court yards, in temples, in village, outside the villages in lonely places as well as in the fields. Their construction is done by individuals personally, by particular families or at the community level. Personal Kundis are constructed in courtyards or on raised platforms of houses, whereas community Kunds are constructed on Panchayat land. Their use is made by one or two villages. There is a door for entry into the enclosure of big Kundis. Mostly there are two open ponds in front of the door. One is big and other is small and their

Figure 16.4

Tanka

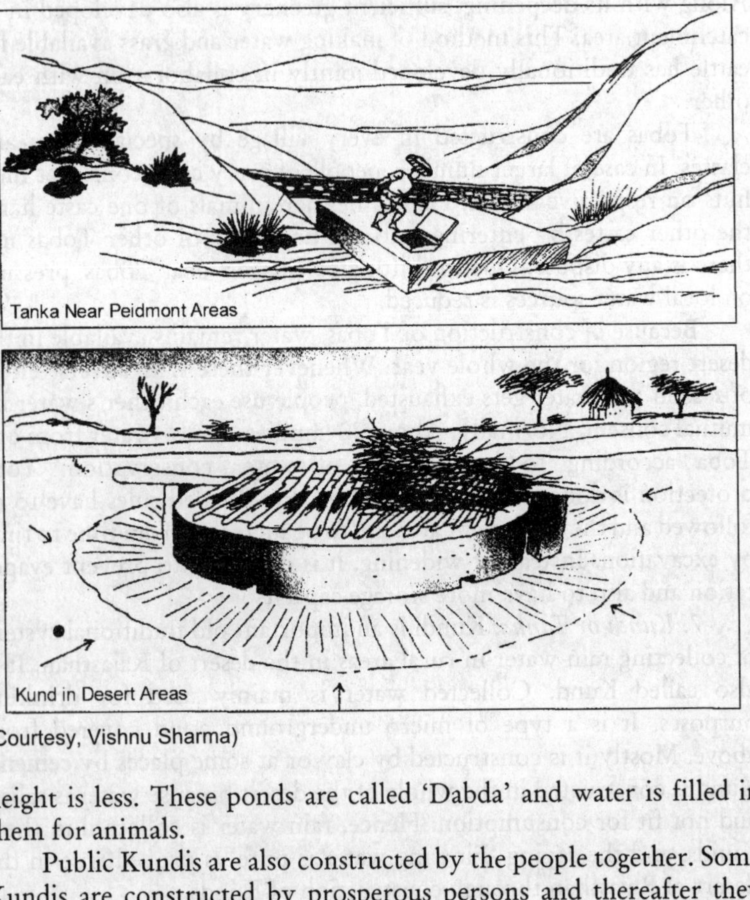

(Courtesy, Vishnu Sharma)

height is less. These ponds are called 'Dabda' and water is filled in them for animals.

Public Kundis are also constructed by the people together. Some Kundis are constructed by prosperous persons and thereafter they make them available for public use. Such type of Kundis are entrusted to a particular family who looks after it while residing in its vicinity.

Construction of Kundi is done according to the place. Depending on the slope of courtyard or the platform, Kundi is constructed in its lower portion, which is besmeared with clay and lime. Kundi is constructed according to the size of the courtyard. Its masonary is done properly from inside to prevent leakage of water. The courtyard where rain water is collected is called '*Agor*' (Catchment area) or '*Payatan*', which means accumulation. The area for collection of water is kept very clean all the year around. Water flows through the area of collection and enters through holes inside the Kundi. Net is fixed on mouth of holes to prevent leaves of trees or other refuse entering into Kundi. In case the size of the Kundi is very big, ventilation for flow of air is also kept. Kundi is always kept covered irrespective of its size. The base of Kundi is constructed with locally available materials like silt, concrete, ash, bajri etc. to prevent the bottom from washing out with water.

Table 16.3

Demand for Water and Sources of Water in Rajasthan

Object	*Sources of Water (in Lakh Acre Feet)*		
	Groundwater	*Surface Water*	*Total Water*
For Drinking	15.05	02.20	17.25 (9.1)
For Industries	0.22	NA	00.22 (0.1)
For Irrigation	44.78	127.42	172.22 (90.8)
Total	60.05	129.62	189.67
Percentage	(31.7)	(68.03)	(100)

Figures in brackets indicate percentage.

Source: Centre for Science and Environment, New Delhi.

Depth of Kundi varies from 40 to 50 feet. A dome is constructed above it and 3-4 steps are built to draw water out of it. Kundis are also built near fields in the village.

8. *Khadeen:* Khadeen is the most multi-purpose method among all traditional systems of water conservation. It is based on indigenous technical knowledge. It was first developed in the 15th century by Paliwal Brahmins in the Jaisalmer district. Rulers used to give land for construction of Khadeen, and farmers used to give one fourth produce in lieu of it to the ruler. Thus, the Paliwals developed about 500 small and big Khadeens in the district of Jaisalmer. Even today, those khadeens are irrigating 12,140 hectares land.

Khadeen is a dam like temporary tank constructed from the earth and it is constructed below slopy land. A raised earthen boundary is erected on two sides and a firm sheet of stone is erected on the third side. This earthen boundary (Pal) is called '*Dhora*'. Length of the 'Dhora' depends on the proportion of water coming during rains. Many Khadeens run for 5 to 7 kms. The earthen boundary (Pal) normally has a height of 1.5 to 3.5 metres. In case of

Figure 16.5
Structure of Khadeen

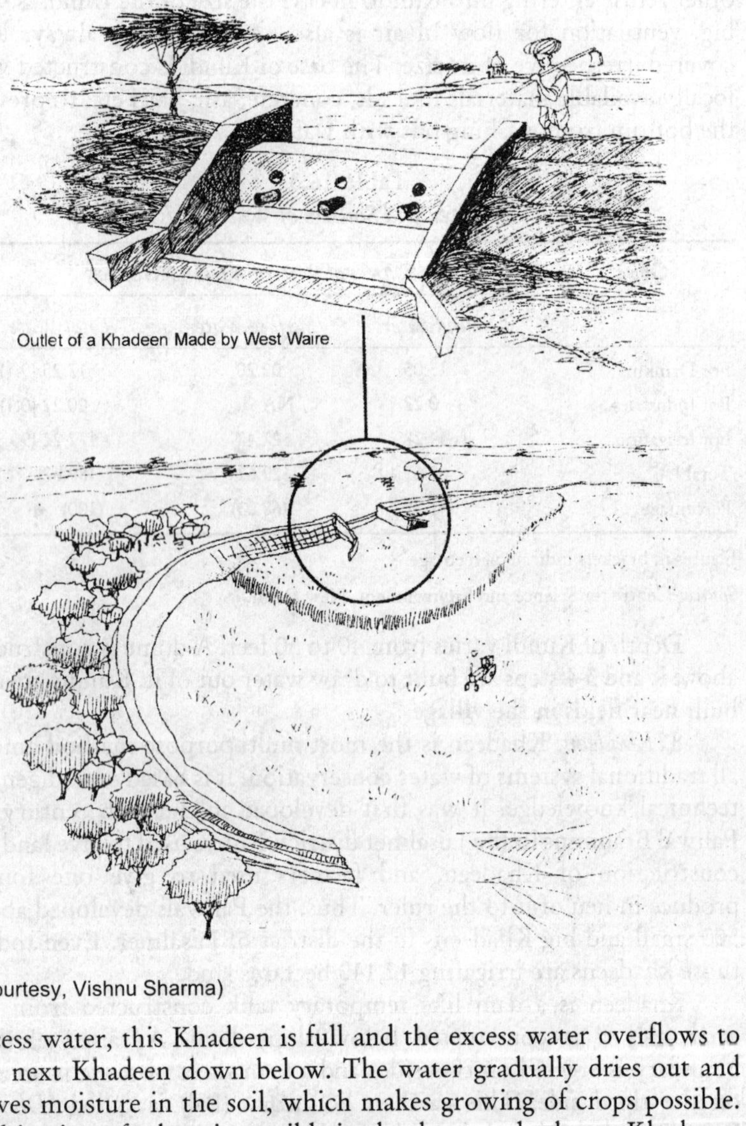

(Courtesy, Vishnu Sharma)

excess water, this Khadeen is full and the excess water overflows to the next Khadeen down below. The water gradually dries out and leaves moisture in the soil, which makes growing of crops possible. Cultivation of wheat is possible in the desert only due to Khadeens. Infertile land can also be made fit for cultivation by the Khadeen technique. Crops are grown by collecting water of Khadeens at lower slopy lands. With the help of Khadeens, at least one crop is grown even in the most arid region. At the time of its construction, quality and structure of the nearby land is kept in mind. Rocky catchment with steep slopes and soil for cultivation of crops in the plain area are considered suitable for construction of Khadeen. The place where water collects is known as Khadeen and the dam for preventing water is called Khadeen dam. The size of Khadeen is decided based on the nature of rainfall, condition of catchment and nature of soil. Khadeen dams are constructed in such a way that they do not breach on account of rainfall. Hence, excess water is allowed to escape ahead. In deeper Khadeens, gates are fitted for exit of water as per need.

Speed of water flow remains slow in arid regions. At the time of intense rainfall, water flows fast on the slopes of Khadeens depending on the slope and nature of catchment. 50 to 60 per cent of flowing water is absorbed into the soil and the remaining water goes waste in infiltration and evaporation. Ratio of Khadeen suitable for agriculture and the catchment area should be 1:15. There is meagre rainfall in desert areas but sometimes there is torrential rain when it flows very fast on slopes. According to one estimate, water up to 1,00,000 cubic metre can collect and come down from 1000 hectare rocky catchment in one rainfall.

Crops can be grown in arid regions with very little labour by the help of Khadeens. It neither needs more weeding, nor is there need for fertilizers or pesticides. As compared to crops grown in the Indira Gandhi Canal area, more production per hectare can be obtained without much labour in Khadeens. This is so because water flowing in Khadeens brings fine and fertile soil with it and after mixing with the previous soil, its composition changes and becomes more fertile. In the catchment area of khadeens, cattle grazing continues and urine and cow dung make the soil fertile. A well is also constructed near the Khadeen and seepage of water from the Khadeen comes into the well, which is used for drinking water.

Due to increase in population, joint families have started disintegrating with the passage of time and lands have been divided. Successful implementation of Khadeens however, depended on vast joint lands. With division of land, greenish catchment of Khadeens has been destroyed due to over grazing. Thus, grazing should be limited in catchment areas, though it is necessary. With density of rainfall, a thick layer of silt also collects in Khadeens. This should be cleared from time to time. Wells are constructed below Khadeens so as to limit the quantity of salts collected in Khadeens.

At present, many Khadeens are neglected. Rajasthan government has realized their importance and planned for construction of new Khadeens since it is not easy to renovate old Khadeens though it is the need of the day. Selection of the place of Khadeen is often disputable. Division of Khadeen lands has not been

Figure 16.6

Kui Constructed in Desert

Anil Agarwal / CSE

done properly and disputes arise in division of water also. Khadeens which are being constructed now are of weak quality. They cannot bear heavy rains and cracks develop in them during rains. Equal distribution of ground surface water has not been encouraged in modern Khadeens. Benefit of a Khadeen is decided by the location of nearby fields. Hence, modern Khadeens should be constructed based on traditional indigenous technical knowledge and scientific maintenance of old Khadeens should be done.

8. *Kui and Dye Keriyan:* Kui (small well) is another example of expertise of the people of western Rajasthan. 'Kui' or 'Beri', as it is called in some places, is mostly constructed near a tank so that water leaking and oozing from the tank collects in it. Thus, wastage of water is minimized. Mostly 'Kui' is 10 to 12 metre deep and remains kacha. Its opening is covered by strips of wood to prevent *men* or cattle falling into it.

'Kui' are still found in quite a large number in Loonkaransar of Bikaner district. They also exist in many villages between Ramgarh and Mohangarh of Jaisalmer district on the Indo-Pak border as well as between Phalodi and Diyatara villages of Phalodi sub-division of Jodhpur district. Water of 'Beri' is kept in reserve for the periods when surface water is exhausted. When severe drought had occurred in 1987 and all the tanks had dried up, even then water was available in the 'Beris' of these areas.

Indigenous knowledge of maximum possible use of water resources is visible in an emergency arrangement named 'Dyekeriyan'. Field boundaries of those fields in which kharif crop has to be grown, are raised high to encircle rain water. This collected water absorbs in the fields. After harvesting, a shallow well is dug in the midst of the field where some part of the percolated water is deposited. Such water is again used.

Relevance of Traditional Water Conservation Systems

System of water conservation and water management in our country is centuries old. Khadeen, Kund and Nadi in Rajasthan, Bandhara and Tal in Maharashtra, Bandhi in Madhya Pradesh and Uttar Pradesh, Aahar and Paen in Bihar, Kuhul in Himachal, Eeri in Tamil Nadu, Surangam in Kerala, Pokhar in Kandi region of Jammu, Katta in Karnataka and so on, were very ancient traditional methods for conserving and transporting water and which are in use even today. Traditional systems are special gifts of ecology and culture of those regions and have developed in local circumstances. They have not only proved their utility on the criterion of times but have also adjusted local needs with environment. Modern systems exploit the environment, whereas old systems emphasize on ecological conservation. Traditional systems possess the advantage of joint human experiences, which is their greatest strength.

Nature of rainfall is seasonal in India. Total average annual rainfall in the country is 1,170 millimetre. Out of this, 90 per cent is received only during the 3 months of monsoon period. Total rain is for 200 hours during the monsoon period and half of it rains in 20-30 hours. As a result of it, most of the rain water flows away waste. Even dams over rivers are not able to conserve more than 20 per cent water flowing through rivers. Remaining 80 per cent water is allowed to flow away to save the dam from damage.

Principle of water conservation says that rain water should be conserved according to local needs and geographical conditions. It also ensures recharging of groundwater storage. Traditional systems of water conservation fulfill needs of irrigation and domestic requirements. Historical and archaeological evidences since the 4th century show that even small communities of different regions of India were running useful systems of water conservation and distribution. During the rule of Nand (363-321 B.C), rulers constructed canals and community based irrigation systems. Gond rulers of Madhya Pradesh not only developed better systems of irrigation and water supply but they also developed necessary administrative and social systems for their maintenance.

All traditional systems were not on a small scale. For catering to the needs of cities, bigger systems were also developed, but they had linkages with smaller systems as was observed in the Chola period (930-1200 A.D) and medieval Vijaynagar kingdom. Without highly praising traditional systems, it can be said that they were useful from the point of view of water supply and cost benefit ratio in those days and even today. These systems are also important because they subsisted communities even in the wake of droughts or famines for long durations. Sometimes, when smaller systems failed due to failure of rains for years together, bigger systems carefully maintained the balance. This cannot happen unless rural as well as urban communities both participate in planning and implementation of these systems. Thus we can learn a lesson from the past.

People want a modern system because if one can get water by turning a pipe in the home itself, who would like to go on foot to fetch water from a well or a tank? Similarly, everyone would like to get water by pressing the button of a pump set or by opening the gate of a dam, but when pipes becomes dry and dams are filled with silt and modern systems start failing, people remember traditional systems. There is a large part of the country where modern systems cannot reach due to heavy investment. In such parts, people are dependent on traditional systems for drinking water as well as for irrigation. Another problem with modern systems is that the rural community has become dependent on the government and the government agencies generally fail to fulfill basic necessities of people. Pipelines are laid when elections are on head but water hardly appears in them.

The object of traditional systems does not mean that they are old and decayed structures. These systems are different from government controlled capitalist systems. Modern systems are not only capital intensive but they also recover heavy price in the context of environment. Use of water received from them is generally made against the principles of basic criterion of agriculture according to season. Groundwater is drawn out from tubewells in excessive quantity. Many government agencies do not understand the ecological importance of the sources from which they are drawing out water. They think that they would continue to get water from these sources forever. Social and political impact of modern systems has also been negative. Community based systems on the other hand, develop social coherence and self dependence. Decision-making power vested in groups or local communities who jointly worked there. It ensured economic independence and full utilization of local resources at the lower level. Traditional systems used cheap and easy technology which could be easily managed by the local people. Modern systems have disassociated the community and increased dependence on government. Problems increased all the more when modern systems based on commercial principles proved weaker on the front of distribution. Corrupt government agencies often fail to maintain modern systems efficiently.

Water is a great source of economic development. If equitable, community based, and practical development of water resources has to be ensured, then traditional systems shall have to be revitalized and developed.

Degradation of Traditional Methods

Traditional water conservation systems in most parts of the country are almost dead. They have been insulted and most of them have been exterminated. Such degradation has been all the more faster after independence. It is clearly visible that the existing centralized system of water supply and management has failed to satisfy the needs of the majority of people. There are about 5 lakh tanks in the country that are mainly used for irrigation, but due to deforestation, silt has collected in catchment areas and tanks, hence they have not remained useful. Groundwater reservoirs have also been affected. In many areas, seepage water tanks have degenerated and encroachments have been made inside tanks, which have resulted in lowering down of water level and drying of drinking water wells.

Reasons responsible for degradation of traditional water conservation systems are as follows:

1. *Increase of population and increasing demand for water,* which could not be fulfilled through traditional systems. Due to government emphasis on more convenient systems of centralized storage arrangements in the form of water reservoirs and canals, extension of traditional systems was stopped and existing systems degenerated due to negligence.
2. *Centralized modern systems* were initially started with the confidence that people would get more water easily through them, but after some years they started taking even more interest in vast, complicated and costly systems which ensured their monopoly over these resources. This led to misuse on a large scale.
3. *Increasing dependence on administrative machinery* even for small matters like maintenance of modern systems and rights vested in the government agencies through laws related to land and water resources, establish complete control on some agencies, which gives rise to misuse of power.
4. *Encouragement to individual beneficiary schemes by the government* simultaneously lead to neglect of active community participation in maintenance of traditional systems.
5. *Commercialization of agriculture* and large scale cultivation of cash crops give short term profit but create grave problems in the long run.

6. *Expectation of instant benefit at the personal level* generally discourages community cooperation.
7. *Change in distribution system* results, and there is concentration of land and community resources in few hands
8. *Sponsoring of land centered organizations by the government,* whereas traditional systems of water conservation looked at land in the context of water, and water entered angle was kept in view.
9. *Investment system of colonization period* did not change, hence small systems of irrigation were neglected.
10. *Inability of government agencies* to consider water in the form of a gift of nature for environment management.

Modern efforts for revitalization of traditional sources of water conservation will have to move forward keeping in view the reasons for downfall of these systems. It will have to be seen whether circumstances exist today for revitalization of traditional systems or not. If there is no community support for them, in that case it would be futile to create structures for this purpose. Hence, first of all, efforts shall have to be made to prepare the community again. Moreover, social contradictions have developed in some systems so that they have lost the coherence. Careful scrutiny of such contradictions will have to be made.

Some systems of water conservation were caste based. Policies of the British provided more strength to the caste system, so that caste based organizations became instruments of oppression and authoritarianism. The British divided the community for filling their coffers. Community based systems were considered adverse to centralized administration of the British model. Perhaps the same reasons were responsible for the downfall of water conservation systems of Tamil Nadu. On the other side, traditional systems like 'Kuhul' were successful in Himachal and Jammu and Kashmir because people themselves contributed labour and capital for their construction and maintenance.

One after the other, governments have encouraged large systems connected with water but their planning and management is looked after by non-responsible bureaucrats. Management of individually owned wells is also difficult because of non-availability of resources. Due to these policies of the government, importance of organizations dependent on the community like the water conservation system, has gone down. It cannot even be understood that in regions having scarcity of water, management of surface water and groundwater should be undertaken jointly by the government and the community. The advantage of big projects has been taken, by the irrigation department, contractors and political leaders. Big farmers have benefited from wells and tubewells, while assistance based programmes have benefited the implementers of the programmes. Community based organizations specially for the poor, have borne the loss.

National policy should encourage small water conservation systems. These policies should be based on community management of natural resources and should encourage them. Government would also benefit by this and basic needs of people would also be fulfilled. Ultimately, they are a part of our legacy.

We should, learn lessons from traditional water management systems, starting with the methods of ensuring community based planning, designing and management of big canals and groundwater systems. For this, the existing process of decision making shall have to be reversed.

In every village or purva (hamlet) of Karnataka, there are old historical temples and mostly every temple has a 'pokhar' (pond). All these tanks are in a neglected state. Management of most of these tanks is in the hands of the irrigation department. Some small tanks are under local trustees of inactive boards. Unfortunately, maintenance of these tanks has become their last priority. Irrigation department or Boards should earmark some part of the income of temples for cleaning and maintenance of these tanks.

In Bangalore, Mysore and many other cities, there were big old tanks. They are being dried and filled for commercial profit. The recent example is the Kormangla lake, which was dried up for the construction of residential premises for the national games.

17

Integrated Basin Planning

Concept of Basin Development

A river basin is considered as a basic unit of water resource from the point of view of planning. Its basin area is known as surface and groundwater flow, which flows from a definite outlet. The National Water Policy has made it clear that water resource planning should be made on the basis of drainage area of a hydrological unit, in which all development projects and working policies are sanctioned. Thus, river basin or basin area is a geographical unit which has been agreed as the main working unit of development for integrated planning. In the decade of the 1990s, basin area was considered as a unit for watershed development programme. Delimitation of basin is decided by water divider. Extension of one basin can be in one or more political units. On the other side, lakes and small seas are also included in the basin besides a river valley in an international context. Arabian Sea and Mediterranean Sea are also considered as hydrological units but the available quantity of water is divided in case of hydrological units coming in international boundaries, since it is also seen in the national context. Cauvery river is such type of hydrological unit. It is clear that a river basin should be considered as one unit and it should be developed in the natural boundaries of a geographical area in the national context.

Need for Integrated Basin Planning

Water resources in India are not equally distributed due to its geographical diversities. Vast differences are found in availability of water in different parts of the country from the point of view of time and place. Due to this, a large part of the country has to face drought, while on the other side, the north-eastern and eastern parts face the problem of floods.

About 51.12 million hectare area of the country is drought affected, whereas 40 million hectare area is drought-prone. Karnataka, Tamil Nadu, Rajasthan and Gujarat are drought-prone states. Similarly, Uttar Pradesh, Bihar, West Bengal, Gujarat, Orissa and Assam have grave problems of floods. According to Central Water Commission, annual average flow of water is about 1,869 cubic kms,

Figure 17.1

Water Areas of India

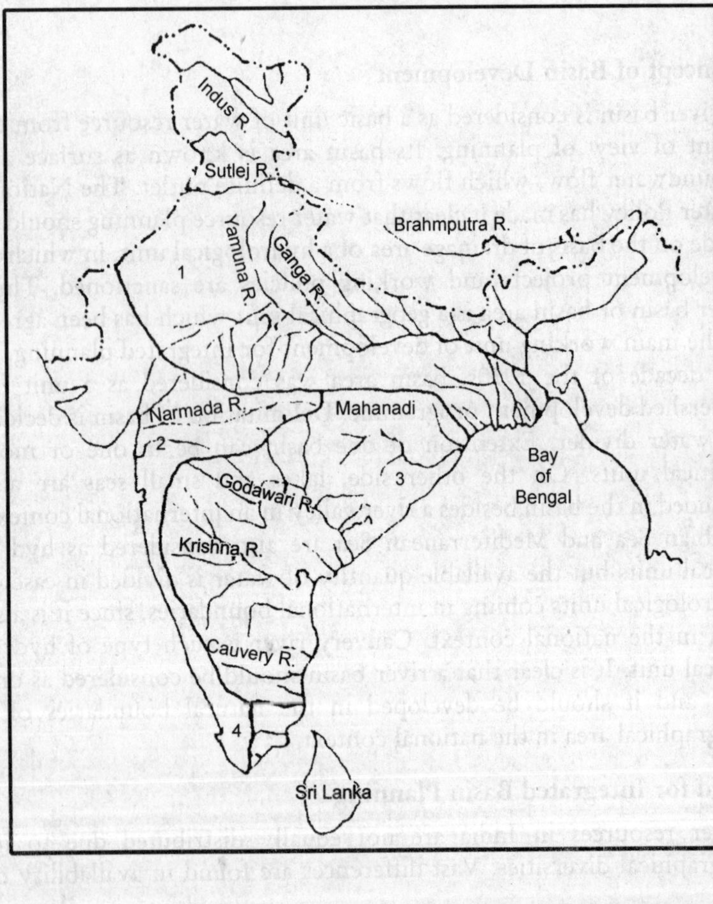

out of which 690 cubic kms water can be used under the existing technology of water development, which is about 37 per cent of the total river water. The remaining 63 per cent water flows away towards seas without any use.

Distribution of water sources in India is very inequitable. There is 60 per cent water capacity available in Ganga-Brahmaputra-Meghna in the north, whereas this capacity is 11 per cent in areas having sufficient rain in the Western Ghat. In the peninsular area, capacity of all rivers including Godavari, Mahanadi, Krishna and Cauvery is about 19 per cent.

The idea of integrated river basin planning was emphasized by the then irrigation minister K.L. Rao after independence during the Fourth Five-Year Plan in the proposed *'National Water Grid'* scheme. It was proposed to link river basins with one another to bring inequitably distributed water in a balanced position through integrated river basin planning. This planning was considered as beneficial from the point of view of increasing population, continuity of calamities of flood and drought, drinking water, and need for sufficiency of food grains.

Integrated river basin or basin planning should be done by developing one basin while linking it with water flow of another basin, so that water is distributed in a balanced way and a balance may be established in the economic context of the whole country. It is necessary to develop inter-basin water transfer plans.

Region	*Extension*	*Agriculture Area (%)*	*Water Probability (%)*
1.	Himalayan River Area	44	64
2.	Western Flow River Area (Including Kerala)	19	9
3.	Eastern Flow River area (Up to Cauvery)	35	19
4.	(a) Southern part of Cauvery	0.8	1
	(b) Kerala	1.2	7

Source: K.L. Rao, 1979.

River basin is treated as one basic unit for inter-basin water transfer under water resource planning, in which excess water is transferred to water scarcity basin and used there. There are many basins like it in India where excess water is found. Inter-basin transfer of water has been thought necessary for transfer of water from river basins having excess water. This hydrological transfer of water would be of a long distance and sufficient water would be required for it. Environmental impact assessment would also be necessary in different regions.

Inter-Basin Transfer: A Historical Perspective

Plan regarding transfer of water under integrated river basin planning is not new, but works have been executed in this area since the last five centuries. Construction of Yamuna canal and Agra canal during the period of Mughal rulers is its excellent example. At that time, water of the Himalayas was being sent to far-off Uttar Pradesh, Punjab and Rajasthan. During the 19th and 20th centuries, inter-basin water transfer was done through the Periyar project, Parambakulam Aliyar project, Kurnool Kuddappa canal and Telugu Ganga project in the south, and Sindhu basin and Rajasthan Canal projects in the north. During the 19th century, a 47.28 metre high masonry gravity dam was constructed in the way of the Periyar river valley flowing towards the west by planning for transfer of water from Periyar basin to Vaigai basin. Similarly, a 1,740 metre long tunnel for disposal of 40.75 cubic metre capacity was constructed in the middle of mountains for diverting the flow of Vaigai river towards the east. This scheme was started in 1895. At present it is irrigating 81,069 hectare area and generating 140 megawatt from a hydroelectric power center established there.

Parambikulam Aliyar project in south India is a plan to transfer water from Chelakudi basin to the Cauvery basin. Water reservoirs constructed in it have been connected by tunnels and water has been taken to drought affected Coimbatoor district of Tamil Nadu and Chittoor district of Kerala. Work of its development was completed after independence during the Third Five-Year Plan. Work of Kurnool Kuddappa canal was started in the year 1863 for transferring the water of Krishna basin to Pennar basin. It was completed in 1882.

Thus, inter-basin water transfer is not a new idea. This work started in India five centuries ago. Western Yamuna canal and Agra canal constructed during the Moghul rule, are its examples. Water of Himalayas was being taken to far-off Uttar Pradesh, Punjab and Rajasthan. In the present century, Himalayan water is being brought to the desert of Rajasthan through the Indira Gandhi Canal Project. It is a multi-purpose project through Harike Barrage on Beas river. Parambikulam Aliyar, Telugu Ganga and Sardar Sarovar etc. are other projects of a similar type.

Projects of Inter-Basin Water Transfer

National Water Grid Scheme

The object of National Water Grid Scheme for inter-water transfer is the maximum use of water sources of different rivers so that water in

Figure 17.2

Ganga Cauvery Link

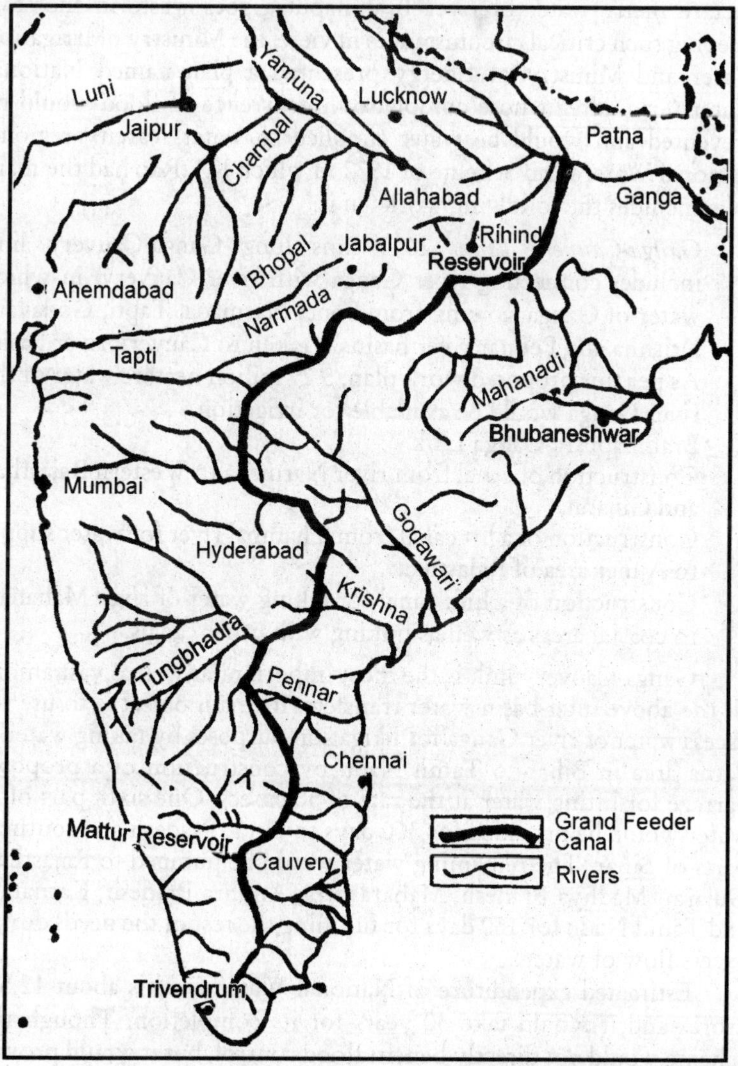

sufficient quantity may be provided to drought affected and water scarcity areas. Himalayan rivers carry huge quantities of water during the monsoon and take it to the seas, whereas peninsular rivers of central India are water scarcity affected for most of the time of the year. It is known that out of 420 lakh hectare metre water in river Ganges, only 125 lakh hectare metre water is used. The remaining water flows away independently towards seas. Similarly, 380 lakh hectare metre water of river Brahmaputra goes waste in the seas. Keeping such critical circumstances in view, the Ministry of Irrigation Water and Ministry of Energy presented a plan named National Water Grid, after whose completion, recurrence of flood would be prevented and would be water supplied to water scarcity regions. National Water Grid Scheme of 1972 in which K.L.Rao had the main role, includes the following works in it:

1. *Ganga-Cauvery Link:* 2640 kms long Ganga-Cauvery link includes connecting river Ganga with river Cauvery, in which water of Ganga flowing from Sone, Narmada, Tapti, Godavari, Krishna and Pennar river basins is taken to Cauvery river basin. As per this proposed work plan, 5.29 milion hectare water of the river Ganga would be available for irrigation.
2. Brahmaputra-Ganga Link
3. Construction of canal from river Narmada to Western Rajasthan and Gujarat.
4. Construction of a lift canal from Chambal river for water supply to Ajmer area of Rajasthan.
5. Construction of a link canal for taking water of river Mahanadi to coastal areas as well as linking with other canals.

Ganga-Cauvery link is the most important as well as vast among all the above inter-basin water transfers. Its main object is to use the excess water of river Ganga for irrigation purposes by taking water of Patna area in Bihar to Tamil Nadu by construction of a proposed barrage for lifting water at the rate of 50 cusecs. One sixth part of its water would be provided for 300 days to Uttar Pradesh and southern parts of Bihar. The remaining water would be pumped to Rajasthan, Gujarat, Madhya Pradesh, Maharashtra, Andhra Pradesh, Karnataka and Tamil Nadu for 150 days for fulfilling the rest of the needs during excess flow of water.

Estimated expenditure of National Water Grid is about 12,500 crores and it would take 30 years for its completion. Though this scheme would not directly benefit flood control, but it would provide irrigation facility for 40 lakh hectares of additional area which has remained scarcity affected since the ancient period. Besides this, drinking water facilities to affected geographical area and water transport would develop. Solution of long pending problem of river water distribution would also be arrived at. Hence, after completion of this project, there would be balanced development by providing green cover to land which has remained thirsty since years. Ultimately it would lead to national unity.

Garland Canal

Captain Dastoor had proposed idea of a garland Canal, which aimed at utilizing the whole of the surplus water of the country and providing 21.9 crore hectare area for irrigation. It is mainly divided in two parts: First, *Himalayan Canal,* which is 4,200 km long and 300 metre wide. It would be between 335 metre to 457 metre above sea level. It would be located in southern slopes of Himalayas, extending from river Ravi in the west to Brahmaputra in the east. Main source of water in it would be from Himalayas. This project would also sustain 50 lakes in the west and 40 lakes of the south. Its storage capacity would be 61.7 million hectares. Second, a canal by the name of *Southern and Central Garland Canal* which would be 9,300 km long and 300 metre wide. It would have a height of 244 metre to 305 metre above sea level. 200 lakes of water storage capacity of 49.7 million hectare metre would be located in the middle of this canal. These Himalayan and Garland Canals would intercept at two places, i.e., Patna and Delhi. According to Captain Dastoor, 219 million hectare unirrigated land would be irrigated from excess water after development of this system. Duration of 3 to 4 years was proposed for it which would provide employment to 16.8 million labourers and would involve an expenditure of Rs 24,095 crores. This estimate by Captain Dastoor was in the year 1974 and it has been analyzed by two committees.

First of all, after analysis by Central Water Commission in 1979, it was told that the proposal of Dr. Dastoor would cost about Rs 12 million crores and it should be implemented. A national scheme was also prepared for it in 1980, which talked about inter-basin transfer. Prior to this, it was studied by IIT, state governments, experts from Central Water Commission and professors of Roorkee University, who said that this project was technically weak and not suitable economically. However after execution of this project, it was

Figure 17.3
Garland Canal

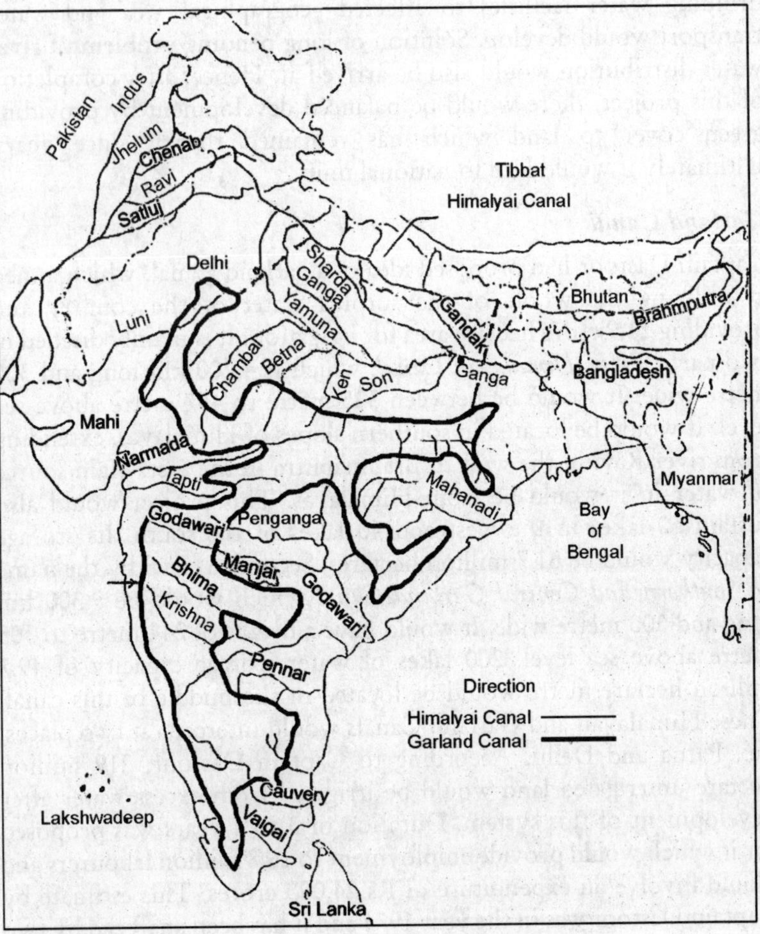

envisaged that a balanced direction would be provided to imbalanced development and improvement would take place in the weaker sides.

Other Projects

1. Executive engineer of technical mission, Gurgaon, R.N. Malik has suggested the following interlinks:

- Ganga with Yamuna
- Yamuna with Ghaghhar

- Ravi with Beas
- Sutlej with Ghaghhar

After construction of these links, 3 districts of Rohtak, Bhivani, Gurgaon of Haryana and Thar desert would be benefited. According to Malik, no other link is proposed any where in the country.

2. S. Dhawan, an engineer-cum-contractor of Delhi has proposed a 3,600 km long and 50 metre wide link extending from Dhateshwari river of Mizo hills in the east to river Ravi, along with foothills of Himalayas in the west. Its width would be 1,448 metres. One other canal is proposed in the south, having length of 9,800 kilometre and width of 247 metres. He has also proposed another canal having length of 400 km extending between Indus and Ganges rivers. About 11,65,800 million cubic metre water is estimated to be collected in the proposed storage areas. Its investment would be approximately Rs 82,000 crores.
3. Subeer Kaur of Indian Institute of Technology (I.I.T.) Mumbai, has proposed construction of water reservoirs of the size of Chilka lake in every village, tehsil, district, state for solution of the problem of drought and flood. They would be connected with big pipelines. She has also given priority to power generation for supporting tubewells and recharging groundwater.
4. Himmat Kumar Maiyar, building engineer, and others, have proposed a 5,000 km long coastal canal having 50*10 metre magnitude, and would be connected with mouths of rivers by contour canal. Total availability of water in it would be 2.5 million cubic metres and it would be costing around Rs 6,000 crores.
5. Haridas Shetty (Mumbai) has proposed a pipeline along the side of the northern railway track where various water reservoirs are located. It would prevent recurrence of droughts and floods.

All the above described proposals are imaginative. Little thought has been given to execute them, even though they can provide some strength to development of the country from the economic point of view.

Projects of National Perspective

Our available water resources should be utilized in such a way that they can be used by us forever. Many working policies have been proposed in this perspective. One main project of linking different river basins in the national perspective has been prepared for optimum utilization of water resources, as a result of which the facility of irrigation would be available for drought affected and water scarcity areas. With completion of this project, the object of the welfare state of supplying drinking water to the entire country, would also be achieved. After redistribution of national water estate as per the National Perspective Plan, regional disparities would also be removed. In the context of the present economic liberalization, possibility of hydro-electricity would increase. As a result of the above described proposals by K.L. Rao, Dastoor and others, a practical plan has been prepared in the form of National Perspective Plan, which has been divided in two parts:

1. Development of Himalayan Rivers
2. Development of Peninsular Rivers

Development of Himalayan Rivers

In the plan for development of Himalayan rivers, dams should be constructed on Ganga, Brahmaputra and their companion rivers for storage of water, and link canals should be constructed among such dams so that surplus water of the eastern companion rivers of Ganga can be transferred to the western parts of the country. In the same development plan, Brahmaputra and its companion rivers should be linked with river Ganges, and through link canals available, surplus water should be connected with rivers Subarnarekha and Mahanadi and thereafter it should be supplied in distant areas of the south. As per one estimate, if the surplus water of Himalayan area rivers is fully exploited, an additional area of about 2.20 crore hectares can be provided irrigation. Besides this, a large area in Ganga and Brahmaputra basins would be relieved from the menace of floods. Development of links between Yamuna and Ganga would give relief to drought affected areas of Rajasthan and Gujarat. Besides ensuring availability of irrigation facilities in the Indo-Nepal valley, 30 million kilowatt hydro electricity would also become generated. Silt would not be deposited in the east of Kolkata port, and an independent water route can develop, which would benefit the whole of India. In planning and execution of this plan, parts of Nepal, Bhutan and Bangladesh would also be included as equal hydrological units.

According to another study, if companion rivers of Brahmaputra are connected with river Ganges, and surplus water of the Ganges is connected with Mahanadi through Subarnarekha and Damodar rivers, there can be an increase in the water of river Mahanadi for irrigation in the south. Excess water of river Mahanadi can be supplied to coastal areas by linking with Cauvery through Godavari. Hence, according to the perspective plan, Himalayan rivers would be connected through canals with rivers of the South and it would develop greenery in the whole of the eastern arid part of the country.

Development of Peninsular Rivers

In the work plan regarding development of rivers of Peninsular India, after inter-connecting Mahanadi, Godavari, Krishna and Cauvery, dams are to be constructed for storage of water. In this plan, rivers flowing in the north of Mumbai and west of Tapti would be connected with the Narmada project and drinking water facility would be provided to Mumbai corporation and Kutch and Saurashtra regions of Gujarat. River Ken, Kalisindh, Betwa and Chambal would be connected to provide irrigation facilities to backward Malwa Bundelkhand areas and other water scarcity areas of Madhya Pradesh. After completion of this plan, additional irrigation would be provided in 1.30 crore hectare area in the states of Orissa, Andhra Pradesh, Maharashtra, Karnataka, Tamil Nadu and Madhya Pradesh, and simultaneously 4,000 megawatt hydro electricity would be generated.

National Water Development Agency was established as an autonomous agency in the Ministry of Water Resources in 1982 for a detailed study of different proposals regarding development of peninsular rivers under the perspective plan on a scientific and practical basis. In 1990-91, work regarding preparation of report for development of Himalayan rivers was entrusted to the committee. At present, the report regarding peninsular rivers is ready, whereas the report regarding Himalayan rivers is in the primary stage.

Studies made by National Water Development Agency regarding peninsular rivers has revealed that after irrigation from Mahanadi and Godavari, sufficient water is saved, whereas water in many small rivers like Krishna, Pennar, Cauvery, Begai and rivers flowing towards the east falls short of requirement. Even though water of some of these rivers is being fully utilized, not more than 60 per cent land of these storage areas can be irrigated and there is possibility of things becoming still worse in future. Surplus water in basins of Ganga, Brahmaputra, Godavari and Mahanadi can be supplied to water scarcity areas to meet increasing demand in future. Various above described water development proposals reveal that they are beneficial not only from the point of view of technical and economic angles but they would also create employment opportunities on a large scale. Simultaneously, social justice would be secured along with improvement in condition of rural areas. Thus, National Water Development Agency has identified 36 water transfer links, out of which 17 links are peninsular components and 19 are Himalayan components.

Role of National Water Development Agency

National Water Development Agency (NWDA) was established in the Ministry of Water Resources in 1982 with the objective of giving strength to the proposal of linking Himalayan and peninsular rivers for transfer of water from basins having surplus water to basins having scarcity of water, so as to have optimum utilization of available water resources. It has studied it in the national perspective as an independent agency. Funds are provided to it by the Government of India. The study of components related to Himalayan rivers was completed in 1990-91. Its main objectives are:

1. To encourage scientific development of optimum utilization of water resources of the country.
2. To study different proposals of Central Water Commission regarding transfer of water of basins in the national perspective and make combined study of Himalayan and peninsular components as well as conduct detailed survey and research of various reservoirs in the country located within different proposed plans.
3. Present detailed report after study of surplus water of Himalayan and peninsular rivers, which can be transferred to other states or basins as per above requirement.
4. To prepare necessary report regarding components related to peninsular and Himalayan rivers.
5. For achievement of the above objectives, prepare such work policy which may be acceptable to the whole society.

Existing Work Policy of the Agency

The agency has so far identified 36 water transfer plans, out of which 19 are Himalayan components and 17 are peninsular components. It has also completed detailed study of the above components. Report on work policy regarding the peninsular components has been presented by the Agency and three reports out of them have been completed. Among the Himalayan rivers, survey has been completed regarding Manas-Santkosh-Tista-Ganga-Sharda and Yamuna components. This work is continued during the Tenth Five-Year Plan also.

Work System Adopted by the Agency Regarding Surplus Water Areas and Scarcity Water Areas

Evaluation of basins and sub-basins having surplus water and those having water scarcity has been done by the Agency. First of all, a study of proposals regarding links between inter-basin water transfer and water balance according to basins was made. Suitable places were studied from different angles. Important guidelines adopted for different studies are as below:

1. *Evaluation of Availability of Surface Water*: Evaluation of availability of surface water is made with 75 and 50 per cent reliability according to basin and sub-basin.
2. *Evaluation of Water Requirement for Future (up to 2025):* Evaluate requirement of surface water for different sectors eg. irrigation, domestic, industrial, hydro electricity and salinity control according to basin and sub-basin.
3. *Requirement of Water in Irrigation:* Evaluation of annual demand in irrigation is done. As per study of the Agency, intensity of irrigation has been proposed to be 15 per cent, 125 per cent and 100 per cent respectively for big, medium and minor irrigation projects.
4. *Requirement of Domestic Water*: As per evaluation up to 2025, it has been told that per capita requirement of water in rural and urban areas would be 70 litre and 200 litre per day respectively. Animals require 50 litre water per capita per day.
5. *Requirement of Water for Industries*: Evaluation of current demand of industries for water is obtained from state agencies. If any type of statistics are not available, it is estimated that requirement of water for domestic purposes and for animals would be equal to demand for industries.
6. *Hydro Electricity:* Agency studies suitable places and availability of water for generation of hydro electricity.
7. *Salinity Control and Environmental and Ecological Studies:* Salinity in river basins is also evaluated. Suitable steps are suggested for bringing it in balanced position. Evaluation study

of environmental and ecological effects is also made. Evaluation of water balancing of selected water collection areas is also done by the Agency.

Present Plan of River Linking

National Water Development Agency (NWDA) has identified 36 links for transfer of water from one basin to other on the basis of the above guidelines and after study of all the river arrangements in the country. Out of them, 17 rivers are of the South India and 19 rivers are from North India. Among Himalayan rivers, it is proposed to connect Yamuna-Sharda and Ghaghra-Yamuna. Surplus water in them would benefit Haryana and Rajasthan. The plan of connecting Gandak-Kosi-Ghaghra with Ganga and Brahmaputra would benefit drought-prone areas of Uttar Pradesh, Madhya Pradesh and Bihar, and the population living in lower parts of Ganga basin would get rid of menace of floods.

Availability of water in peninsular rivers of Southern India is comparatively lesser than that of the Himalayan river system, because their source is not snow bound mountains. Hence, the quantity of water in them keeps on changing according to seasons. In many parts of Maharashtra, Karnataka and Tamil Nadu situated in east of Western Ghat (Sahyadri Hills), situation of drought remains due to shadow rain area. Similarly, Rayalaseema of Andhra Pradesh and Telangana also face scarcity of water.

Due to the inequality of water distribution in the country, K.L. Rao had first proposed the "National Water Grid" scheme in 1972. After him, various schemes related to connecting different rivers of the country were made continuously for three decades, as has been mentioned in the previous sections. But due to dispute with Karnataka regarding supply from Cauvery waters to Tamil Nadu, dispute between Punjab and Haryana on construction of Sutlej-Yamuna link canal and severe devastating drought in 14 the States before green belt, this work became more highlighted. This situation was created because in spite of availability of water resources in the country, the inequitable distribution of water allowed 50 per cent of the available water to be exploited by western states during the past three decades. As a result, the only alternative before the country for getting rid of water crisis in the 21st century was the linking of rivers.

Figure 17.4

Present Scheme for Connecting Rivers
(Radhakant Bharati, Rivers of India)

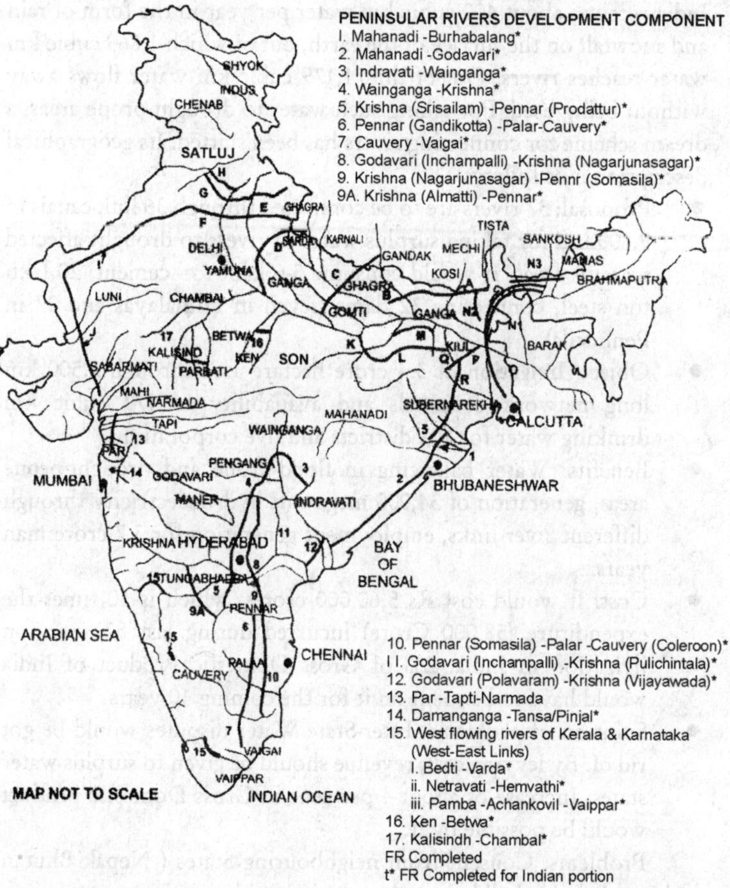

A-Kansi-Mechi (Alternative II)
B-Kosi-Ghaghra (Alternative III)
C-Gandak-Ganga
D-Karnali-Yamuna
E-Sharda-Yamuna
F-Yamuna-Sirsa Branch of WYC
G-Ganga-Sirhind Canal
H-Tajwala-Bhakra
I-Iraic Tailand of Rajasthan Canal
J-Extension of Rajasthan Canal
K-Chunar-Sone
L-Sone-Baage-Cule
M-Sone Dam-Southern Sub-River
N-Brahmputra-Ganga (Alternative I)
N2-Brahmputra-Ganga
N3-Brahmputra-Ganga
O-Farraka-Sundarban
P-Farraka-Durgapur
Q-Durgapur-Dwarkeshwar
R-Durgapur-Subernarekha
S-Dwarkeshwar-Mahanadi

Box 17.1

(Jal Swapna Yajana for Linking of Rivers

India receives about 400 cubic km water per year in the form of rain and snowfall on the surface of the earth, out of which 1,869 cubic km water reaches rivers. Out of this, 1,179 cubic km water flows away without being used. For taking such water to drought-prone areas, a dream scheme for connecting rivers has been started. Its geographical description is as follows:

- Proposal: 37 rivers are to be connected through 36 link canals in 9,600 km for taking surplus water of rivers to drought affected scarcity areas. It would consume 5-6 lakh ton cement, 20 lakh ton steel, connecting 32 dams (seven in Himalayas and 27 in Peninsula)
- Object: Irrigation of 3.4 crore hectare area through 2,500 km long network of canals and availability of 173 cubic km drinking water for 101 districts and five corporations.
- Benefits: Water balancing in flood-prone and drought-prone areas, generation of 34,000 megawatt hydro-electricity through different river links, employment generation for 3.7 crore man years.
- Cost: It would cost Rs 5,60,000 crores, which is 10 times the expenditure (58,000 Crore) incurred during last 52 years on irrigation in India. 2% of Gross Domestic Product of India would have to be spent on it for the coming 10 years.
- Solution of problems: Inter-State Water disputes would be got rid of. By levying cess, revenue should be given to surplus water states. Increase of about 4 per cent in Gross Domestic product would be possible by it.

Problems: Consent from neighbouring States (Nepal, Bhutan and Bangladesh) shall have to be taken since they may not cooperate. Arrangement of Rs 5,60,000 crore may not be an easy task. Increase in cost may create crisis for the scheme in future. Problem of up rooting 4.5 lakh persons would be there. Submergence of 79,292 hectare forest cover would reduce biodiversity, which would create ecological problems.

A public interest litigation petition based on a speech delivered by Dr. A.P.J. Abdul Kalam, President of India, on the eve of independence day on 31st October 2002, was filed. A Supreme Court Bench consisting of Chief Justice B.N. Kirpal, Justice Y.K. Sabbarwal and Justice Arijit Pasayat issued directive to the central government of India to complete the work of connecting rivers of India within a period of 10 years. It suggested the Parliament to ensure for passing a law according to entry 56 of the first list in the Constitution that any state government should object to this work. The central government clarified that the centre would connect peninsular rivers by the year 2035 and Himalayan rivers by the year 2043. Thus, it has now been clear that during the coming 37 years this work would be completed.

The preliminary feasibility report regarding connecting rivers has been prepared. National Water Development Agency (NWDA) has prepared the final feasibility report for development of peninsular rivers of India. Mahanadi, Godavari, Krishna, Pennar and Cauvery rivers would be connected in this work. Along with these big rivers,

Box 17.2

Outline of Water Dream Scheme

Before starting this scheme for construction of 36 river links to solve the water crisis of the country, former Prime Minister Atal Behari Vajpai had named it as *Jal Swapna Yojna*. Optimum utilization of yearly surplus 1,869 cubic km surface water deposit would be possible by it. India was divided in two geographical divisions for this purpose. Their comparative description is as follows:

Construction Work	*Himalayan River System*	*Peninsular River System*
Length of Canal	6,099 km	4,777 km (Including 94 km tunnel)
Links	19	17
Big Dams	9	27
Water to be Diverted	3,298.3 cubic km	14,128 cubic km
Generation of Hydro Electricity	30,000 megawatt	4,000 megawatt
Total Investment (At 2002 price level)	Rs 1,84,929 crore*	Rs 1,05,745 crore*

* Total cost of the plan also includes Rs 2,69,326 crore cost of electricity projects.

small rivers like Ken, Parvati, Par, Tapti, Damanganha, Panba, Achankovil, Netrvati and Bedti. Surplus water would be diverted towards dry basins of river Krishna, Penar, Cauvery, Vaigai and Vaipar. In this system, 17 links, 27 big dams along with 4,777 km long canal would be constructed, including a 94 km long tunnel.

This canal would be diverting 14,128.8 crore cubic metre water which would be used for drinking as well as for generation of 4,000 megawatt hydro-electricity. Total investment on development of peninsular water system would be Rs 1,05,745 crores. Maximum benefit of connecting rivers of the south would go to the states of Orissa, Andhra Pradesh, Karnataka and Tamil Nadu.

Among Himalayan rivers, Ganga, Brahmaputra and its tributaries would be connected in Ganga in India, Nepal and Bhutan and Ganga would be connected with Mahanadi. In this system, the water of Sharda, Ghaghra, Gandak, Manas, Aaee, Redak, Torsa and Jal Dhaaka would be taken up to Yamuna (Okhla), Ganga (Narora), Son Basin (Up to Indrapuri Barrage and Kudwan Dam) and sub-basin of Ram Ganga. 19 links, 9 big dams, and 6,099 a metre long canal system would be constructed for it and 30,000 megawatt hydro electricity would be generated. National Water Development Agency (NWDA) has already started development works of Himalayan river system. Water of Himalaya area would be taken to Assam, West Bengal, Bihar, Jharkhand and Orissa. Water of Ghaghra, Gandak, Sharda and Yamuna would be taken to Rajasthan and Gujarat by linking with Sabarmati river. This canal would make water available for large parts of Rajasthan which remains dry for most of the part of the year. The development of the total river system in the national perspective would be costing Rs 5,60,000 crores, which includes cost of Rs 2,69,326 crore involved on generation of hydro-electricity. This cost is estimated on 2002 price level. This cost seems justifiable looking at the increase in gross domestic product on completion of the plan. China is also preparing a plan for use of water resources of river Hwang ho which is the most devastating river of the world. It also includes construction of 660 km long water reservoir which it has been named as "Three Georges Dam". Along with flood control, it would also generate hydro-electricity which would cost Rs 3,00,000 crores.

Water Resources Department had initially prepared a two-phase programme for connecting rivers but the special task force considers it more practicable and more organized to complete it in three phases.

In the first phase, water basins would be connected, in the second phase, activities connected with rivers flowing downwards would be integrated, and in the third phase, small rivers would be connected with big rivers. Arrangements have also been made for review of progress of implementation of this whole plan for which quarterly working report will have to be presented to the Supreme Court of India.

Physical, economic and cultural changes would be coming after connecting rivers of the country and different parts of India would be benefiting in different forms. After completion of this plan, following benefits are likely to accrue:

1. Flood Control
2. End of Drought
3. Irrigation facilities for 3.4 crore hectare un-irrigated area
4. Cheap hydro-electricity due to generation of 34,000 megawatt cheap power generation
5. Supply of drinking water facility to 101 districts and metropolitan towns
6. Increase in agricultural products
7. Navigation
8. Fisheries development
9. Freedom from inter-state water disputes
10. Employment opportunities, generation of 3.7 crore man years
11. Strengthening of national economy due to increase in gross domestic product

Problems of the Plan

Though the multipurpose plan for connecting of rivers of India is being called water dream or river garland dream or nectar revolution *Amrit Kranti*, but many problems might be faced in its completion. The size of the plan, area being spread over in the country, flow route through difficult hills, desert, dense forest, uneven surface, and different types of cultural environments, may result in opposition which may come from local voluntary organizations to the state government. Till completion of this plan, the following problems will have to be faced:

1. Technical Problems
2. Political Problems
3. Economic Problems
4. Environmental Problems

Technical Problems

Engineers working in this project have said that though it is technically feasible but there has been no unanimity in it. According to engineers, water from one basin can be transferred to another basin but the proposed water links have been indicated by lines delineated on the map of India in a traditional and dogmatic fashion, which show only length and width. The third dimension of height or depth has not been indicated. Surface ground of India is plain at some places, while at some places it has plateaus or hills, the due to which water has to be lifted with pressure at some places. In northern plains, from east to the whole of the southern boundary, there is a plateau bank, where maximum rate of water lift is more than the water inlet during floods. It is clear that even after development of this system, a large part of water still goes to the sea.

Political Problems

Political thinking of different states will also create problems in linking of rivers. Issues regarding distribution of water and opposition from voluntary organizations might create problems. At present, no state is ready to agree that any quantity of water is surplus with it. The states are giving priority to inland basins rather than inter basins. Hence, creation of an environment of consensus among states is the main political function. Consultation with people is also essential for it. Nature of some rivers is international for which it is necessary to think about international consequences. It has become evident from the experience of earlier multipurpose projects (Sardar Sarovar and Tehri Project) that many social issues arise in the process, in which uprooting and rehabilitation of people are the main ones. Solutions to all these problems are not yet in sight. Water is in the concurrent list of the Constitution and without amending the Constitution for making water a national resource, the Centre cannot compel the states to agree to it.

Economic Problems

An investment of Rs 5,60,000 crore at the price level of 2002 is required for completion of the plan, which includes canal system, dams and generation of 34,000 megawatt hydro electricity. According to experts, such an amount for a developing country like India is not a small sum. It is not an easy job to arrange so much funds. The cost is

Box 17.3

Climatic Changes and Plan for Linking of Rivers

Basic aspect of this plan is related to linking of rivers having surplus water with water scarcity rivers, for which it is necessary that sufficient quantity of water should be maintained in the rivers. If water in the rivers goes on drying, what would be the importance of this vast structure? During last decades of past century, climatic balance has been disturbed because of fast economic activities. Consequently, snow of Himalayas has been melting, and its extension is reducing. During winters, peaks of Himalayas should remain covered with thick sheet of snow. According to environmentalists, snow of Himalayas would be reduced to a large extent during the coming 20 to 50 years and water coming from melting snow of Himalayas would cease. Because of increasing temperature in the world, glaciers of Himalayas are being converted into snow lakes. According to an international magazine "New Scientist" all the glaciers of Himalayas would be destroyed by the year 2025. In such a situation, existence of glaciers in changing phase of climate will also have to be kept in view. During last four decades, deforestation has taken place on a very large scale in mountainous region of Himalayas and in the South. Its adverse effect has also its impact on the plans.

likely to increase with the passage of time. Nearly 150 projects are already lying incomplete for want of funds.

Environmental Problems

For such a vast plan on the national scale, it is necessary to have an environmental impact assessment. When a tunnel is constructed for transfer of water from one basin to another, the forest cover has to be cleaned and the population of the area under submergence has to be shifted from the existing habitation. If the route of the canal passes through a sensitive forest area which has been reserved for wild animals, in such a case, by destruction of the forest, ecological imbalance will be created. A forest cover of 79,292 hectares will be submerged in this scheme which is a great ecological problem. According to Medha Patkar, the force behind the "Narmada Bachao Campaign", it is a political molestation with ecology. By diverting the water flowing to seas , lands of deltas would become barren and wetlands would be destroyed. By disturbing the drainage area of Sar Dariya and Amu Dariya in west Asia, the capacity of water storage of Aral Sagar has reduced by 60 per cent because the water was diverted..

Even after completion of the scheme, many environmental problems may arise. Such a situation is being experienced in projects implemented in irrigation areas during the past. Land which has remained dry for years, cannot bear more water. Problems like waterlogging, salinisation and alkalisation have been observed in the Indira Gandhi Canal region where barren land has extended. The main reason behind such problem has been the unevenness in distribution of water.

Because of the problems related with linking of rivers, many voluntary agencies, non-government organizations, state governments, political parties, panchayats, municipalities etc., shall have to become united on a common platform for the completion of this plan, otherwise it will not be possible to get rid of the water crisis.

Today India is also facing grave water crisis along with the whole world. Due to continuous shortfall in rains during the last decade, surface water is reducing. On account of this 50 per cent of groundwater is being exploited in Punjab, Haryana, Gujarat, Uttar Pradesh and Tamil Nadu. When groundwater is also at the verge of extinction due to scanty rainfall, linking of river basins remains the only alternative.

Progress of Linking of Rivers

River linking plan has been passing through disputes and opposition since long. It was finally sanctioned in August 2005. According to an agreement between Chief Ministers of Rajasthan and Madhya Pradesh, it was agreed to link Parvati, Nivaj and Kalisindh rivers with the Chambal river. In additional as a result of an agreement between the Chief Ministers of Madhya Pradesh and Uttar Pradesh, both the states would jointly develop the union of Ken and Betwa rivers.

1. *Linkage of Parvati, Nivaj and Kalisindh with Chambal river:* Consent regarding linkage of Parvati, Nivaj and Kalisindh rivers with Chambal was obtained on 25th August 2005, between Rajasthan and Madhya Pradesh. Madhya Pradesh agreed with the proposal of Rajasthan regarding transfer of water of river Chambal into Banas River. This whole work would cost Rs 4,000 crores. Madhya Pradesh and Rajasthan have agreed to make full use of 12 MAF (Million Acre

Foot) water of the Chambal basin and prepare a master plan for it within one year. At present, both the states were using only 3.5 MAF water of this river. A plan would be prepared for use of the remaining 8.5 MAF water. Madhya Pradesh has given consent for implementation of nine schemes. After their completion, the problem of drinking water would be solved along with that irrigation.

Since 2.21 lakh hectares of land would be irrigated in both the states. This facility would be given to 6 districts of Madhya Pradesh and one district Jhalawar in Rajasthan. Additionally, 1.12 lakh hectare land would be irrigated in districts of Ujjain, Dhar, Ratlam and Shahjahanpur, whereas another canal would irrigate 1.09 lakh hectares land Rajgarh, Shajapur and Jhalawar. Farmers of Mandsaur and Neemuch would also benefit from it. Madhya Pradesh has permitted Rajasthan to implement Kalisindh, Parvan, Manohar Thana, Andheri, Lhasi, Hathiya Deh, Peeplad, Rajgarh and Hindlor projects below the Chambal barrage. These nine projects can irrigate 1.67 lakh hectare land.

Both the states have agreed to jointly repair the right main canal starting from Kota barrage so that cultivators at the tail end are also able to get water for irrigation. Environmental sanction has also been accorded to Rau Ghat Hydroelectric power station.

2. *Ken-Betwa Link:* Chief Ministers of Madhya Pradesh and Uttar Pradesh sanctioned the union of Ken-Betwa rivers in August 2005. This would cost Rs 4,000 crores. The oustees of this project would get all possible assistance, for which an outline has been prepared. A dam and reservoir would be built on river Ken and power generation would also be done. The balance water of Ken river would be flown to river Betwa through a 231 km long canal. 9,000 hectare land would be submerged by this project and it would irrigate areas in eight districts of Uttar Pradesh including Jhansi, Lalitpur, Banda etc., and Raisen, Chhatarpur, Panna, Vidisha and other districts of Madhya Pradesh.

18

Watershed Management

Watershed

Watershed is a geographical unit of area from which rain water after falling over it, collects through a river or connected small rivers and flows through one outlet.[1] This place is known as exit or water outlet point. Thus, it is a unit of area that includes all land and water areas which contribute to run off to a common point. It is possible to control all the rain water received in that area. According to Mahnot, "A watershed is that area from which all precipitation water flows to a single stream".[2] He has considered it as a synonym for 'Catchment Area' or 'Drainage Basin'.

The boundaries of watershed are known as drainage divide. Rain falling on the opposite sides of a drainage divide falls into different watersheds. Naturally, there are hierarchies of streams or tributaries which are called macro watersheds and mini/micro watersheds. All India Soil and Land Use Survey has serially delineated watershed areas at the national level and has classified them in five sequential classes:

1. Region
2. Basin
3. Catchment
4. Sub-catchment
5. Watershed

Table 18.1
Classification of Hydrological Unit

Class of Hydrological Unit	*Size (lakh hectares)*
Region	More than 300
Basin	30-300
Catchment	10-30
Sub-catchment	2-10
Watershed	0.5-2
Sub-watershed	0.1-0.5
Milli-watershed	0.01-0.1
Micro-watershed	0.001-0.01
Mini-watershed	0.0001-0.001

The most suitable size of watershed is 500 to 1000 hectares as a unit of planning. In the recommendations of Hanumant Rao Committee report (1994), size of an ideal watershed was considered to be 500 hectares and this is still being followed. The structure of sustainable development is prepared with the 500 hectare, which is considered as the unit of planning. For this purpose, appropriate steps are taken for physical and ecological restoration. Thus, a watershed has been defined as a geographical unit in which the base for sustainable development can be prepared through different physical, economic and social activities.

Simultaneously, the objective of resources development and management is also achieved. The physical and biotic activities executed in any watershed are affected and controlled by the geographical peculiarities of that basin. The main peculiarities of a watershed are as follows:

Drainage

Boundaries of a watershed area are spread on both sides of water streams. Layout of streams of any watershed area can be known from the maps of Survey of India. In those maps, seasonal drains are shown in black colour. Serial of drainage, and its form and density affect flow of watershed and its infiltration.

Box 18.1

Delineation of Boundaries of a Watershed

Every watershed area is attached to some water stream and generally it is known by the name of such stream. The whole area which supplies water to this stream, it is covered within the limits of the watershed. The line connecting the hills and high places on four sides in the area of the stream normally determines the boundary of the watershed area. Every watershed area of a definite class covers a certain area, hence one place cannot come in the area of two or more watershed areas of the same class. Before starting development programmes in any watershed area, decision regarding problems of that watershed area is very essential. It is a very easy task and the following things would be helpful in it:

- First of all, find out concerned watershed area in the morphological site map by Survey of India.
- Make a study of streams of this area and select a place from where the whole water of the concerned watershed area comes out and supplies water to the streams of higher class.
- Identify higher points on four sides of the selected place where water of the other side does not flow towards the concerned stream.
- By joining these points, we shall get boundaries of such watershed area in which we want to execute work.
- If the desired area is not fully covered by the concerned watershed area, a watershed of higher class than it may be made.
- Even after making many micro watersheds on both sides of some main river, some areas may not be covered within the boundaries of any watershed area. Such areas would be covered by the macro watershed of that area.

Size and Shape

Watershed may be macro, medium and micro according to size, and by shape it may be elongated like a pear, triangular, or circular. Its size determines the quantity and flow of rainfall while its shape determines the quantum and time of flow. Besides them, its physiography is also an important factor.

Figure 18.1

Some parts of a watershed area are not covered in any area near main river

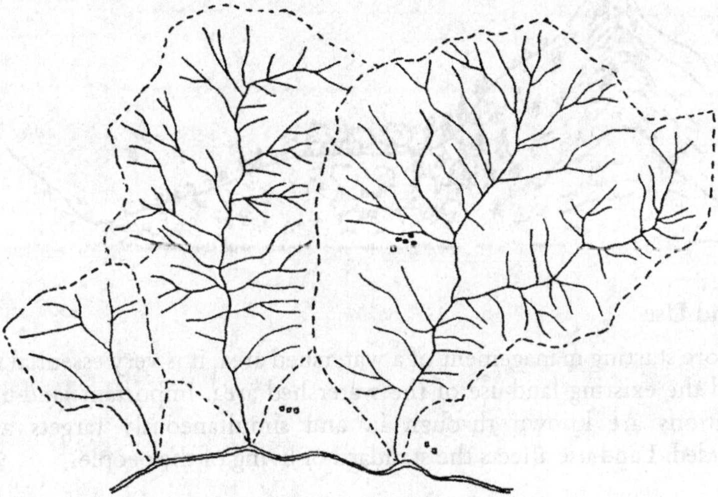

Figure 18.2

Layout of water streams

Figure 18.3
Relation between Slope and Land-Use

Land Use

Before starting management of a watershed area, it is very essential to find the existing land-use of the watershed area. Important land-use relations are known through it, and simultaneously targets are decided. Land-use affects the standard of living of the people.

Slope

Slope affects movement and distribution of rainfall as well as land-use in the watershed area. Land having steep slope is mostly barren, whereas land having light slope, is cultivable. (Figure 18.3) Surface flow of water is decided by the slope of the land.

Water does not stay for long on land having a steep slope and flows away downwards fast. Since the sun's heat is not similar on both sides of the slopes of hills on a height, hence slope on one side is normally barren. Land-slides depend on composition and slope of the rocks.

Soil and Water Relationship

Land and water are principal resources of a watershed area. Activities between both of them mutually affect each other's nature and availability. Both these resources decide land-use of the watershed area.

Climate

Precipitation, temperature, speed of winds, humidity, evaporation and other seasonal standards decide availability of water in a watershed area. Boundary of every watershed area is spread from the

Figure 18.4
Vegetative Filter Strip

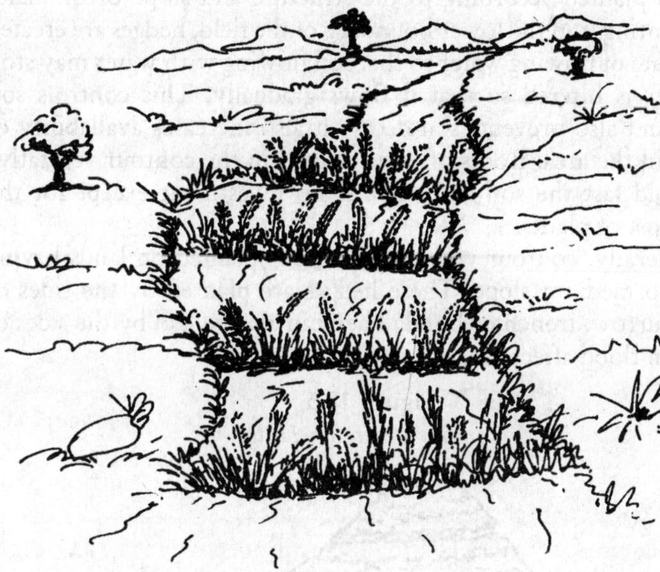

reduces formation of sand dunes. For stabilization of fine sand dunes, mulching of dry shrubs is also done in desert areas.

Vegetative Filter Strip

An irregular and continuously changing vegetative filter strip is prepared by planting local shrubs . Its plantation is done in such a way that it develops in shape of a net and the flowing water passes through it after being filtered. This technique is known as vegetative filter strip. Because of filtering of water through the strip, not only does the force of water flow reduces but silt is also prevented from going farther due to filtering. Hence, soil erosion is controlled. Moreover, availability of biomass also increases in the form of vegetative filter strip in the watershed area. Such type of structure is developed mainly at places where soil erosion is faster due to changing flow of water and water does not stop flowing. After development of a vegetative filter strip, flowing water goes under the ground slowly and increases the groundwater. This is suitable for cultivable as well as un-cultivable lands.

Contour Vegetative Hedge

For establishing Contour Vegetative Hedge, vegetative hedge of local species is planted according to the structure and slope of the land. After planting it in the lower boundary of the field, hedges are erected in the route of flowing water so that soil flowing with water may stop and water is filtered so that it flows gradually. This controls soil erosion and also prevents soil decay. It also increases availability of biomass in the area. Roots of trees planted in the contour vegetative hedge hold fast the soil and there is no investment except for the normal cost of plants.

Generally, contour vegetative hedge is planted on lands having normal to medium slope. These hedges are planted by the sides of narrow furrow, trench or bund. They can also be laid by the sides of streams in flood affected areas.

Figure 18.5
Contour Vegetative Hedge

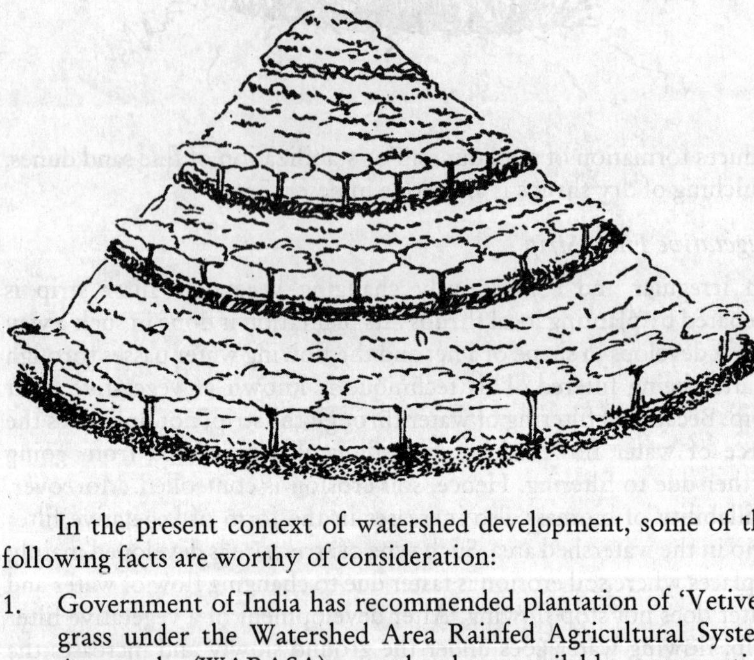

In the present context of watershed development, some of the following facts are worthy of consideration:

1. Government of India has recommended plantation of 'Vetiver' grass under the Watershed Area Rainfed Agricultural System Approach (WARASA), as the best available conservation vegetation.

2. In Rajasthan, vegetative species like Ber (Zizyphus), Mehandi, Moonj, Dhaman grass, Khas, Hemta, Lemon grass and other local brands of vegetation can be used.

Gully Control

Gully control is an important conservative measure in uncultivable land reform programmes. It controls soil erosion from being caused in watershed areas. The upper layer of soil is destroyed by gully erosion and the levelled ground becomes a wide and deep valley after that.

Soil erosion can be controlled by constructing check dams and developing vegetative covers. The form of gully erosion depends on the hydrological condition, quantity of silt, and requirement of

Figure 18.6
Gully Control

vegetation by soils and people. Gully erosion mainly depends on necessity of vegetation by the structure of the check dam and gully erosion can be mainly controlled by check dam structures. Check dams are of two types: (1) Pervious Check Dam (2) Impervious Check Dam. Classification of both these types depends on the structure of flow of water from any tank or gap. Check dam is constructed on the upper parts of a river. The number of check dams that may be constructed over a river would depend on the flow of water, slope and force of the flow. This number may vary from one to one hundred. This whole system can be constructed with the participation of the local people.

Pervious Check Dam

The main objective of constructing such a check dam is to filter the flow of river water after controlling the eroding power of water so that silt deposits in it, and the water moves ahead after being filtered. Its construction is done with logs of wood and pieces of rocks. Water remains stored for local use in it, but it is constructed for some future object. Its construction can also be done by trees of thick wood, loose stones, and fences.

Impervious Check Dam

This dam is constructed on natural ground parts or concrete, and stone is used for strong cover. It is usually constructed in loamy soil or barren land where gullies are formed of shallow and loose soil.

The objective of construction of impervious check dams is to use water after its storage as well as to check soil erosion since water does not pass through these structures.

Conservation measures of check dams have following advantages:

Figure 18.7
Live Fencing

(a) Un-cultivable Area

(b) Cultivable Area

- Force of flowing water is reduced by check dams.
- Power of eroding capacity of water is reduced by them.
- There is improvement in moisture condition of soil also.
- There is reduction in severity of floods.
- There is storage of water in loamy soil.

Live Fencing

For protection of crops and grasses during watershed development programme in preliminary stages, live fencing is done around the fields. The following vegetation is considered more suitable for live fencing:

1. Thorny shrubs like jhadber, gokhru
2. Thor plants and fast growing trees like subabool etc.

The above plants form a type of wall which then becomes a protective arc. In case there is risk of over grazing, a ditch is constructed by its side and with its help, bund fencing is done on both its edges. Thus, live fencing is constructed by this method.

Drainage Line Treatment

Drainage line treatment programme of low cost has been developed in watershed areas to collect and store water of natural drains. This is called drainage line treatment. It is mostly constructed in hilly areas and barren lands where there is very fast speed of hydrological flow. When hydrological flow enters from hilly and uncultivable lands into land of normal slope, its speed increases all the more. Hence, its treatment is started in upper parts itself. In such areas, very little water can be stored in ground levels in both the cultivable as well as the uncultivable lands. Hence, most of the water flows away very fast through the drainage line, soil erosion increases due to deep furrows and drains, and finger-like gullies are developed in cultivable as well as uncultivable lands. Soil and water conservation method of drainage line treatment is adopted for it since it is useful for both cultivable as well as uncultivable lands. Rain water starts conserving in fields automatically by this method.

Activities Undertaken for Drainage Line Treatment

Drainage line treatment includes the following activities for treatment of watershed area:

1. Bank Stabilization
2. Activities in Upper Reaches Area:
   (i) Live Check Dam
   (ii) Bush Wood Check Dam
   (iii) Loose Stone Check Dam
   (iv) Small Dug Out Sunken Ponds
3. Treatment in Middle Reaches Area
   (i) Loose Boulder Structure
   (ii) Earthen Structure
   (iii) Run-off Management of Dug-out Ponds
4. Activities in Lower Reaches Area:
   (i) Small Dug-out Sunken Structure
   (ii) Gabion Polybag Structures
   (iii) Removal of Nalla Congestion

Bush Wood Check Dam

Construction of Bush Wood Check Dam is done with logs of wood and bushes. Logs are fixed in the ground and branches of bushes and branches of trees and plants are attached to them. This construction is affected by availability of local plants. Standard local techniques and standard local materials are used in its construction. This method has been considered more suitable in case of small gullies. Logs bound with bushes and plants are fixed in the ground for firmness and bushes and grasses are also bound with wires. It is constructed in two forms:

Figure 18.8
Check Dam of Dense Bushes

Figure 18.9
Loose Stone Constructed Dam

(i) Single Line Check Dam
(ii) Double Line Check Dam

Loose Stone Check Dam

Such type of structure is constructed in watershed development areas for effective control of flow rate of water over steep slopes and in scattered gullies. Big size stones (each stone weighing at least 40 kg.) are required for its construction so that it requires lesser care and it may last longer. To keep the stones together firmly for a long time, they are bound by iron wires. They are also called as wire crates or Gabion dam.

The place where the dam is constructed should be clean and the slope of banks should be 1:1.Its foundation should be dug up to 30 cms. uniformly. Dug out hollow places should be filled with stones. From the top it is half moon shaped so that excess drainage water overflows from it.

To prevent erosion from permanent parts of the banks of gully, stones of 30cm. to 60 cm. height should be filled. A cover of stones of suitable length and width is constructed in the lower part. Thickness of this cover should not be more than 45 cms. Big stones are used to give a permanent form to this structure. Rectangular stones having a levelled base are considered more suitable for its construction.

Small Dug Out Sunken Ponds

In the upper portions of the drainage area, pits are made in the route of water flowing with great force. Ponds are dug in circular shapes.

Figure 18.10
Small Dug Out Sunken Ponds

No pakka work is done on the flooring so that water inside ponds can be soaked by the earth. A part of water flowing in them is absorbed. This increases groundwater and the water in ponds can also be used for local consumption . This structure is made in upper as well as lower parts.

Earthen Structure

Earthen structures are constructed in areas of medium flow of water where flow of water is not very much spread over. Its construction is done by earth and stones/pebbles. Non-porous clay soil is more suitable for its construction. Saline, alkaline or peat soils are not suitable for it. This man made structure is built at suitable points in suitable conditions where it can be most useful. The storage water is used for animals and for irrigation. Wild animals are also dependent on it. However, regular maintenance of such structures is required.

Depth is also very important in earthen structures. If water stays throughout the year, it is essential to keep more depth so that water remains there even after evaporation. Moreover, water is also required for irrigation. Soil of its side walls should be free from white ants.

Figure 18.11
Earthen Structure

Earthen structures should be constructed at such places where maximum water can be stored. This is possible where the valley is narrow and slopy and slope of the base of the valley is deep. The tank to be used for the purpose of irrigation should be as far as possible, near the point of use. The capacity of the tank should be equal to water for use during a year. Source of water is mainly rain water. This is the best method of water conservation from the local point of view.

Run off Management of Dug Out Ponds

At present, water conservation is being done in an important form through management of flowing water. In this method, tanks are constructed in those areas where water of the whole habitation collects outside the habitation at one place and earth is excavated on the flow route of the water. They are constructed at the Panchayat level and public participation plays an important role in them. Their size depends on two situations - first, the quantity of flow of water and second, the local demand for use. Their construction is done after study of both these situations. Tanks are constructed either kacha or pakka from cement and concrete. Their floor should have clay soil so that water does not percolate easily and water is used for a long time for fulfilling local needs.

Soil otherwise eroding in large quantities is stabilized by regularizing flowing water by digging of tanks. Because water collects at a small deep place, evaporation reduces and a part of the stored water percolates underground, thereby increasing groundwater level.

Polybag Check Dam

Polybag check dam is constructed for controlling flow of water where the flow is at a medium speed and water is reachable in lower portions. Empty cement bags are required for its construction. The check dam is constructed by placing bags filled with sand in the route of the flowing water. Only plastic thread is used for stitching polythene bags. Each one is tied with the same plastic thread so that they are placed as one unit. This structure is cheaper than the loose stone check dam. All the polybags are tied together for giving force to them. Rain water is collected by it and used for local purposes. Its main objective to prevent soil erosion by reducing the force of flow of

Figure 18.12
Polybag Check Dam

water and store the water for use by local population as well as for use by wild animals. This type of construction is made at medium as well as lower portions of the route of the flow of water.

Methods of Land and Water Management based on Traditional Knowledge

Farmers have developed many traditional methods of soil and water conservation according to local needs through the 'Forget and Learn Principle'. Groundwater resources of the state can be conserved easily by following such methods. Contour farming, minimum cultivation, ridging and strip cultivation etc., are important among such methods. Contour line farming is done along the contour against the slope rather than towards the slope. By cultivation on contour line, rain water stops in the furrows itself. It neither erodes the soil of the field nor takes the fertilizer farther from the field. It maintains moisture of the soil and increases groundwater. In areas having sandy soil, strip cultivation restricts the process of wind erosion and strengthens land and water management. Farmers ply patta (roller) behind the cultivator so as to maintain moisture in the soil.

Land and Water Management through Biotic Controls

Land and water conservation is also done through biotic controllers. Important biotic controllers are horticulture, agro forestry, dry gardening, pasture development and forestry, along with green manure, vermin culture and cow dung manure etc.

Horticulture

In horticulture, maximum trees can be planted with minimum water. This would convert the barren ecological system of arid regions into greenery and also yield higher per hectare benefit as compared to other crops. It has been proved by different experiments that drip irrigation or pot irrigation methods can give maximum benefit in commercial crops like myrobalan (anvla), grapes, kinoo, guava, palm, papaya etc. Hence, farmers should be encouraged to practice horticulture and prepare them for adopting nursery culture.

In agro forestry, khejdi, local acasia (babool), rosewood (sheesham), ardoo and other trees can be planted on field boundaries, and in arid regions plum and palm trees can be grown.

Pasture Development

To decrease the pressure of increasing cattle grazing in watershed areas, improved varieties of grass seeds are sown to develop pasture lands. For its protection, fencing of thorny plants (Thor) is done around the boundary. For greater protection, a ditch is also dug to erect a raised earth boundary along the side of the Thor fencing. Grasses, and fuel and fodder trees of species like Israeli acacia (babool), local acacia, Jhad beri (Plum) and Kumat etc., are planted. Thus after such a scientific use of pasture, land and fodder would be available for a long time and ecological balance would be encouraged.

Crop Demonstration

Improved varieties of hybrid seeds and plants are made available to farmers in a watershed area. Here, crop demonstrations are organized under the supervision of technical personnel. Most successful demonstrations are also shown to neighbouring farmers on the site so that they also feel encouraged to adopt improved seeds and improved techniques to get higher benefits.

Cow dung manure, green manure and vermi culture are also adopted in watershed development programme to discourage use of chemical fertilizers and for sustainable development.

Wasteland Development

Barren land or waste land is that part of the land on the ground level where production of biomass is either less than its productivity, or as much productive activities are not done in that land as compared to its existing productivity. This land can be brought under vegetative cover by suitable management. According to National Wasteland Development Board (1990), lands having the following symptoms can be included in wasteland:

1. Land whose structure is in the form of gullies or ravines or upland.
2. Coastal land including saline and alkaline lands.
3. Degraded forest area, degraded pastures and degraded plantation.
4. Sandy soils including coastal or desert soils. Along with stony land, ground level with steep slope is also included in it.

Tree plantation should be done with people's participation on roadside, so that they can easily get fodder and fuel wood also. Roots of trees would keep the earth fastened to them and the trees would also attract monsoon.

Draft for Implementation of Watershed Development Programme

Experts have suggested a special management system for implementation of watershed area development programme. Organizations involved in carring out such programmes can be successful through ensuring its implementation by people's participation using the participatory rural appraisal method.

Implementation of Programme

These programmes will be implemented mainly through the Zilla Parishads (ZPs) of the District Rural Development Agencies (DRDA). However, the projects can be implemented through any department of the atate government or autonomous agencies of central government or state governments with the approval of the Department of Land Resources, Government of India, wherever they are expedient in the interest of Watershed Development Programmes.

Sanction of Projects

The projects will be sanctioned by the Government of India as per the procedure in practice at the time of adoption of these revised guidelines. The Department of Land Resources, Government of India, may amend or relax this procedure from time to time. In case of interpretation of any of the provisions of these guidelines, the Department of Land Resources will be the final authority. However, the Department of Land Resources may sanction special projects for treatment of wastelands in Special Problem Areas such as high altitude regions, land slide areas, slopes having more than 30 degree gradient and so on or for any other specified technical reason. These projects need not necessarily be implemented through the participatory mode but may be implemented on the intensive treatment specific departmental approach.

Preparation of Perspective Plan

The Department of Land Resources brought out the 'Wastelands of India' in May 2000, in collaboration with NRSA, Hyderabad. The state governments are expected to prepare a long-term perspective plan for treatment of wastelands/degraded lands, drought-prone and desert areas over a period of 15 years. It is suggested the plans keep in view the availability of funds under IWDP, DPAF, DDP and other such programmes being implemented through international cooperation projects and other agencies. In most of the states, State Remote Sensing Application Centres have developed watershed-wise atlas showing code numbers. These code numbers should be specified in the perspective plan indicating already treated watershed areas, ongoing project areas, and new project areas to be taken up in a phased manner.

Criteria for Selection of Watersheds

The following criteria may broadly be used in selection of the watersheds:

1. Watershed area may be about 500 hectares. However, if on actual survey, a watershed is found to have slightly less or more area, the total area may be taken up for development as a project. Even small contiguous watersheds with an approximate total area of 500 hectares may be taken up for development.

2. In case a watershed falls in two villages, it should be divided into two sub watershed areas confined to the designated villages. Care should be taken to treat both the sub watershed areas simultaneously.
3. Watershed, which has acute shortage of drinking water.
4. Watershed, which has a large population of Scheduled Castes/Scheduled Tribes dependent on it.
5. Watershed that has a preponderance of non-forest wastelands/degraded lands.
6. Watershed, which has a preponderance of common lands. However, in view of the fact that watershed development aims at poverty alleviation by improving productivity of land and generation of employment, projects not having preponderance of common lands may also be considered for sanction, provided there is adequate justification.
7. Watersheds where actual wages are significantly lower than the minimum wages.
8. Watershed, which is contiguous to another watershed that has already been developed/treated.
9. Watersheds where people's participation is assured for raw materials, cash, contribution on labour etc., for its development, as well as for the operation and maintenance of the assets created.

Implementing Agency

The following governmental and non-governmental agencies can execute watershed development programes.

1. Watershed Development and Soil Conservation Department
2. Forest Department
3. Panchayat Samitis and Gram Panchayats
4. Voluntary Organizations

According to the latest policy, all the activities pertaining to agriculture development, animal husbandry, land and water conservation, pasture development and agricultural engineering would by executed by a single agency. Forest development works would be executed only by the forest department. Project implementation agencies would provide technical guidance to villagers. All other works would be executed by the villagers themselves.

Project Commencement

The date of sanction of the project shall be the date of commencement of the project for all purposes. The project shall be implemented over a period of five years from the date of sanction.

State Watershed Development Committee

To ensure coordination among various government departments/institutions and voluntary agencies, a State Watershed Development Committee (SWDC) shall be constituted under the Chairmanship of Additional Chief Secretary/Agricultural Production Commissioner/Development Commissioner. Secretaries and Heads of Departments of Rural Development, Panchayati Raj, Forests, Soil Conservation, Horticulture, Agriculture, State Remote Sensing Centre, Harijan and Tribal Welfare, two representatives of prominent NGOs working in the field of watershed development, two prominent women representatives, and one representative of State Training Institutions in the related field will be the members of the committee. A designated department in the state government shall be the nodal department to service this committee and to supervise the implementation of watershed development programmes.

This committee may meet twice a year to monitor, review and evaluate the progress of implementation of the watershed development programmes. If there is adequate justification, the committee may recommend relaxation of the cost norms for a specific project, especially in difficult areas.

The ZP/district Rural Development Agency shall normally be the authority competent to decide on the suitability or otherwise of the Project Implementation Agency for taking up watershed development projects. However, the state government will be competent to change the Project Implementation Agency (PIA). In case of change of PIA in the projects under IWDP, prior concurrence of the Department of Land Resources, Government of India, will be necessary.

District Watershed Development Committee

To ensure coordination at the district level, a District Watershed Development Committee (DWDC) shall be constituted under the Chairman, Zilla Parishad or DRDA, as the case may be. It shall consist of CEO/PD, ZP/DRDA and district level officers of the line departments associated with the implementation of watershed development projects, one representative of the State Remote Sensing Centre, one member from the relevant Research and Training institutions in the district, one NGO representative, at least two prominent women workers and one prominent social worker. The District Watershed Development Committee will advise and assist the ZP/DRDA on matters regarding selection of PIAs, members of Watershed Development Teams, training, community organization, publicity campaigns and such other items/activities. The committee will also approve the detailed action plan for watershed development projects in the district. It should meet at least once in a quarter and review the progress of the watershed development projects, assist in resolving management and administrative problems, guide in implementation, and identify policy issues, if any, for reference to the state government/Government of India.

ZPs/DRDAs or any other institution in whose favour the watershed development projects have been sanctioned, shall be fully responsible for their implementation. The CEO, ZP/PD, DRDA will be the Member Secretary of the DWDC. The ZP/DRDA at the district level shall have the right and responsibility to monitor and review the implementation of the programme. The CEO, ZP/PD, DRDA, shall maintain the accounts of watershed development projects and shall sign all the statutory papers such as UCs, Audited Statements of Accounts, Progress Reports, Bonds etc.

Role of Panchayati Raj Institutions

The Zilla Parishads and other Panchayati Raj Institutions (PRIs) shall have very important roles to play in Watershed Development Programmes. Wherever the DRDA is made responsible for implementation of the watershed programmes, the Chief Executive Officer of the Zilla Parishad shall be a member of the DWDC. The PRIs shall have the right to monitor and review the implementation of the programme and provide guidance for improvements in the administrative arrangements an procedures, with a view to ensure convergence of other programmes of Ministry of Rural Development such as JGSY, SGSY, IAY, CRSP, Rural Drinking Water Supply etc.

At the village level, the Gram Panchayat shall be fully involved in the implementation of the programme, specially in community organization and training programmes. It may use its administrative authority and financial resources to support and encourage the formation of SHGs/UGs for two or more cluster of projects provided it has the requisite capacity and capability. However, one NGO may not normally be given more than a total of 12,000 hectare area to develop in a district out of all the programmes of similar nature, and 25,000 hectare in the state. This will encourage more NGOs in the field of watershed development. Due consideration and preference should be given to those institutions, which have done their basic homework in identifying villages/watersheds/areas where they would prefer to work. Past experience in the village or in a nearby village, and quantum of funds handled by the PIA in the last three years may be taken into account for their selection by the ZP/DRDA. The non-governmental organizations blacklisted by CAPART or other departments of state government and Government of India should not be appointed as PIAs.

The Project Implementation Agencies (PIAs) will motivate the Gram Panchayats to pass necessary resolutions to make public contributions, conduct Participatory Rural Appraisal (PRA) exercises, prepare the development plans for the watershed, undertake community organization and training for the village communities, provide technical guidance and supervision of watershed development activities, inspect and authenticate project accounts, undertake action research to adapt low-cost technologies and/or validate and build upon indigenous technical knowledge, monitor and review the overall project implementation, and set up institutional arrangements for post-project operation and maintenance and further development of the assets created during the project period.

Watershed Development Team

Each PIA shall carry out its duties through a multi-disciplinary team designated as the Watershed Development Team (WDT). Each WDT may handle 10-12 watershed development projects and may have at least four members, one each from the disciplines of forestry/plant science, animal sciences, civil/agricultural engineering and social sciences. At least one member of the WDT should be a woman. Preferable qualification for a WDT member would be a professional degree. However, qualification can be relaxed by the ZP/DRDA in deserving cases keeping in view the practical field experience in the respective discipline. One of them shall be designated as the Project Leader. The PIA will be at liberty to either earmark its own staff exclusively for this work, or engage fresh candidates including retired personnel, or take people on deputation from government or other organizations. The establishment charges for the WDT shall be subject to the limits prescribed in Annexure-I and debited to the Watershed Development Projects. The WDT shall be located at the PIA or Block headquarters or any other small town nearest to the cluster of selected villages. The appointment of the WDT must be completed within a period of two months from the date of nomination of PIA, failing which the ZP/DRDA shall have the right to cancel the nomination of the PIA and attach the villages to some other PIA.

Self-Help Groups

The PIA shall constitute Self-Help Groups (SHGs) in the watershed area with the help of WDT. These Groups shall be homogenous groups having a common identity who are dependent on the watershed area, such as agricultural labourers, landless persons, women, shepherds, or scheduled caste/scheduled tribe persons. Around 50 per cent of villagers who are directly or indirectly dependent on the watershed, should generally be enrolled as members of at least one Self-Help Group. Separate Self-Help Groups should be organized for women, scheduled castes, scheduled tribes etc.

User Group

The PIA shall also constitute User Groups (UGS) in the watershed area the help of WDT. These Groups shall be homogenous groups, who may be the most affected by each work/activity and shall include the persons having land holding within the watershed areas. Each UG shall consist of the persons who are likely to derive direct benefits from a particular watershed work or activity. The UGs should actually take over the operation and maintenance of the completed community works or activities on common property resources.

Watershed Associations

Where a watershed is coterminous with a Village Panchayat or its area is confined within the boundaries of a Village Panchayat, the Gram Sabha of the Panchayat concerned will be designated as the Watershed Association (WA). However, where a watershed comprises of areas coming under the jurisdiction of more than one Panchayat, members of the community who are directly or indirectly dependent upon the watershed area, will be organized into a Watershed Association. Such a Watershed Association should be registered as a Society under the Registration of Societies Act, 1860. The Watershed Association will meet at least twice a year to evolve/improve the watershed development plan, monitor and review its progress, approve the statement of accounts, form User Groups/Self-Help Groups, resolve differences or disputes between different user groups/Self-Help Groups or amongst members of the User Groups/Self-Help Groups, approve the arrangements for the collection of public/voluntary donations and contributions from the community and individual members, lay down procedures for the operation and maintenance of assets created, approve the activities that can be taken up with money available in the Watershed Development Fund, nominate members of the Watershed Committee from amongst the User Groups/Self-Help Groups by a system of rotation, and take disciplinary action like remove membership from the Watershed Committee or User Groups or impose whatever other disciplinary action it deems fit. The WA will elect its own President who shall also be the Chairman of the Watershed Committee. The Watershed Secretary shall assist the President of WA in discharging of the responsibilities entrusted to the WA.

Watershed Committee

Subject to the overall supervision and control of the Watershed Association, a Watershed Committee (WC) shall carry out the day-to-day activities of the Watershed Development Project. The Watershed Committee may consist of 10-12 members who will be nominated by the Watershed Association from amongst the User Groups (4-5), Self-Help Groups (3-4), Gram Panchayat (2-3) and a member of the Watershed Development Team. While making nominations, it may be ensured that the Committee has at least one-third representation of women. There should be adequate representation of members from the Scheduled Castes/Scheduled Tribes etc. Needless to stress, the Watershed Committee will be responsible for coordination and liaison with the Gram Panchayat, the Watershed Development Team, the ZP/DRDA and Government Agencies concerned, to ensure smooth implementation of the Watershed Development Project. The WC shall meet at least once in a month on predecided date. It shall be responsible for undertaking watershed developmental works and to make payments for the same.

Watershed Secretary and Volunteers

Each Watershed Development Project shall have a Watershed Secretary (WS) engaged by the Watershed Association. He should preferably be a matriculate from the same village or at least from a nearby village and agree to live in the watershed village during the project period. In case of non-availability of such a person, a moderately educated person of the village with good writing and reading skills sufficient for maintaining records and accounts of the project, may be engaged to function as WS. He will work under the direct supervision of the Chairman of the WC and will be responsible for convening meetings of the WA/WC and for carrying out all their decisions. He will maintain all the records of project activities and the proceedings of the meetings of the WC and the WA. He will also maintain the accounts. If the PIA feels it is necessary that volunteers should assist the Secretary, the WC may be permitted to provide not more than two volunteers to assist the Watershed Secretary. Watershed Secretary as well as watershed volunteers shall be engaged on an honorarium basis subject to the ceiling prescribed in Annexure-1. They shall not be treated as employees of Watershed Committee/PIA/ZP/DRDA, state government/Government of India.

Community Organization and Entry Point Activities

Participatory Rural Appraisal (PRA) is one of the most important exercises in watershed development projects before finalization of the Action Plan. In fact the watersheds should be selected after due PRA and the watersheds with greater participatory response should be preferred. After sanction of the project, the WDT should tour the watershed area extensively and organize various Self-Help Groups, UGs and other groups. As a part of the confidence building exercise, some community benefiting entry point activities can be taken up by the PIA directly. The treatment plan and interventions should be decided after elaborate PRA exercise. These activities can be renovation of village level school, Panchayat buildings, community houses, common places, drinking water sources/wells, bathing ghats, approach roads to water tanks, village roads, village sanitation improvement works etc. Water harvesting measures and improvement of drinking water sources should be preferred over other activities. The entry point activities can be undertaken out of the grant available for community organization. Religious activities and activities for individual benefits should not be permitted.

Capacity Building and Training

Capacity Building is an important aspect for the successful implementation of watershed development programmes. ZP/DRDA shall ensure that relevant training programmes are organized for all the functionaries involved in watershed development. PIA may ensure that a majority of the members of SHGs/UGs are given basic training involving skill upgradation and orientation on the technical and organizational aspects of low-cost locally available technology, that is simple and easy to operate, and to maintain works and activities. The items, *inter alia* that can be included in the Watershed Development Plan are:

1. Land development including in-situ soil and moisture conservation measures like contour and graded bunds fortified by plantation, bench terracing in hilly terrain, nursery raising for fodder, timber, fuel wood, horticulture and non timber forest product species.
2. Afforestation including block plantations, agro-forestry and horticultural development, shelterbelt plantations, sand dune stabilization, etc.
3. Drainage line treatment with a combination of vegetative and engineering structures.
4. Development of small water harvesting structures such as low-cost farm ponds, nalla bunds, checkdams and percolation tanks and groundwater recharge measures.
5. Renovation and augmentation of water resources, desiltation of tanks for drinking water/irrigation.
6. Pasture development, either by itself or in conjunction with plantations.
7. Repair, restoration and upgradation of existing common property assets and structures in the watershed to obtain optimum and sustained benefits from previous public investments.
8. Crop demonstrations for popularising new crops/varieties or innovative management practices.

9. Promotion and propagation of non-conventional energy saving devices and energy conservation measures.

Project Approach for Watershed Development

Initial sanction of Watershed Development Project is only indicative of location of watershed development projects, physical target and financial outlay. The detailed action plan in the form of an integrated project has to be prepared by the WDT in consultation with the Watershed Community. Community Organization is an important component; at the same time technical requirement and feasibility of appropriate biophysical measures are to be carefully worked out for long-term sustainable interventions for the entire area of the watershed. The action plan should specify among others, the following:

1. Pre-set deliverable output;
2. Elaborate road map with definite milestones;
3. Definite time frame for each activity;
4. Technological interventions;
5. Specific success criteria; and a
6. Clear exit protocol worked out as per Standard Schedule of Rates (SSR) approved by the state governments in representative areas.

Funding Pattern

Depending upon the eco-system and major problems faced by different districts/blocks, each Watershed Development Project shall be eligible for funds as per the funding pattern prescribed in Annexure-II for the whole project period. Department of Land Resources, Government of India may extend the project period in deserving cases without enhancement of cost and reduction of work components. This amount shall be divided amongst the following project components subject to the percentage ceiling mentioned against each:

1.	Watershed Treatment/Development Works/Activities	80%
2.	Watershed Community Organization including entry point activities	5%
3.	Training	5%
4.	Administrative Overheads	10%
	Total	100%

Under the Watershed Development Programme, sizeable amounts of funds are made available for capacity building and training. However, it is observed that in several cases, these funds are not being utilized fully. While it should be our endeavour to attach due importance to this crucial activity, it should also be ensured that funds allocated for capacity building and training are spent meaningfully and to the fullest extent.

Credit Facility

The general cost norms for watershed development projects shall remain as per Annexure-I. However, the PIA/ZP/DRDA shall explore and encourage availing the credit facilities to WA, SHGs, UGs, Panchayats and individuals provided by banks or other Financial Institutions for further developmental activities in watershed areas.

Revolving Fund for SHGs

The PlAs will set up a revolving fund not exceeding Rs one lakh per watershed to be given as seed money for vocational development by the Self-Help Groups (SHG) at the rate not exceeding Rs 10,000 per SHG for undertaking income generating activities. This seed money must be recovered from the SHG members in a maximum of six installments on monthly basis. This could be reinvested in the same or other SHGs.

Fund Flow

Funds for the projects from Department of Land Resources shall be as per procedure in practice at the time of implementation of these guidelines. However, various instalments for ongoing projects shall be as per Annexure-II. Release of next instalment will be made when the unutilized balance is not more than 50% of the previous instalment released and on receipt of satisfactory progress report and audited statement of accounts of the previous year after taking into consideration the interest accrued on the previous releases.

Funds from Government of India shall be released to ZP/DRDAs. They may retain their share of the administrative cost from each project and release the funds to PIA for administrative cost, community organization, training and works components. PIA should release the work component to the Watershed Committee Account and the Secretary, Watershed Committee should withdraw the funds and make the payment for the works etc. The Secretary should give details of expenditure to the Gram Panchayat/Gram Sabha. ZP/DRDA should endeavour to release the funds directly to Watershed Committees on recommendations/consultations of PIAs.

Watershed Project Account

The WC will take necessary action to open a bank account in its name in the local branch of any Nationalized Bank/Cooperative Bank. This account shall be operated upon jointly by the Chairman of WC, one member of the WDT, and the Watershed Secretary. Application to the ZP/DRDA for release of funds to the Watershed Project account shall be jointly signed by the Chairman of WC and the Watershed Secretary and shall be duly recommended by the Project Leader of the WDT. The Watershed Secretary shall maintain the necessary records of income and expenditure from this account in the manner prescribed by the ZP/DRDA.

Watershed Development Fund

One of the mandatory conditions for selection of villages under the Watershed Development Programme is the contribution towards watershed development fund as per the activities being undertaken in individual lands as well as community lands. The contributions to (WDF) shall be at least at the rate of 10 per cent of the cost of works done on individual lands and 5 per cent of the works on community lands. However, in case of SC/ST and persons identified below the poverty line, the minimum contribution for works on all lands including private lands shall be at the rate of 5 per cent. Care should be taken that the contribution comes from the farmers and not from the labourers engaged to treat the private lands. These contributions would be acceptable either in cash or voluntary labour or materials. A sum equivalent to the monetary value of the voluntary labour and materials would be taken from the watershed project account and deposited in this fund. All such contributions shall be deposited in Watershed Development Fund, which shall be maintained separately, to be operated after the project is completed.

The Watershed Development Fund shall not be a one time measure but should be organic in nature. Development Fund Contribution in respect of community property may come from the Village Panchayat also. Contribution to this fund will keep on accruing even after the project period is over, by way of charges paid by the community for use of assets generated in the project, e.g., water for irrigation, fuel wood, fodder etc. Contribution to this fund can be in cash or in the form of labour.

In case no separate institutional arrangement is done by the ZP/DRDA for post project management, Chairman, WC, and the Secretary will operate this account jointly. The individuals as well as charitable institutions should be encouraged to contribute generously in the Watershed Development Fund. The proceeds of this fund shall be utilized in maintenance of assets created on community land for common use after completion of the project period. Works taken up for individual benefits shall not be eligible for repair/maintenance out of this fund.

1. The watershed development projects may be sanctioned at the rate prescribed by the central government from time to time. The prevalent rate with effect from 1st April 2000, is Rs 6,000 per hectare.
2. Ceiling on Administrative Overheads

1.	*At ZP/DRD, A Level*	*Rs*
	WDT Members Training (For 10 WDP's)	30,000
	(i) Proportionate Expenditure for one WDP	3,000
	(ii) Miscellaneous Expenditure/WDP	3,000
	(A) Total for one Watershed Project	**6,000**
2.	*At PIA WDT Level (For 10 WDP's)*	
	(i) WDT members honorarium	7,50,000
	(ii) TA/DA	4,50,000
	(iii) Office Staff/Contingencies	3,00,000
	Total for 10 WDPs	15,00,000
	(B) Expenditure for one WDP	**1,50,000**
3.	*At Village Level*	
	(i) Watershed Secretary & Volunteers Honorarium	1,20,000
	(ii) TA/DA	12,000
	(iii) Office Contingencies	12,000
	(C) Total for each Watershed	**1,44,000**
	Grand Total Cost Ceiling on Administrative Overheads (A+B+C) Per Watershed Of 500 Ha Area	**3,00,000**

Chart Showing the Release of Project Funds by ZP/DRDA

Year	Installment	%	Agency	%	Component Breakup	% Breakup
1st	1st	15%	PIA	9%	Administrative Cost	2%
					Community Organization	4%
					Training	3%
			WC	6%	Works	6%
2nd	2nd	15%	PIA	4%	Administrative Cost	2%
					Community organization	1 %
					Training	1 %
			WC	11%	Works	11%
	3rd	15%	WC	15%	Works	15%
3rd	4th	15%	PIA	3%	Administrative Cost	2%
					Training	1 %
			WC	12%	Works	12%
	5th	15%	WC	15%	Works	15%
4th	6th	15%	PIA	2%	Administrative Cost	2%
					Training	1%
			WC	13%	Works	13%
5th	7th	10%	PIA	2%	Administrative Cost	2%
			WC	8%	Works	8%

Monitoring and Review

The PIA shall submit progress reports on each of the Watershed Development Projects once in every quarter to the ZP/DRDA. Similarly, each Watershed Committee shall submit a quarterly report to the PIA after it is scrutinized and approved by the WDT. The state governments shall be responsible for regular monitoring and evaluation of Watershed Development Programmes through independent evaluators. The Ministry of Rural Development may also appoint independent institutions/individuals to carry out concurrent as well as post-project evaluations/impact studies of the Watershed Development Projects.

The services of identified district-level and state-level institutes may also be utilized for establishing a partnership relationship with district authorities/state governments. These identified institutes will assist in generating information/feedback based on 'observations' made of the projects at prescribed points of time. They will also provide timely and precise inputs regarding the pace and the quality of the implementation of the project on permanent and dedicated basis. These institutions are expected to provide the services of personnel trained for performing the expected tasks such as identifying the lead activity in each of the watershed projects, visiting the project areas at pre-determined events in the life cycle of the project to obtain readings in the form of pre-set deliverable quantified outputs, and assisting and ensuring the implementation of the Exit Protocol in the "Watershed Plus" phase to ensure that adequate attention being paid to considerations of sustainability and equity.

Projects on Probation

Under the Watershed Development Programme a watershed project is taken up for a period of five years, which includes an initial phase of 9-12 months for establishing the necessary institutional mechanism for execution of the project. While this process has been working satisfactorily by and large, there seems to be instances where the PIA is not in a position to grant the required amount to village level institutions due to conflicting interest groups in the project area. It is, therefore, desirable to put a project on probation for a period of one year. During the year, if it is considered that the project cannot be implemented successfully for certain unavoidable circumstances and reasons, the ZP/DRDA shall recommend for its foreclosure to the state government, which will consider such request on merits and send its recommendations to the Department of Land Resources. Before considering any such request, the state government shall ensure that the amount already spent in the project area has been duly accounted for. The projects will be formally closed only after approval by the Department of Land Resources. In such cases, the unspent amount should be refunded to the Department of Land Resources. This review may take into account the sustainability of the project, equity, and other related issues. However, such requests shall not cover the cases of financial embezzlement, defalcation and other deliberate irregularities, for which responsibility is to be fixed.

Training

It would be responsibility of the state government to ensure effective training arrangement at every level:

S.No.	*Name of Training and System*	*Duration*	*Training Venue*
1.	Specialists of Watershed Development Team (Engineer, Botanical Specialist, Veterinary Science Specialist, Social Science Specialist)	One Month	Agriculture University or State Level Organizations
2.	Watershed Development Committee Members, Secretary and Volunteers	5 Days	Krishi Vigyan Kendra and Block Level Organizations
3.	Consumers/Self-Help Groups	2 Days	Block and Village Level
4.	Participatory Meetings of Beneficiaries and Awareness Camps	3 Days	Village Level

Krishi Vigyan Kendra, team specialists and local officers would be associated with the training. The honorarium of the resource persons would be decided by the concerned District Rural Development Agency.

Village Development Fund

Financial provision for maintenance after the project period is met from the Village Development Fund. During the 5 year period of implementation, 10 per cent amount is deposited by beneficiaries as contribution and this amount is deposited in the Development Fund. Scheduled caste, scheduled tribe and poor persons deposit only 5 per cent. Similarly, only 5 per cent contribution is deposited for work done on common lands. Interest would also accumulate during the 5 years. After the project period, maintenance expenses and salary of secretary is met out of this fund.

Opportunity of Rural Employment

Labour employment would be, available to the villagers for construction of earthen check dams, stone dams and vegetative hedges, growing of grass, digging trenches and pits, doing plantation etc. Hence, pakka construction is discouraged in watershed development programme.

Progress, Review and Evaluation

An Advisory Committee at the district level and a State Coordination Committee at the state level would review the progress every month and every quarter. Provision also exists for evaluation by an independent agency.

Thus, it is an integrated programme for creation of maximum employment opportunities in rural areas, increasing productivity and income through optimum utilization of land and water resources, and thereby taking the country on the path of prosperity. It gives a new direction to development based on people's participation and creates a democratic system for development of greenery in agriculture, pasture development, forest and hill development, increased production of fuel, fodder and building wood, increased production of grass, of development cottage industries, and creating thrift schemes for women and special schemes for water management.

People's Participation

The main object of the new policy is to associate farmers, breeders, villagers and people's representatives with the programme. Village Panchayats have to pass a resolution for executing pasture development in watershed development programmes and for bearing the responsibility for maintenance of project works after the project is over.

Villagers would get the scheme prepared according to local problems and circumstances. They have to contribute for 10 per cent for development work on individual fields and 5 per cent for works on common lands, pasture, government land. The contribution can be in the form of cash, kind or labour. Scheduled caste, scheduled tribe and poor would contribute only 5 per cent of work on individual fields. The Executive Committee of beneficiaries would implement the works, which would be executed according to their opinion. About 10 Self-Help Groups would be formed in every watershed area. Initial assistance of Rs 5000 to Rs 10,000 can be given to each group for increasing the sources of income and this loan amount would be repayable without interest in 6 months.

Since the development works are executed by the committees of beneficiaries they would maintain their interest in the maintenance of those works also, They would not consider the programme as a government programme but as their own, and would perform their roles willingly.

Public representatives, social workers and educationists should also be concerned to think about the programme and wherever there is problem of soil erosion due to rains, rain water flows away in a nalla out of village, pasture land and hilly area is available in sufficient quantity, they should advise the people's representatives to approach the concerned District Rural Development Agency for sanction of watershed area development programme. Thereby rain water would become available for the village and productivity would also increase. In a watershed development programme, falling rain water can be checked by contour bunding, by planting vegetative hedge on field boundary, or by construction of dams, or ponds tanks for storage of water. The villagers would get the following benefits by successful execution of this programme:

1. Water conservation
2. Soil conservation
3. Increase in groundwater level
4. Increase in moisture in fields
5. Availability of fruits, fodder and fuel wood
6. Economy of chemical fertilizers due to availability of organic manure
7. Protection from ill-effects of chemical fertilizers
8. Saving of electric bill because of lesser requirement for irrigation.

If villagers are made aware about such benefits they would definitely welcome it. If works are executed by villagers for their own selves, they would be durable and sustainable. Hence, agencies engaged in watershed development programmes should adopt convergent community action methods and participatory rural appraisal so that rural development programmes become lively and fast. So far there has been absence of PRA in rural development programmes, hence the proportionate benefit has also been lesser. Hence, it seems necessary to give detailed knowledge about the process of PRA.

Participatory Rural Appraisal (PRA)

To know about the people from the people themselves, is participatory rural appraisal. Rural people themselves prepare the plan for their village on the basis of available resources in this technique. PRA is a vision as well as a system. It helps the outsiders to understand about the form of the village. Effort is made to understand the systems in the village through different methods such as time line, seasonality diagram, trend analysis, mathematical and statistical methods, and observation and discussion. This technique is very effective in preparing a firm plan for the village. Often we are not able to get participation of villagers while preparing a plan for them. Thus, its implementation becomes unsuccessful because the programme is not useful from the point of view of villagers. Many a times, when the question of people's participation arises while preparing government plan, opinions of some influential persons of the village are taken. Requirements, resources and deficiencies pertaining to the beneficiaries of the plan are not evaluated. Many a times, it is not possible to reach the needy persons even if we want to do so. These deficiencies in planning and implementing the plans can be removed through different methods used in PRA technique. Since rural people are experts of their villages, if any suggestion has to be given to them, their systems and work methods have to understood. Only then will it be possible to run a programme successfully in their village.

Use of Maps and Symbols

To attract the attention of villagers during PRA. discussions, local maps and symbols must be used. For example, leaves, pebbles, seeds, small pieces of wood etc. may be used for counting and measurement, so that literate as well as illiterate both can actively participate in the discussions and the analysis. These symbols and impressions can be drawn on the ground or on a big drawing sheet so that many persons can see the information being filled. This will enable an open discussion among all of them.

Seasonality Diagram

As a part of the programme, a chart is prepared for seasonality diagramming which denotes what things are produced, when and in how much quantity. Ups and downs during the year are also indicated in it. Before making entries in it, it is necessary to decide the definition of time according to the villagers. For example, villagers generally do not use the Gregorian Calendar, whereas city dwellers understand only English months. Thus, the city and village time does not match. Hence, it is essential that time is linked with festivals or seasons or local months as per the Vikram Calendar so that outsiders and local persons follow in the same context.

On the ground or on a big drawing sheet, months are indicated in the horizontal column, while crops are written in the vertical column. The villagers are then asked about production in individual fields. Production is expressed in the form of symbols such as one seed for one bag. Thus, it can be known when other cereals can be produced. Total calendar regarding variety of concerned crops, irrigation management, agricultural labour, fodder for cattle, land exploitation, migration, cattle wealth, income generation, indebtedness, sickness, crop diseases etc. can be prepared by such a method. This process can be repeated with different groups at different places and thereafter it can be finalized.

Hindrances in Watershed Development Programme

There are many natural and man made situations, which affect the normal process of development. The main among them are as under:

- Lack of awareness due to illiteracy.
- Lack of participation and local politics.
- Difficult geographical boundary , difficult accessibility.
- Sensitive environment.
- Lack of technical knowledge and suitable human capacity.
- Lack of correct information and statistics.
- Lack of useful and good developmental structure.
- Lack of funds.
- Feeling of mutual suspicion between villagers and project implementation agency.

Watershed Management

Watershed is that geographical area in which falling water collects through a river.

Watershed area can be identified in hilly and slopy regions even by visual inspection. The area above the confluence of rivers where water falling in it flows through the river, would decide the boundary of the watershed area of the river. As there is a geographical boundary of a district, development block or village etc. and mostly it is decided by nature and no changes can be made on account of administrative needs.

Normally, some indications are found in every small or big watershed, for example,

1. Complete water of one watershed area crosses the boundary of watershed area only through a single exit.
2. No area falls within two watersheds of the same class.

Since the last few years, the word watershed has become very popular among development organizations. Various developmental projects, which have been affected by the concept of watershed, are being re-defined.

Based on past experiments of development specially in drought and hilly regions, it was realized that mutual convergence of different development programmes is necessary for optimum sustainable development. Excessive exploitation of resources has been done during the past years in the name of development. This has resulted in, scarcity of resources as compared to their availability adversely affecting infrastructural development.

The demand of the day is that fast depleting resources should be conserved and renovated. Renovation of old resources is difficult without conserving them. The beginning of the process of conservation can be made by better management of primary resources of land and water. Land and water conservation are strongly connected with each other and their relationship affects the quality of life. This relationship can be sustained by concentrating on a limited area. Improvement in mutual relations between primary resources and their suitable management is most essential for working in a watershed area decided by land and water.

Various processes executed within the boundaries of a watershed area affect available resources. Sustainability of any development process depends on deep understanding of mutual relationship and working according to situations.

Watershed development programme is an integrated effort for increasing production through better utilization of primary resources without causing any adverse effect on the natural balance. The technique of watershed development concentrates on solving local problems by taking advantage of traditional knowledge and cooperation of local people. These solutions found by the cooperation of local people can become easily popular in society. Since centuries utilization of resources has been indirectly done on the basis of boundaries of watershed. In fact whatever utilization of fallow lands that can be seen in arid regions today, is based on better utilization of natural resources and is with the conducted boundaries of a watershed area.

Unfortunately government departments, non-government organizations and rural people have not yet been able to expedite watershed development programme according to the suggestions given by Hanumant Rao because most of the development works have been done in individual sectors like land, water, agriculture, forests and so on. Most of the expenditure is directly done by the government or non-government agencies resulting in a number of malpractices, as per one estimate only 12 per cent of development fund actually reaches the beneficiary. According to the existing practice of watershed development programme, the entire amount is transferred to the project account of the Gram Sabha of the village. The beneficiaries themselves maintain the accounts of total receipt and expenditure. The government or non-government agency only looks after the management and control aspect of it.

For facing the challenges of the present times it is the demand of the day to ensure that the development policies implemented in drought, hilly and desert areas according to geographical circumstances, are carried out in true spirit. However, although at the village level the workers connected with programme implementation are capable of executing development activities, but there is absence of effective plan formulation and required technical assistance from expert bodies.

Principles of Watershed Management

The main principles of watershed management are as follows:

1. Land should be utilized according to its capacity.
2. Protection of vegetative cover should be done in sufficient quantity for control of soil erosion.
3. Storage of rain water as far as possible.
4. Conservation of excess water through various methods.
5. Gully control for controlling soil erosion.
6. Ecological restoration methods should be adopted like pasture development, horticulture, agro forestry and reforestation
7. Renovation of traditional methods of water conservation etc.
8. Development.
9. Sustained community action for the operation and maintenance of assets created and further development of the potential of the natural resources in the watershed.
10. Employment generation, poverty alleviation, community empowerment and development of human and other economic resources of the village.

Community participation and technical knowledge are two pillars of watershed area development, and coordination between the two is necessary. Many a times, the organizations engaged in watershed development cannot pay heed to them, due to which the pace of development is affected. Planning of different activities and their execution should be done only after taking local people into confidence and after proper discussions. Local people should be acquainted with different aspects of development and awareness generation should be done regarding better utilization of resources. Simultaneously, they should be made competent regarding agriculture, horticulture, water-management, animal husbandry and forestry. Availability of information for planning and execution as well as people's participation should be ensured at the village level and the watershed level.

Notes

1. Jat. B.C., *Watershed Management*, Pointer Publications, p. 15.
2. Mahnot. S.C. and Singh. R.K., *Soil and Water Conservation*, p. 7.

19

River Water Disputes

Rivers flow in their natural form according to geomorphic conditions. They are not controlled by political units. Flow of water is distributed according to the relief and size of the watershed area. If a country is situated in an uneven relief where a river is in its primary stage, its distribution will not be in proportion though there may be sufficient quantity of water. Instead it would be distributed to the succeeding drainage area in definite proportion. Today, water is considered as the heart of an economy of any country and all economic activities depend on it. However at the national as well as international level, river water disputes are being created because of continuously increasing demand for water, its decreasing availability, and its deteriorating quality. Geographical descriptions of important national and international river water disputes and agreements are given below:

National River Water Disputes

Krishna River Water Dispute

This dispute is linked with the three states of Karnataka, Andhra Pradesh and Maharashtra. The main basis of the dispute is associated with the Almati Dam constructed on river Krishna in the district Bijapur of Karnataka. Its foundation stone was laid by the then Prime Minister of India, Lal Bahadur Shastri, in 1968. Thereafter, the Central Water Commission changed it into the form of a multipurpose project having a proposed height of 528 metres. This was objected to by Andhra Pradesh who said that by increasing the height of the dam, the natural water flow of river Krishna would be reduced, resulting in non-availability of water for Jurala, Shri Selam, Nagarjuna Sagar and other projects of Andhra Pradesh. According to the distribution of water decided by Bachhawat Tribunal, Andhra Pradesh was to get 811 thousand million cubic feet (TMC), Karnataka would use 734 TMC water and Maharashtra could use 656 TMC water, although only 173 TMC feet water had been allotted for the north Krishna project.

Issue of Dispute: Karnataka wants to raise the height of the dam up to 528 metres. Andhra Pradeshs objection is that with 524.45 metre height, Karnataka would be able to stop 227 TMC feet water, whereas only 173 TMC water has been allotted for north Krishna project. Andhra Pradesh holds that Jurala, Shri Selam, Nagarjuna Sagar and other projects constructed in the Krishna river basin would not be able to get water. This matter is still pending in the Supreme Court.

Cauvery River Water Dispute

The dispute over Cauvery, famous as the Ganga of South India, is again in the limelight. The problem between Tamil Nadu, Karnataka, the central government and the Supreme Court is at its climax. The origin of river Cauvery is in Brahmgiri hills (1341 m) near Sahyadri (Western ghat) in district Coorg of Karnataka. It flows through Karnataka and Tamil Nadu, finally merging into the Bay of Bengal. Total length of the river is 802 kms. At present, the problem of distribution of water is pending due to the intervention of central government and Supreme Court. In spite of orders of the Supreme Court dated 5 October 2002, the Karnataka government decided not to give water to Tamil Nadu. Hence J. Jayalalitha, Chief Minister of Tamil Nadu, filed a contempt petition against the decision of Karnataka government in the Supreme Court of India. The Supreme Court had ordered as per its judgment on 4th November 2002, that Karnataka would release 9000 cusec water daily for Tamil Nadu. Even after the order of the Supreme Court, it was decided in a cabinet meeting under the chairmanship of S.M.Krishna, Chief Minister of Karnataka, not to give water to Tamil Nadu. This decision was unanimously supported by the all party meeting of Karnataka government. Thereafter J. Jayalalitha, Chief Minister of Tamil Nadu, approached the Supreme Court under Article 355 and 356 of the Constitution for action against Karnataka government. Although Karnataka government has started releasing 1000 cusec water in view of the stern

Box 19.1

Cauvery: A Cultural Heritage

Cauvery river is known as Ganga of the South India. The name of Cauvery is worth mentioning among the seven holy rivers of India. The origin of this river is in the Brahmgiri hills in district Coorg of Kerala. After flowing for 802 kms from Karnataka through Karnataka, Tamil Nadu and Pondicherry, it falls into the Bay of Bengal. Many dams have been constructed in Karnataka over Cauvery river and "Shiv Samundram" and "Rangpatanam Dam", two among them, are considered as holy places of the Hindus. At Shiv Samundram, river Cauvery forms any 100 metre high waterfall. Hydroelectricity is also generated from this waterfall. Tanjavur district (Tanjavur plains), famous as gardens of the south, is situated in the delta portion of this river. It is an important tourist place. The famous Srirangapattam city is also located on the bank of river Cauvery. Harivansh Puran gives a clear description of the origin of Cauvery, which is also known as Hal Ganga. According to it, due to curse of Yuvanashva, Ganga was born with a half body as the daughter of Yuvanashya, who was named Cauvery. Cauvery has been named as the daughter of some sage in the Skand Puran also.

attitude of the supreme Court, but even then a permanent solution of Cauvery river water dispute is still due.

The dispute regarding the distribution of the water of river Cauvery among Kerala, Karnataka and Tamil Nadu is a very complex problem. The real dispute started between Karnataka and Tamil Nadu in 1889 when Karnataka prepared a plan to construct a dam on river Cauvery. Looking to the seriousness of the dispute between Mysore State and Madras Presidency, Mysore state entrusted the dispute to the central government in 1892. One member of the Mediation Commission (Justice H.D. Griffin, Allahabad) constituted by the central government, gave its decision dated 12th May 1919, but the Madras Presidency refused to follow it. Thereby the first effort of the central government failed. Another effort for agreement started on 13th April 1920, which was partly successful on 18th February

1924. Both the states got sanction for construction of one dam each (Krishna Sagar in Karnataka and Mysore dam in Tamil Nadu) as per the agreement. The agreement was valid for 50 years or up to 1974, and neither state could construct any dam/water reservoir without mutual consent. However, the Karnataka Government presented a plan for construction of three water reservoirs in violation of the agreement and inspite of opposition from Tamil Nadu. In response, the Farmer's Union of Tamil Nadu filed a petition in the High Court opposing the plan, which was withdrawn in 1971 after an other agreement.

After the agreement, a committee under the Chairmanship of C.C.Patel presented two reports in September and December 1972, but it did not suit any of the Governments. The Farmer's Union again filed a petition and the dispute continued till 1979. After termination of the duration of the agreement in 1974, the Chief Ministers of all the three states, in a meeting dated 26th November 1974, temporarily agreed on the "Cauvery Valley Authority", but it has again got into a dispute due to mutual differences. The issue of water dispute has adopted political dimensions by now by the hide and seek method. As per the insistence by the Tamil Nadu Government in 1986 under International Water Dispute Act, 1956, the Supreme Court constituted a one-man Cauvery Water Dispute Tribunal under the Chairmanship of Justice Chitosh Mukherji on 2 June 1990, with the objective of finding a permanent solution of the dispute among all the four states. The Tribunal announced its judgment on 25 June 1991, according to which:

1. Karnataka will have to release 205 TMC (Thousand Million Cubic feet) water to Tamil Nadu on monthly and weekly basis for Mysore water reservoir.
2. Karnataka will not extend its Cauvery basin (11.2 Lakh).

But the Karnataka government refused to abide by it, saying that it was a unilateral decision.

The central government then called a meeting of all the four states for common consent for it but without any concrete result. The Chief Minister of Tamil Nadu went on a hunger strike in July 1993 and the matter again went to the Supreme court. The then Prime Minster Narsimha Rao announced a three point package on 1st June 1996 for solution of this problem.

1. Release of 600 crore cubic feet water to save crops of Tamil Nadu.

2. Constitution of an expert group.
3. Early meeting of Water Resources Council.

A four-point agreement was arrived at among the Chief Ministers of all the four states on 7 August 1998 after long discussions but without a common consent. Treating the above agreement as a formality, the central government issued a notification on 11th August 1998 for implementation of the interim order and other related orders of the Tribunal constituted for solving the Cauvery Water dispute. According to the Government of India Gazette 675(E), a Cauvery River Water Dispute Authority would be constituted under the Chairmanship of the Prime Minister of India with its headquarters at Delhi. Chief ministers of all the four states would be its members. It would prepare its own regulations and a vigilance committee would be constituted for assistance and implementation of decisions of the Authority. A committee under the chairmanship of the secretary, Central Water Resources, and the Chief Secretaries of all the four states would be its members. In spite of the intervention of the central government and the Supreme Court, the problem of water dispute is still unsolved. In violation of the orders of the Supreme Court, Karnataka government has decided not to give water

Box 19.2

Interstate Water Dispute and Indian Constitution

Vast provisions have been made in the Constitution of India regarding deciding inter-state water disputes under Centre-State relations. According to Article 262 of the Indian constitution, the Parliament can frame law for judicial decision in such disputes. According to Article 262 (Part-II), the Parliament can also provide by law that no type of court shall have jurisdiction in matter of such dispute. Parliament had legislated "River Board Act and Inter-State Water Dispute Act" by using such provision in 1956. It may also be stated that River Board Act, 1956 was passed on insistence of the States. This Board advises the Government with reference to exchange and development of inter-state rivers and river valleys. Along with it, "Inter-State Water Disputes Act, 1956" empowers the central government for establishing Tribunals/Arbitrators for settlement of such water disputes. Supreme court or any other court has no jurisdiction on decisions taken by such tribunal.

to Tamil Nadu from 5 October 2002. As a reaction, Jayalalitha, Chief Minister of Tamil Nadu, has filed a contempt petition in the Supreme Court. Thus, this problem is still continuing and looking to its history and complexity, there is little possibility of its permanent solution in future. The main reason is that the Cauvery river water dispute has been strongly politicized. Tamil Nadu believes that Karnataka is intentionally doing the malpractice of withholding Cauvery waters to keep Tamil Nadu as an arid region. Karnataka's version is that the water of Karnataka is meant for Karnataka and if it is given to Tamil Nadu, scarcity of water would be faced in Karnataka. In such circumstances, any unanimous solution seems a difficult task unless both the states sympathetically follow the decision of the Supreme Court.

Sutlej-Ravi-Beas Dispute

The dispute over the waters of Ravi, Beas and Sutlej is mainly between the states of Punjab and Haryana. At the time of partition, an agreement had been signed between India and Pakistan regarding distribution of water of Indus river. According to it, waters of river Indus, Jhelum and Chenab were to go to Pakistan and waters of rivers Sutlej, Ravi and Beas would be for India. India had made a payment of Rs 110 crores to Pakistan in lieu of it. The dispute started with the division of Punjab. Haryana demanded a share of the waters of their rivers in 1966 and the dispute started from there. A committee was constituted for it in 1970 and according to its recommendation, Haryana was to get 37.8 lakh acre feet water. Immediately after this, The Planning Commission recommended giving 37.8 lakh acre feet water to Haryana, 32.6 lakh acre feet water to Punjab and 2.00 lakh acre feet water to Delhi, but the dispute has not yet been settled.

The Punjab government filed a petition in the Supreme Court in 1976 against this award, while Haryana filed a petition in support of it. A new agreement was signed in 1981, according to which, Rajasthan was also included along with these two states. In 1981, the Rajiv Longowal Agreement was signed, which maintained the quantity of water as already agreed upon.

In 1986, the Ravi-Beas Tribunal (Iradi Commission) recommended the allotment of 37.30 lakh acre feet water to Haryana and 50.00 lakh acre feet water to Punjab, Haryana however challenged the recommendation, demanding 56.00 lakh acre feet water, but the Iradi Commission could not be challenged in the Supreme Court also because the recommendations of tribunals regarding water disputes referred in it, do not fall within the jurisdiction of any court including the Supreme Court according to Section 11 of Inter-State Water Dispute Act, 1956.

The dispute still continues and the Sutlej-Yamuna link Canal has not yet been completed due to the dispute. Thousands of acres of water of Ravi and Beas rivers is going to Pakistan, while the lands of farmers of south Haryana are becoming barren land due to non-completion of the link canal.

Yamuna Water Dispute

This dispute is connected with five states of north India, i.e., Haryana, Uttar Pradesh, Rajasthan, Delhi and Himachal Pradesh. An agreement regarding Yamuna water was signed between Haryana and Uttar Pradesh in 1954 according to which 77 per cent water of Yamuna was to go to Haryana and 23 per cent to Utttar Pradesh. After 20-25 years, Delhi, Rajasthan and Himachal Pradesh also started demanding a share in the water of Yamuna.

On 12th May 1994, the then Water Resources Minister Vidya Charan Shukla arranged an agreement among all the five states. According to it, Haryana was to get 47.82 per cent water, Uttar Pradesh was to get 33.65 per cent, Rajasthan to get 9.34 per cent, Delhi 6.24 per cent and Himachal was to get 3.15 per cent. This agreement would continue till 2025. In the agremeant it was provided in Section 7(3) that whenever there is water supply in Yamuna, water supply would be given to Delhi on priority.

Dispute arose on this point many times among these states, so on 22nd March 1995, the Upper Yamuna River Board was constituted.

Son-Rihand Water Dispute

The dispute over the waters of rivers Son and Rihand is between the three states of Bihar, Uttar Pradesh and Madhya Pradesh. In 1973, the Baan Sagar agreement was signed between them for sharing the waters but Bihar accused the other two states later for violating thc agreement. According to this agreement, all the water of river Rihand was allotted to Bihar, but the Central Government Undertaking, National Thermal Power Corporation and Uttar Pradesh Government are all the using water of the river Rihand. Hence, the farmers of Bihar are not able to get water for irrigating crops, which are dyring in 22 lakh acre land.

Mahi Water Dispute

Agreement for distribution of waters of Mahi river was entered in 1996 between Rajasthan and Gujarat. According to it, part of the water of Kadana dam and all the water of Banswara dam (Mahi Bajaj Sagar) would be released for use of Rajasthan when the area irrigated by Mahi river would be interchanged for the Narmada Project. However, the Gujarat government is hesitating to fully implement conditions as yet. On insistence from the Government of Rajasthan, the Central Agriculture and Irrigation Minister organized a meeting of chief ministers of both the states on 24 January, 1980 for solving the dispute. Many meetings have been held thereafter at the Ministerial level and official level both at the mutual level as well as under the auspices of the central government. The Chief Minister of Rajasthan addressed the Gujarat Chief Minister on 16 January 1996 also for early disposal of the matter.

Narmada Project

According to the final order and decision of Narmada Water Dispute Tribunal, the share of Rajasthan is 0.50 MAF in the waters of river Narmada. This water is proposed to be brought to Rajasthan through a 460 km long canal from the Sardar Sarovar dam in Gujarat. This water is proposed to benefit 74 villages of tehsil Sanchore of district Jalore and 1.35 lakh hectare area of tehsil Guda Malani of Barmer district, and also provide drinking water to 124 villages falling near the canal. The estimated cost of the project on 1989-90 price index was Rs 467.53 crores, out of which the cost of works to be carried out in Rajasthan was Rs 187.39 crores and share of cost payable to Gujarat was 280.14 crores. The Planning Commission had sanctioned the expenditure of such cost on 23 January, 1996.

Availability of water on the Rajasthan-Gujarat border would be from the Narmada main canal. The estimated period for its construction in Rajasthan was eight years. Out of the 74 km length of the canal in Rajasthan, earth excavation, bridge construction, regulator construction and other pakka works are in progress in the first 30 km length. Besides it, earth work of 'Vaak' distributory to be dug at 8 km is also in progress. A 30 km long road work along the canal is also in progress. Residential and non-residential houses at Jalore and Sanchore have already been completed. Siphon work on river Sukdi is also proposed to start during this year.

By the end of the year 1995-96, Rs 7,464 lakhs were spent on the project, which includes expenditure on the Sardar Project in Gujarat. The system established by the central government to monitor the work was to set up two committees, namely the Narmada Control Authority and the Sardar Sarovar Construction Advisory Committee. During 1996-97, a provision of Rs 13 crores was made for works to be executed by the Rajasthan state government.

Ravi-Beas Dispute

Both in January 1955 and December 1981 agreements have been made for distribution of excess water of Ravi-Beas between the states through which they flow. However, some inter-state disputes are going in relation to the distribution of the waters, which are as follows:

1. Regarding Transfer of Control of Ropad, Harikke and Ferozepur Headworks to Bhakra Beas Management Board according to Punjab Reorganization Act:

According to the provisions of the Punjab Reorganization Act, 1996, control of Ropad, Harikke and Ferozepur Headworks should be with Bhakra Beas Management Board (B.B.M.B). However, since Punjab government has not yet transferred the control of head works to B.B.M.B., hence the central government had to issue directions to the state government in this regard. In meetings dated 29-30 July and 6 August, 1992 consultations were done by Central Water Resources Minister with the chief ministers of Punjab, Haryana and Rajasthan for transfer of control of headworks to the Board including headwork in Rajasthan. After these meetings, the Central Water Resources Minister sent a draft of the decision to all the states. According to it, all the states were to be appointed as members of the Board by rotation. Hence, it would not be necessary to transfer control of headworks to B.B.M.B. The Rajasthan government then sent a letter to the Central Water Resources Minister on 23 January, 1993 that transfer of control of headworks was included in Punjab Reorganization Act, 1996, hence it was necessary to change it. The issue has not yet been solved and the state government has requested the central government to arrange another meeting for this purpose.

2. Share of State in Power Projects over Ravi-Beas in Punjab:

According to an agreement dated 10 May 1984, the central government has not referred the matter regarding share of Rajasthan and Haryana in 11 power projects at Thein, Anandpur Sahib, Mukeriyan, Shahpur, Kandi and U.B.C.D. etc. to the Supreme Court.

On the frequent insistence of the Rajasthan government, the Central Minister organized meetings on 29-30 July and 6 August 1992. Thereafter, a draft decision was circulated, according to which it was proposed not to refer the matter to the Supreme Court. This proposal however was not agreeable to the Rajasthan government.

3. Implementation of distribution of waters of Ravi-Beas rivers among participating states:

According to an agreement made in 1981, the share of Rajasthan was 8.6 M.A.F. in the surplus Ravi-Beas water, which is 52.70 per cent. This water has to be distributed among Punjab, Haryana and Rajasthan. However, by wrong interpretation of articles in agreement of 1981 no decision can be reached. After the meetings held on 29-30 July and 6 August 1992, only 49 per cent water is being given, the Central Minister sent a draft of the decisions to the state governments which were accepted.

Gang Canal System and its Modernization

Gang Canal system is one among the three main canal systems of Ganganagar. Total length of Gang Canal is 129 km out of which the first 112 km falls within the boundaries of Punjab. The canal was made by lime lining. The authorized capacity at its top place is 2,720 cusec and total irrigated area is 3.08 lakh hectares (7.60 lakh acre). The estimated irrigation intensity of this canal is 60 per cent (Kharif 24% and Rabi 36%).

Availability of Water: According to division of water done before partition, share of this canal was 1.11 M.A.F. waters of Ravi-Beas rivers. After partition, 0.33 M.A.F. additional water was allotted for this canal. Thus, the total water of Ravi-Beas available for this canal is 1.44 per cent at present.

Existing State of Gang Canal System: The lining of Bikaner Canal, which was constructed with lime, has been damaged at many places and its lifetime has expired. Consequently, the quantity of water available on the Punjab-Rajasthan border has reduced and additional loss of about 440 cusec water per day is being incurred on this account. Bikaner Canal within the boundaries of Punjab is to cost Rs 98 crores for re-lining of the the canal, as estimated by Punjab. The matter is under consideration of the Rajasthan government.

Modernization of Gang Canal System: After starting re-lining work of Bikaner Canal, it is proposed to supply water to the area irrigated by Gang canal through Indira canal feeder. A link canal has already been constructed between Gang canal and Indira canal feeder and 1,100 cusec additional water was provided for crops from the link canal during the year 1994.

After the modernization plan, water allowance of the canals is proposed to be the same as 3.00 cusec water for 1,000 acres irrigated area. Irrigation intensity would thereby increase from 60 per cent to 70 per cent from the 1.44 M.A.F. water allocated for Gang Canal. But this would be possible only if the lining of all the canals of the system is made pakka. Additional irrigation capacity would be thereby available for about 50,000 acres.

International River Water Disputes

India-Bangladesh Dispute on River Ganga Water

Ganga water dispute is connected with the Kolkata port. The flow of water of river Ganga from the steep slope of Himalayas is very fast and within the boundaries of Uttar Pradesh and Bihar, many companion rivers merge into it. After its entry into Bangladesh, flow of its water becomes slow and it divides into two rivers named Padma and Ganga. Padma merges into the Brahmaputra and becomes Meghna in Bangadesh. It then falls in the Bay of Bengal. Ganges divides itself into many streams in the delta of Sunderban. In 1974, Farakka barrage was constructed on this flow of water in 1974 to make water available for Kolkata port, so that ships can sail up to the dock even during dry season and the dock remains free of silt. Bangladesh had objections to this saying that (1) India diverts the whole water during dry season towards Kolkata port, which results in insufficient quantity of water in Meghna-Padma river, and (2) the gates of Farrakka barrage are opened during the rainy season, which results in flood in Bangladesh.

Opposition of Bangladesh

1. Maulana Abdul Hamid Khan Bhasni organized a long march to Farakka with three lakh persons on 16 May, 1976.
2. Bangladesh raised this issue in the General Assembly of United Nations Organization on 15 November 1996. Begum Khalida Zia delivered a speech against India but no country supported her except China and Pakistan.

A treaty regarding distribution of water of Ganges-Farrakka was entered into between India and Bangladesh on 5 November 1977. According to it, 34,500 cusec out of the distributable 55,000 cusec in Farrakka was allocated for Bangladesh. The remaining 20,500 cusec would remain for Kolkata. In the agreement it was also decided that out of the total water available in Farrakka barrage, 25 per cent would remain stored while 60 per cent would go to Bangladesh and 40 per cent to West Bengal. This treaty continued up to 31 May, 1988.

On 11 December, 1996 river water treaty was again signed. It is valid for 30 years having a clause for review after every five years. It can also be reviewed after one year on unilateral demand.

Among the 12 clauses of the agreement, some important ones are as under:

1. Both the countries would distribute the water on equal basis if water in Farakka remains 70,000 cubic feet.
2. In case water remains between 70,000 to 75,000, Bangladesh would get 35,000, and the balance water would remain with India.
3. If the quantity of water exceeds 75,000, India would get 40,000 and the balance would go to Bangladesh.

According to the treaty, as per formula prepared for five months, during the dry season starting from January, water would be divided between the two countries as per its availability up to 40 years (1949-89). These five months have been divided into 10 fortnights. The share of India would be a maximum of 40,000 cusec and a minimum of 25,992 cusec per day, while the share of Bangladesh would be a maximum of 67,516 cusec and a minimum of 27,633 cusec per day.

Mahakali River Dispute

This dispute regarding the water of river Mahakali (called Sharda in India) is between India and Nepal. The latest dispute relates to Ranakpur and Pancheshwar projects, but the dispute regarding water of Mahakali stopped after the Nepal visit of the Foreign Minister of India, Pranab Mukherji, in February 1996. In 1998, the Indo-Nepal Sub-Commission was established with the objective of taking action on all the aspects related with the development of water resources.

Indus Valley Treaty

After the partition of India, an agreement for distribution of waters of river Indus was signed between India and Pakistan. As per the treaty, the waters of rivers Indus, Jhelum and Chenab were earmarked for India. India for which paid Rs 110 crores as a one time amount to Pakistan. On 19 September 1960, both the countries entered into the Indus Valley Treaty and a permanent Indus Commission of both the countries was constituted which meets every year.

Tulbul Navigation Project Dispute

This dispute between India and Pakistan relates to the construction of a barrage on Wular lake by Pakistan. About 20 km route of river Jhelum passing through Wular lake in Jammu Kashmir to Baramula via Sopor is to be made navigable. A barrage is proposed to be constructed over the lake so that even during water scarcity 4,000 cusec flow of water is ensured. This will make the river remain navigable in India throughout the year. Work on this project was started by India in 1984 but when Pakistan raised objections in view of the Indus Water Treaty 1960, it was stopped since 1987.

Allegations of Pakistan

1. Tulbul project is really a water storage project and it would affect the flow of water of river Jhelum towards Pakistan.
2. This project is against the spirit of Indus Water Treaty of 1960, according to which India has no right to store the water of river Jhelum.
3. India cannot create any hindrance in the flow of the water of river Jhelum river water as per terms of the treaty.

Discussions have taken place between India and Pakistan on this issue 21 times, but without any success.

20

Water Management through Remote Sensing Technology

Remote Sensing

Remote Sensing means getting information regarding any place located at a distance using artificial satellites and space skills. This is an advanced technique, which gives information regarding changes occuring on the earth from time to time. Electro-Magnetic energy obtained from sun or auto-abandonment is used as a source in it. Transmission of this energy into the atmosphere is caused on the ground as a source through absorption or expansion. Reflection and abandonment of electro-magnetic energy from the ground is important for getting information from the earth. They cause interaction, which transmits energy from the earth to remote sensors. The remote sensors located in satellites, collect different types of information and send it to receiving stations on the earth, which is used after digital image processing. The resolution of remote sensors is the most important fact in this technique. For example, the resolution of remote sensors LISS III of Indian Satellite IRSIC is 23.5 metre whereas the resolution of Pancromatic (pan) is 5.8 metre.

There are three types of platforms in the remote sensing technique. Platform is a rectangular position of the camera which obtains information regarding the target. According to height, they are of three types. These are:

1. Ground Borne
2. Air Borne
3. Space Borne

Ground-borne remote sensing system is used for study of land resources for which detailed information is received with the help of space technology and satellites. Airborne technique is normally used for getting air photo pictures for photo interpretation and detailed description can be obtained at any time through them. Spaceborne platforms are generally not affected by the environment of the earth and they move independently in their orbit. Vast statistics can be obtained from them though they depend on the extension of the sphere of the satellite.

For the last 30 years, many geographers had been thinking about a system by which spatial information could be organized and stored using computer. During the last decade, this growing technique came to be known as 'Geographical Information System'. Geographical Information System is mainly a system of computer hardware and software that organizes and stores geographical data for counting, analysis and individual planning. In other words, it is an information technology which analyzes spatial and non-spatial data after its collection. Many areas have been included in this Information Management Technology. They include mainly computer science, cartography, information management, telecommunication, geology, photo-geometry, remote sensing etc. Geographical Information System has developed in many forms. The first publication of the word 'Geographical Information System' was made in 1965 by Michael Decy and Deven Marbel of Northwest University in a paper written by them. This name is related with the following techniques:

1. Automated Mapping
2. Computer-Aided Mapping
3. Computer-Aided Design
4. Computer-Aided Drafting
5. Geographical Information System
6. Geo-processing and Network Analysis
7. Land Information System

Elements of Geographical Information System

The main elements of Geographical Information Systems (GIS) include hardware, software, scanners, digitizers, C.D., key board, graphics, monitor, plotter and printer.

Computer Hardware

The computer hardware has a hard disc for collection of data and programs. Digital tape cassette, C.D.Rom etc., are also its parts.

Scanner digitizer is used for converting maps and data in digital form. Plotter or printer is used for demonstrating the result.

Figure 20.1

Important Hardware Constituents of Hardware of GIS

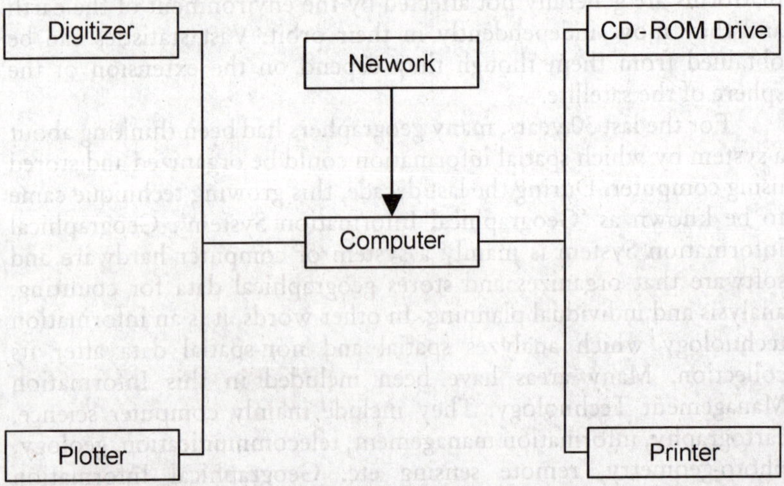

GIS Software

Geographical Information System is divided in five functional classes. They are:

1. Presentation and Verification of data
2. Storage and Management of data
3. Exit and Presentation of data
4. Transfer of data
5. Interaction with user

Utility of Geographical Information System and Remote Sensing

Use of Geographical Information System and Remote Sensing is possible in the following areas:

1. Agricultural development
2. Land valuation analysis
3. Study of changed reflection capacity of vegetative regions
4. Analysis of deforestation and environmental crisis
5. Supervision of vegetative health

Picture 20.2

Important Constituents of GIS Software

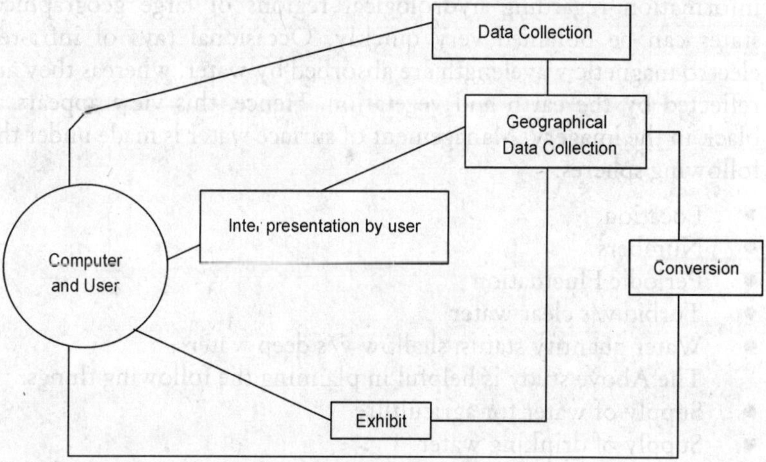

6. Management of land degradation
7. Estimation of area of crops and production
8. Mapping of waste lands
9. Soil Resource Mapping
10. Groundwater Probability Mapping
11. Discovery of geological minerals
12. Supervision of forest fire
13. Supervision of Oceanic products
14. Water Resources Management etc.

Geographical Information System and Remote Sensing technique has performed an important role in water resources management since suitable and sufficient quantity of information can be obtained from satellites. The following aspects of water resources management can be studied with the help of Geographical Information System and Remote Sensing:

1. Survey and search of surface water
2. Hydrological studies
3. Watershed conservation, planning and management
4. Management of flood affected areas
5. Water management in irrigated areas
6. Groundwater management

Surface Water Inventory and Surveying

Satellite capacity is important for surface water management because information regarding hydrological regions of large geographical states can be obtained very quickly. Occasional rays of infra-red electro magnetic wavelength are absorbed by water, whereas they are reflected by the earth and vegetation. Hence, this view appears as black in the imagery. Management of surface water is made under the following spheres:

- Location
- Numbers
- Periodic Fluctuation
- Turbid v/s clear water
- Water quantity status: shallow v/s deep water

The Above study is helpful in planning the following things:

- Supply of water for agriculture
- Supply of drinking water
- Development of probable sources for recharging of groundwater

Thus, constant use of surface water would be possible in these spheres.

Hydrological Studies

Study of different aspects of hydrology can be done easily with the help of Geographical Information System and Remote Sensing technique. Modelling of rain water, study of water balance, flow forecast and estimate etc. are important among them. In modelling of rain water flow first of all, contour map is prepared. After that an elevation map is prepared. This is superimposed on the land utilization map and a soil map is then prepared. Knowledge of rainfall, evaporation, evapo-transpiration, infiltration etc. is essential for study of water balance. Hydrological balance is important in certain conditions when hydrological cycle is affected. Hydrological cycle affects climatic changes. Forecast regarding flow of water and its estimate can also be easily done with the help of Geographical Information System and Remote Sensing technique.

Watershed Conservation, Planning and Management

Work plan of watershed management can be prepared with the help of Geographical Information System and Remote Sensing technique. After scanning the toposheet, drainage system maps and contour maps can be prepared through digitization. A digital elevation model can then be prepared after counting from the contour bund map. Slope map and three dimensional view are prepared thereafter. Soil map, classified slope map and land utilization maps are superimposed on the basis of facts collected during topographical survey. Soil map is cross checked from the conservation work plan. Water management has become the necessity of the day. Optimum utilization of land and water resources is possible only through this technique. Watershed management is resorted to with the following objectives:

1. Control over degradation of flow water.
2. Management and use of flowing water for useful purposes.
3. Control on soil erosion and reduction in quantity of silting.
4. Diversion of water towards suitable direction in flood affected areas.
5. Making provision for groundwater recharging wherever possible.
6. Suitable use of land resource.

For meeting the above objectives, Geographical Information System and Remote Sensing technique is used and prioritization of methods is done for watershed development. Land utilization maps are prepared from statistics of remote sensing. Micro watershed area maps are drawn from contour maps and drainage basin maps are prepared from toposheets.

The map obtained by cross tallying of watershed priority map, soil map, slope map, and land utilization map, is cross tallied with the micro watershed maps to obtain the final micro watershed map.

Work plan of watershed management is affected by the following features of watershed area:

1. Size and shape of watershed
2. Morphology of the site
3. Soil and its specialities
4. Rainfall
5. Land utilization
6. Vegetative cover

Essential statistics required for watershed planning can be divided in three categories:

1. Hydrological Information
2. Soil and land use data
3. Economic and social data

Hydrological information includes information regarding rainfall, requirement of water for crops on climatic basis, and data regarding flow of river and silting.

Soil and land use data include data regarding existing land, data related to soil, morphological data, geological map and types of vegetation. Economic and social data includes calculation of investment on the project. Thus, the process of planning, conservation and management of watershed can be completed easily by Remote Sensing and Geographical Information System and a suitable conservation plan can be implemented.

Management of Flood-Prone Areas

Survey of flood-prone areas is essential for economic assessment of measures to be adopted for flood control and to assess intensity as well as extension of floods. Remote sensing technique has provided praiseworthy methods, with the help of which maps of flood affected areas can be prepared through Remote Sensing and Geographical Information System for managing floods.

Maps of flood affected areas are special types of morphological maps indicating special features of floods by graphs prepared from an imaginative point of view. Trends of previous floods are also indicated in it. This map is a type of basic instrument on the basis of which a strong future strategy is chalked out. These maps are prepared for fulfilling the following objectives:

- Basic primary information regarding land utilization plans are provided by them.
- They provide permission for correct development of new urban areas
- Assessment of loss due to floods can be done with the help of such maps.
- They provide help in controlling unstructured floods.
- Public awareness is spread by them.

Indian remote sensing satellites are now collecting information regarding various peculiarities of floods. In the G.I.S. series, micro sensors LISS-I and LISS-II located in IRS-1A, IRS-2B and IRS-P2 satellites having resolutions of 36 metres and 72 metres arc collecting information every 24 days.

IRS-IC is the latest modern satellite in which three sensors, PEN, LISS-III and WIFS, possessing local resolution capacity of 5.8 metre, 23.5 metre and 188 metre respectively are located. WIFS sensor is established in IRS-P3, which reaches the ground again after five days.

The regional coverage of WIFS sensor is 810 km NOAA (AVHRR) and Landsat (TM) satellites, established by Indian National Remote Sensing Agency (NRSA) which are in the space for climatic study, have local resolution capacity of 30 metres and repetition duration of 16 days. European Remote Sensing Satellites (ERS) are also collecting data about flood affected areas.

Effective use of such data generated by remote sensing system are being made in the following forms:

- Obtaining information about duration, extension and size of flood
- Getting information regarding construction of dams etc. to control floods
- Obtaining knowledge about areas preventing drainage
- Making assessment about losses on account of floods
- Studying morphology found in flood affected areas
- Studying erosion of banks and migration of rivers

Geographical information system provides description of vast areas affected by floods and gives estimate about future data. A digital elevation model is prepared by using local data of flood affected areas. It can provide suitable direction for construction of structures in the context of water received, height of water and slope of the area, so that the adverse effects of flood are prevented.

Water Management in Irrigation Command Areas

Up-to-date pictures of irrigated areas and crops grown on them can be taken by remote sensing satellites, and assessment regarding requirement of water for irrigation can be made after analysis through Geographical Information System. Sensors like LISS-III and thematic Mupper (TM) in the latest satellites IRS-1C and Landsat, have proved very effective. Their inverse colour combination (ICC) provides correct and timely analysis of crops. ICC is a technique involving little time and nominal expenditure for management of irrigation in command areas through Geographical Information System. The following data is required for it:

1. Crop Areas
2. Crop Calendar
3. Soil Map
4. Transpiration
6. Crop Coefficient
6. Canal Network

7. Rainfall Data
8. Discharge
9. Other Sources of Irrigation

For water management in irrigated areas, first of all the cropped area is assessed through Geographical Information System. This information is received from visual interpretation of imageries of satellites. Thereafter, polygons are digitized and a supervized classification is done.

Optimum utilization of irrigation water is essential for proper water management. Some crops require less water, whereas other crops need more water. Computation of water requirement is done on the basis of the above data through Geographical Information System. Soil map is required for this purpose to calculate loss from percolation. This map is digitized and converted into the polygon mode. Infromation about crop calendar is also required. This information is important for assessment of requirement of water. These figures are kept in the form of a table. Study of sources of irrigation is also very important because quantity of water is decided on this basis, for example, to decide how much quantity of water is lost if it is carried through the canal system. Hence, computation of the following morphometric elements is important:

- Calculation of cropped area
- Transpiration
- Stage of crops
- Percolation loss
- Conveyance loss

During present times, water management is being done by analyzing data regarding timely agricultural crops furnished by Sensors LISS-III and Thematic Mupper (TM) in latest satellites IRS-1C and Landsat through remote sensing by Geographical Information System.

Management of Groundwater

In India, more than 90 per cent rural population and more than 30 per cent population in urban areas is dependent on groundwater for drinking water and other domestic requirements. 60 per cent irrigation in the country is done by groundwater. Demand for water is 3-4 times more than its availability, hence by proper management, this gap between demand and availability can be balanced. Different countries of the world have been making use of satellite imageries for management of groundwater since the last two decades. In the beginning, regional maps were developed on a 1:250,000 scale from imageries received from Landsat and IRS series satellites having lesser resolution. Now, the National Drinking Water Technical Mission, Space Department, has developed maps on 1:2,50,000 scale regarding probable availability of groundwater in 446 districts of India during 1987 to 1992, with the help of some non-governmental organizations. More information is now possible through IRS-IC that has a high resolution capacity. Recently, Rajiv Gandhi National Drinking Water Mission in the National Remote Sensing Agency, has prepared groundwater probability maps on 1:50,000 scale for the states of Rajasthan, Madhya Pradesh, Andhra Pradesh, Karnataka and Kerala. Companion places are being developed as sources for drinking water and a scheme for artificial recharging of groundwater is also being resorted to at places wherever it is feasible. Maps regarding surface sources and groundwater in highly irrigated and canal areas are being prepared on 1:15,000 scale with the help of LISS-III and PAN data in IRS IC/ID satellites.

The following integrated information can be obtained regarding groundwater management from Remote Sensing technique and Geographical Information System that is directly or indirectly related to geology, morphology and hydrology:

1. Assessment of groundwater resources.
2. Assessment of surface water flow.
3. Computation of available groundwater balancing for macro use.
4. Identification of underground areas based on groundwater development.
5. Identification of overexploited groundwater areas.
6. Delineating areas having increased groundwater resource.
7. Identifying problems created in the context of water conservation.
8. Planning for combined use of surface water and groundwater resources.
9. Making optimum management and use of groundwater.
10. Maintaining constant vigil on development and use of groundwater.

Thus, a detailed outline for constant use of groundwater in future is being prepared on the basis of imageries received from satellites through remote sensing and analysis through geographical information system.

Indian remote sensing agency is performing a very important task in the direction of water resources management in the country.

IRS P4 (Oceansat) was launched in May 1999 for management of the area of oceans. Its sensor is Ocean Colour Monitor (OCM). On the basis of information received from it, suitable management of oceanic resources like fish, quality of oceanic water and management of coastal areas would be possible. Its repetition is once in two days.

India is constantly progressing in the direction of launching satellites with higher resolution and improving quantity and quality of data. The following satellites are important from this point of view:

CARTOSAT-1, IRS-P 5: It is a mission of cartographic methods in which there are two panchromatic cameras. One camera is in the front side and other is in the back side. They draw three-dimensional pictures daily. Its resolution capacity is 2.5 metres. Its main features are: planet route is 618 km, local time is 10.30 A.M., weight is 1450 kg, mission duration is five years and time of re-visit is five days.

India launched *METSAT* stationary satellite on the earth in September, 2001. It was renamed as *KALPNA-1.* This satellite has a very fast resolution and has been sending pictures of water vapour and clouds in three bands of radio metre.

RESOURCESAT, IRS-P 5: This satellite is in the continuous series of IRS – IC/ID. Its sensor is LISS-III having resolution of 23.5 metre with three bands (Green, Red, Near IR). Pan camera of IRS – IC/ID will soon be converted into multi-spectral camera which will also have 3-bands, resolution capacity 5.8 metre and WIFS sensor having resolution capacity of 70 metre.

Its height is 817 km, equator crossing time is 10.30 A.M., weight is 1250 kg, mission duration is five years. Data received from PSLV –1 RESOURCESAT is very important for soil agriculture, water resources etc.

RESOURCESAT–1 was launched on 17 October 2003 from PSLV. Its next series CARTOSAT-1 and CARTOSAT–2 would be launched very soon.

CLIMATSAT: Knowledge about water vapour, snow, clouds etc. in tropical regions is being sent by this satellite.

Thus, in the programmes pertaining to space. India has made revolutionary changes from the point of view of resources management and climatic data during the last two decades. It has developed the most modern techniques and the information obtained through them can be analyzed using Geographical Information System. Thereby quantitative availability and use of resources can be balanced.

21

Environmental Disasters and Water Crisis

Many environmental disasters have been created during the last century because of human activities. Man has affected ecology everywhere by pursuing fast economic development and by adopting a physical life and a materialistic culture. The deforestation of Congo basin and Amazon basis on the equator as well as the increased use of fossil fuels in Europe and America, has awakened our sensitivity of nature. In the latter half of twentieth century, there has also been deterioration of green house effect and the ozone layer, resulting in increase in temperature. Additionally global warming has enhanced the process of snow melting from Kilimanjaro Mountain on the equator to Antarctica.

The maximum impact of climatic change has been on water resources. Out of the total water resources in nature, two per cent water is frozen in the form of snow and only little less than one per cent water is available for use of mankind. Even this water is suffering from qualitative and quantitative deterioration because of environmental crisis and human activities. Fresh water is being polluted by acid rain. Thus, water resource available in nature is being affected quantitatively and qualitatively by climatic changes, global warming, acid rain, melting of snow and other such activities in these days. Due to such factors, the availability of water is continuously decreasing, creating water crisis. Natural disasters responsible for water crisis are as under:

Climate Change

Balanced climate has become imbalanced due to selfish teasing of nature by man. The earth is becoming hotter day by day due to fast industrialiszation and tremendous increase in transport vehicles. Snow is melting in the Arctics due to climate changes. To study this situation, the United Nations Ecological Programme and World Meteorological Organization together formed a group of scientists called International Inter-Governmental Panel on Climate Change (IPCC) in 1988. Its research has pointed out that a mere increase of 0.3 to 0.6 degree celsius in average temperature during the last century, has caused nature to become imbalanced, resulting in dangerous consequences. Water level of oceans has risen by 10.25 cm during this period while a 2.7 cm rise is considered to be due to spread of water on account of rise in temperature.

Climate is a complex system. A change in climate affects the whole atmosphere including oceans, snow, land, rivers, lakes and mountains, as well as groundwater. A change in these factors affects vegetation and the biotic world. The colourful vegetation on rocks of coral called rain forest of sea, is being heavily affected. Droughts are caused by climate change, which directly affect food production. It is predicted that availability of water would also decrease in future because 50 per cent of fresh water is being used for human consumption. Serious water crisis would be faced by Kuwait, Jordan, Israel, Rawanda and Somalia and other countries already suffering from scarcity of water. American Security Agency has estimated that the annual supply of water in California can reduce between 7 to 16 per cent due to generation of heat by doubling of quantity of carbon dioxide there. Along with agriculture, the natural formation of forests can also change due to climate change. Temperature and moisture right from micro vegetation to big trees remains favourable only up to a certain limit. Any change in them either shift the vegetation from their places or extinguishes them for ever. It is considered that some other means will have to be adopted due to increasing population and urbanization. One-third forests of the world have to face danger on account of climate change. The number of forest fires can also increase due to higher temperature and the quantity of carbon dioxide can increase by forest fires.

Climate Change and Water Crisis

The initial water crisis had started due to uneven distribution of rainfall in nature. Today, it has become more serious due to increase in demand. For example, 60 per cent population of the world lives in Asia, whereas there is only 36 per cent flow of water in rivers here. On the other side, only 6 per cent population of the world resides in South America but there is 20 per cent flow of total surface water of the world there.

Most of the water resources available for human consumption are found in the form of rivers and lakes on the surface of the earth. Both of them are elements of climate and are important in the process of hydrological cycle. They also play important roles in redistribution of water. Increasing human intervention during the last decades of the 20th century has changed the nature of hydrological cycle to a great extent. Continuous deforestation in arid regions of India and related economic activities have expedited the process of desertification, resulting in reduction in quantity of rainfall. Annual availability of water in the Indian sub-continent depends on balanced activity of monsoon. Its gravity and centre of low pressure are located in the Thar Desert.

India is divided in three geographical divisions from the point of view of natural composition. They are: Himalayan mountain in the north, Plain areas of the north and the Plateau part of the south. The Himalayan mountains having vast glaciers from where rivers originate, is the source of important Indian rivers like Indus, Ganga, Yamuna, and Brahmaputra. Hundreds of other small and big rivers also start from Himalayas, which supply water to small areas. Indus river starts from the west of Mansarovar lake and reaches Pakistan after crossing through Karakoram hills. The main rivers Ganga and Yamuna originate from Gangotri and Yamnotri glaciers, located at a height of 6,000 metres respectively in the Himalayas. The water flowing in these rivers is available throughout the year because of melting of frozen snow as per the hydrological process. On the other hand, the Brahmaputra originating from the eastern end of Kailash mountain and flowing to the eastern India, depends on rainfall rather than snow as its main source of water.

During the last decades of the previous last century, glaciers of the Himalayas have started melting due to fast rise in temperature. Vast glaciers are converting into snow lakes. According to the international magazine 'New Scientist', all the glaciers of Himalayas would be destroyed by the year 2025. During a certain period, there would be a dangerous flood situation in rivers subsisting on them but after that a serious water crisis would be faced. Climate change affects not only glaciers but also lakes and groundwater. Change in pattern of quantity of rainfall is also affecting its distribution pattern. On one side, excess water is found in eastern India because quantity of rainfall is increasing. On the other side, there is continuous decline of quantity of rainfall in western and south central portions of the country. In the 'Johannesburg Earth Summit' the main reason for continuous drought was considered to be fast climate change. In India, local and regional climate changes are occuring due to fast depletion of forests, and the rain and the increase in industrialization and urbanization.

Besides drought, climate change is also causing deterioration in quality of water through acid rain. Effect of acid rain has been noticed in India also like that in other countries of the world. Mathura Oil Refinery, located on the bank of river Yamuna emits out sulphur dioxide, which results in acid rain by mixing in rainfall. This is polluting water of river Yamuna. Climate change and water crisis in the Indian sub-continent are becoming jointly visible in the form of changing nature of monsoon. Effective area of monsoon has changed due to climate change during the last century, and the quantity and timing of rainfall has also changed. During the year 2002, the main reason for failure of monsoon was considered to be smoky clouds of Asia. These clouds were formed due to fast industrialization and they had a important role in climate change.

Water resource of India is controlled by the nature of rainfall and climatic balance. Scarcity of water is created by even a partial change in it. Hence, necessary proportionate forest cover should be developed for a balanced climate. Along with planned urbanization, a tight control on increasing population should also be exercised. Since change in climate would also affect change in distribution of water, hence sufficient quantity of water can become available only with healthy nature. The process of climate change is regional and is a long term process, rather than being local and a short term one. Hence, for meeting the water crisis, we shall have to maintain a qualitative level of available surface water sources and also emphasize on conscientious exploitation of groundwater. A proper crop rotation system should be adopted as per the climatic conditions specified by agriculture climatic regions at the national level. Availability of water and crop planning should be given priority in it.

The following hydrological areas would be affected by climate changes:

1. Hydrological area of rivers and lakes
2. Effect on glaciers by melting of permafrost
3. Successive action of snow peaks
4. Change in sea level
5. Effect on coastal water portion and backwater area, delta and humid land on account of rise in sea level

Green House Effect

Due to policies and activities of man against nature, the balanced climate has become imbalanced. Quantity of some particular gases has increased in the atmosphere of the earth to such an extent that warmth or heat of earth is not able to come out. Such an effect is called Green House Effect. According to Oxford dictionary –

"As a result of blanketing effect of the man made carbon dioxide in atmosphere, progressive warming of surface of earth is called Green House Effect".

The main gases causing Green House Effect are carbon dioxide, water vapour, methane and nitrous oxide. Due to absorbing long waves coming out of the earth, a halo created by gases and chlorofluorocarbon (CFC) are also included in the category of green house gases. Maximum contribution is due to carbon dioxide (CO_2).

CO_2 is released in atmosphere by burning fossil fuels like coal and petroleum for different purposes. Besides this, 30 per cent CO_2 has increased due to electricity generation centers, emission from chimneys (human volcanoes) of industries, burning coal and minerals, burning of fuels in means of transport and burning of fuel wood for domestic use. 65 per cent of the Green House Effect has increased through them. Quantity of CO_2 was only 320 ppm (Parts per 10 lakh) in the atmosphere during 1965 but it increased to 360 ppm in 1990. At present, concentration of carbon dioxide in the atmosphere is about 380 ppm. Increase of CO_2 in the atmosphere has been at the rate of about 0.5 per cent per annum. Scientists have estimated on such basis that quantity of CO_2 will become double by the year 2050.

Carbon dioxide (CO_2) is really a natural constituent of the atmosphere. It is not a polluting part in the lower portion of the atmosphere but there is possibility of climate change and change in warmth balance due to its higher concentration. Scientists think that an adverse effect is caused on biotic life on account of change in temperature of the world during climate change. Rise in temperature would reduce rainfall and the extent of soil moisture, thereby affecting agriculture. Many places would be affected by drought. Forest area would also reduce due to increase in temperature and decrease of rainfall. As a result of decrease of forest cover, there would be loss of wild life and other creatures. Coastal areas of seas would also be submerged under water due to melting of ice on account of rise in temperature. According to one estimate, sea water level would rise by one metre by 2050 because of rise in temperature due to Green House Effect. In case man generated concentration of carbon dioxide CO_2 is continuous in atmosphere, oceans would have to do maximum absorption. If absorption of carbon dioxide (CO_2) and decomposition in oceanic water is more than normal, then acidity would increase, resulting in decrease in productivity of maritime ecology system. Because of green house gases, temperature would increase due to decrease of ozone in stratosphere.

Scientists have discovered a new green house gas named trifloromethyl (F_3) in recent years. As compared to CO_2, this gas possesses about 20,000 times the capacity of absorbing infra-red rays for a duration of 100 years.

Global Warming

The quantity of carbon dioxide has increased in the atmosphere on account of increasing pace of industrialiszation, thereby resulting in green house effect. Due to increase in quantity of carbon dioxide found on the surface of the earth, thermal energy released by reflected rays is concentrated in the atmosphere, resulting in increase of average temperature of the earth. This is called global warming.

Scientists believe that climate of the earth would change on account of catastrophic rise in temperature of the world, resulting in decrease of rainfall. Direct effect of decrease in rainfall would come on agriculture, incidence creating of drought. Forest area would reduce in the world due to rise in temperature and it would lead to decrease of rainfall. Wildlife and other creatures would also reduce because of this. Besides, carbon dioxide, methane, chlorofluorocarbon (CFC) compound and nitrous oxide are also responsible for rise in temperature. Immediate and far reaching consequences of warming of ionosphere would do severe harm to human health as well as environment. Immediate effects mainly include death, drought, storm, flood and environmental degradation, while far reaching effects will be various kinds of infection and related diseases, food problem, famine, and danger to biotic life. Besides these, many cities located in coastal regions of seas may submerge in water due to melting of snow at the arctics and high mountains due to rise in temperature.

According to meteorological scientists, temperature has increased by 0.3 to 0.6 degree Celsius during last century. Intergovernmental Panel on Climate Change (IPCC) scientists have remarked after estimating the severity of Green House Effect in view of the population increase and economic and technical progress, that the quantity of carbon dioxide would be doubled as compared to the industrial era in the earlier period. As a result of this, average temperature of the earth would increase at the rate of 0.3°C every century. It would this increase by 1°C by 2025 and 2°C by the year 2100. The increase of temperature in atmosphere by 0.3 to 0.6 degree Celsius has raised water level of seas by 10.25 cms. Scientists have estimated on

Figure 21.1

Probable Chronological Chain of Effects of Global Warming, Water being a Main Constituent

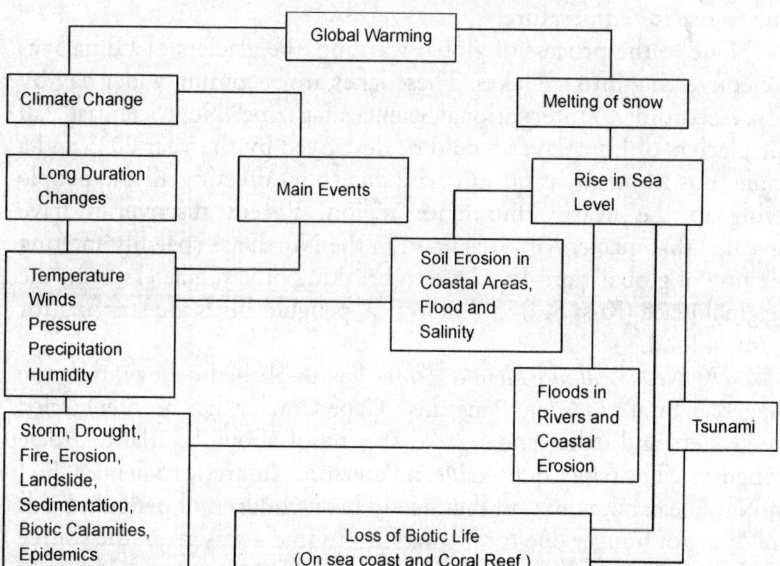

such a basis that water level in oceans can rise from 15 to 95 cms during the next century. Due to the rising height of oceanic water, many towns and metropolitan cities situated beside seashores of Indian Ocean, Mediterranean Sea, African Atlantic Ocean and Caribbean Sea can submerge under water in the future. According to the famous 'Time' magazine, Maldives would submerge under water by 2025. According to scientists of Meryland University, the life of about 9.40 crore people of Egypt, Bangladesh, China, Nigeria, would be in danger if water level of oceans rises by one metre. Fifteen per cent agricultural land on coasts of Mediterranean Sea of Egypt alone would be destroyed. Similarly, the famous 'Sunderbans' of West Bengal in India and Bangladesh would drown in water. It would destroy the ecological marshy system and Peninsular India may also face danger.

Climate as a natural element, possesses a complex composition which is interconnected with the atmosphere, oceans, snow, land, rivers, lakes, mountains etc. Climate produced by their interaction harms the existing vegetation and biotic life. Change in composition of climate becomes visible in the form of famine. According to a study by American Environmental Security Agency, production of wheat, rice and other food grains would be reduced by 1.2 to 7.6 per cent by the year 2060. Global water crisis would also become acute due to rise in temperature.

Due to the process of global warning, the glaciers of Himalayas are converting into ice lakes. These lakes are becoming wider day by day. According to International Science magazine "New Scientist", all the glaciers of Himalayas would be destroyed by the year 2025 and a dangerous flood situation can arise due to it, affecting all the people living in the nearby mountain region. Recent discoveries have revealed that glaciers of Greenland in the North are speedily melting because of global warming. Due to breaking of vast glaciers from the original place (Ross Sea) in Antarctica, penguin birds are starving for want of food.

The New Zealand Herald Times has published a news that the biggest habitat of 'Adele Penguins', Cape Crazier, has been encircled by glaciers and it has endangered the life of about 1.30 lakh 'Adele Penguins'. The paper quoted Dean Paterson, Antarctica Science Chief of New Zealand, who said that newly born children of penguin birds are dying of hunger due to the inability to find a way from the source of feeding place to the habitat. This situation has been created because of the coming of 37 km wide and 55 km long glaciers in March 2001. These glaciers have created an obstacle between Ross Sea area and the open sea.

160 countries signed a resolution regarding climate change in the 'Earth Conference' held at Rio city in 1992. In December 1997, many countries consented to reduce the level of quantity of carbon dioxide at the 'Kyoto Conference' held in Japan. 300 non-governmental organizations also participated in this conference along with the representatives of 160 countries.

Ice Melting

The biggest source of fresh water of the world, which is in the form of ice, has been melting due to rise in temperature on account of climate change. After researches done in recent years, scientists have concluded that deterioration is being caused in snow cover of Antarctica and Greenland. In March 2002, scientists of London, on the basis of data received from remote sensing satellite, said that Larsen B Iceberg connected with the eastern peninsular part of Antarctica has broken. Breaking of ice rock of 1,250 sq mile area and 650 feet thickness is considered as a grave danger for the world. Larsen B Iceberg, which is 1500 year old, it was broken and scattered in the space of one month during March 2002, and serious note of this incident was taken by scientists who consider it as the main reason for rise in temperature in the world. British Arctic Survey had also warned four years back that some icy rocks of peninsular part are melting fast.

Box 21.1

Antarctica

Antarctica is located on 66° 30′ southern latitude in the middle of Antarctica convergence zone, having an area of 1,40,00,000 sq km. 90 per cent fresh water of the world is found there. It is an important place from the point of view of habitat where lakhs of various creatures like whale, seal and creel fish, and penguins and many other kinds of birds are found. It plays an important role in oceanic ecology system and in controlling world climate. It provides the best facility for research related to changes occurring on the earth.

A strange fact is that temperature is reducing in other parts of Antarctica, whereas it is fast rising in the peninsular part. Research is being conducted, to establish whether temperature is really reducing or rising. Director, National Centre for Antarctica and Ocean Research, Goa, told on 1st May 2002, in connection with change in global warming, that snow is not melting in South Pole but it is actually increasing, whereas temperature is continuously rising in the world. This doubt is the result of local variation only.

Apart from Antarctica and Arctic region, indications about melting of vast snow reservoirs of high mountains have been received from other parts of the world also. Gangotri glacier, which is the main source of river Ganga, is also retreating fast. If ice of the world continues melting, the sea level will also rise. IPCC had indicated about such a possibility one decade back, and it had declared that sea level would rise from 30-100 cms, which would destroy the natural ecological system including agricultural land of coastal areas, mangrove land, as fossil walls.

Acid Rain

Water is important for various uses of humans. This water is mainly received from rainfall. During present times, sulphur dioxide (SO_2) emitted from man-generated sources, reaches the atmosphere and forms sulphate and sulphuric acid (H_2SO_4) after mixing with water. When it reaches the ground along with rain, it is called acid rain. Acidity of water is measured in terms of pH. When pH of water is 7, it is called neutral water, pH of normal rainfall is 5, but when this standard is less than 4, it is harmful for biotic community, pH of rain water would reduce in proportion to increase in concentration of sulphur dioxide and nitrogen oxide in the atmosphere.

Scientists believe that the effect of acid rain is not limited only to areas where sulphur dioxide and nitrogen oxide are emited by industrial and transportation sources, but it affects areas far and wide because pollutant factors (eg. sulphur dioxide) are in gaseous form which spread with air and clouds in far off areas. For example, sulphur dioxide and nitrogen oxides emitted by mills located in England and Germany cause acid rain in Norway and Sweden where biotic life has vanished in most of the lakes. They are called 'Dead Lakes' from the biotic point of view. Oxide emitted by industries of United States of America is visible in the form of acid rain in Mexico, while smoke emitted from the fire ignited in natural oil wells by Iraq during the Gulf War, had an effect on Iran, and Jammu and Kashmir in India. CO_2 existing in carbon smoke came along with rain water in Jammu and Kashmir, and it remained frozen in the midst of ice layers.

Acid rain has assumed dangerous dimensions in industry dominated countries like United States of America and Western Europe. Acid rain is also sometimes called 'Lake Killer' because acid rain is the main factor responsible for the death of many water creatures in different water sources like lakes and other big and small water reservoirs. Thousands of lakes of Ontario province of Canada are affected by acid rain and 140 lakes have been declared as unsuitable for biotic life and have been declared 'dead lakes', pH value of their water has become extremely low. According to Hamilton of United States of America, 7,500 to 12,000 people die and lakhs of people suffer from skin cancer and related diseases in the world on account of acid sulphate generated by burning of fossil fuels like coal and petroleum.

Soil productivity is also affected by acid rain. Minerals and other nutritive elements of the soil are destroyed due to acidity. Acid rain has its impact on forests also. It has destroyed many forests of North America (United States of America, Canada, Mexico, and the Western Island group), and European countries (Norway, Sweden, Germany, and Central European Countries). The growth of leaves and roots of trees is hindered by the acid. Ancient historical buildings constructed of marble or red stone are also affected by acid rain. Harm is caused to these buildings due to corrosion on account of acid rain. In India, Taj Mahal, Moti Masjid, Red Fort etc., are facing damage on account of acid rain. Effect of sulphur dioxide emitted from Mathura Oil Refinery is visible in the form of yellowness on the marble walls of Taj Mahal.

Box 21.2

pH Scale

Acidity is measured by pH scale. pH of pH denotes hydrogen potential (pH= Potential Hydrogen). The value lies is found between 0 and 14. Maximum acidity is at 0 and maximum alkalinity is at 14. Normal pH is less than 5-6 pH or somewhat more than that. When pH of rain water is less than 5-6, it is called acid rain. Its measurement is done on geometrical basis. For example, ph-5 is 10 times more alkaline than pH-6 whereas pH-4 is 100 times more alkaline.

The problem of acid rain is not so serious in India as compared to that in America or in other western countries. Researches at Bhabha Atomic Research Centre and World Meteorological Organization have found that acidity of rain water in most of the Indian cities is less than the safe limit. In a few big cities (Delhi, Bombay, Chennai and Kolkata), increase in level of sulphur has been found in atmosphere.

Water Pollution

Problem of water pollution is assuming dangerous dimensions. Harmful chemicals of chemical, textile, sugar, metal, medicine, and fertilizer industries are being disposed off in rivers. Water pollution is caused mainly by the following sources: (1) Industrial effluents (2) Domestic effluents (3) Sewage (4) Agricultural effluents (5) Oil pollution (6) Thermal pollution (7) Social sources (8) Radioactive disposal. Their detailed description has been given in the Chapter on water pollution.

Natural form of water has deteriorated due to various natural and human sources. Some pollutants enter into water, causing great harm to the environment . The main among them are mercury, lead, zinc, fluoride, sodium, phenol, nitride and armenizm.

Water pollution in India has increased to a great extent. According to the estimate by researchers of National Environmental Engineering Research Institute (NEERI), Nagpur, about 70 per cent groundwater in the country has become unfit for human consumption. According to scientists, there is satisfactory sewerage arrangement only in eight out of 142 big cities, partial sewerage system in 52 cities and in remaining 82 cities there is no sewerage, resulting in absence of disposal of waste and leading to water pollution.

Various sacred rivers of India are losing their quality of water. 37 per cent population of India lives in the drainage basin of river Ganga and 47 big cities and 66 small towns located on its sides, are polluting its water. Apart from rivers of India, lakes and other important water reservoirs are also becoming polluted.

Glossary

Acre inch: Amount of water (3630 cubic feet) needed for one inch water depth on the entire surface of an acre area.

Aggressive Water: Water which is capable of dissolving rocks, is called aggressive water. In karst areas, there is carbon-dioxide mixed water which dissolves limestone rocks.

Agricultural Drought: When rainfall is so less that it is insufficient for growth of crops, then it is called agricultural drought. Growth of crops becomes slow in this condition. Quantity of moisture at the time of lesser rainfall depends on the soil and nature of plants, because plants demanding more quantity of water feel it very early. Drought is also caused by change of crop rotation.

Agro-climatology: It is the study of interaction between climatological and hydrological factors, including forestry and animal husbandry. Its main aim is to apply climatological information for the purpose of improving farming practices and increasing agricultural productivity.

Agro-meteorology: The science concerned with the application of meteorology to the measurement and analysis of the physical environment in agricultural systems. Physical factors affecting agriculture are studied with the help of meteorology. It includes meteorological effects on agriculture over a long duration in any area. The object, marketing area and process of working of agro-climatology and agro-meteorology are similar, but there is difference in their use. Agro-climatology makes use of climatic data for long range planning, whereas agro-meteorology uses standard weather records to analyse and predict requirement of water by plants and animals and makes use of it for planning needs of irrigation.

Apposed Glacier: A unitary glacier formed by combination of two mutual glaciers, is called apposed glacier.

Aquiclude: An impermeable rock incapable of transmitting significant amounts of water for supply of water to a well.

Aquifer: Permeable or porous group of rocks that hold water and can transmit water in sufficient quantity for wells or falls.

Aquifuge: An impermeable rock incapable of absorbing or transmitting water because it has no pores or voids.

Artesian Well: Artesian wells are such natural water sources from which water automatically appears on the surface of the earth. For their construction, two types of favourable structures are required. First, synclinal or dome shaped inverse structure, and second, monocline structure. Region from where water of artesian well moves upwards is called aquifer. There should be non-permeable rocks below as well as above aquifer whose nature is similar to Aquiclude. The lower level prevents seepage of water, whereas upper level prevents water from evaporation. Part of aquifer opening on the surface from where water entering through the permeable layer; reaches the central portion is called drainage area. By and by, as soon as permeable layer is full, 'Hydrostatic Pressure Head' is formed. If the well is dug from the surface to the depth of aquifer in this situation, then water of aquifer would appear automatically on the surface due to hydrostatic pressure. This forms an artesian well. Maximum number of artesian wells are found in Australia. Artesian wells are also found in the London area of Great Britain, artesian basin of Moscow in Russia and near dotes in Ukraine.

Artificial Rainfall: Scientific method by which water existing in vapour is converted into water drops by creating an obstacle in motion of clouds. Such rainfall is called artificial rain. In this method, vapour of carbon oxide and silver iodide is sprayed on the layer of clouds. These chemical elements by their coldness, convert vaporous position of clouds into water drops and create rain. In India, Tropical Zone Meteorological Institute, Pune has conducted many successful researches related to artificial rain and this institute has arranged artificial rain on experimental basis at Baramati and Siruk in Maharashtra, Rehan dam in Uttar Pradesh and in coastal areas of Kerala.

Autogenic Stream: It has been derived from the Greek word 'Autos' which means 'Self'. It is a stream whose flow is sustained by a continuously positive water balance along its course. Most of the rivers in humid regions are of such type. A distinction is made between a river which traverses through negative water balance and a autogenic stream which is sustained by a water source in wetter headwaters.

Available Soil Moisture (ASM): It is the amount of water in the soil between FC and WP (FC-WP).

Back Swamp: Low-lying marshy or swampy area on a flood plain where overbank flood or tributary drainage water may collect in the form of a pond between river levees and valley sides, is called back swamp.

Back Water: Such a stationary water area connected with a river but not affected by such a stream, is known as back water. It occurs in natural circumstances, but it also develops during construction of dam in flood-prone areas. Back water area develops fast in such places where a river is divided in two parts and one river forms a very narrow meander where back water collects. Thus back water is water which has been separated from the main river but is mostly attached to the river. Its flow is very slow and almost stationary.

Bai-U-Rains: Due to cyclones during the months of June and July in north-eastern Japan, temperature rises very high resulting in heavy rains. Such rain is known as Bai-U-rain in Japan.

Banner Cloud: Banner cloud is a cloud which seems to extend downwards from isolated peaks looking like a flag. Its classic example is found in the Matter Horn area.

Basin Cultivation: In regions having heavy rains, basins are formed on land by cutting with pickaxe and spade. Soil erosion is prevented through basin formation. Cultivation is done in basins.

Basin Irrigation: There is a traditional system of limited agriculture which is done by controlling flood water of river Nile in Egypt and Sudan. Flood water is stored in basins of different sizes on both the sides of the river. These basins are separated by earthen walls and cultivation is done after drying of water in them. This kind of irrigation is called basin Irrigation.

Bayou: This is a shallow marshy back water area which is also called swampy investor. It is mainly found by the side of the Gulf Coast, and Mississippi and its delta in United States of America.

Bitter Lake: Water of such lakes is bitter and pungent. Great Bitter Lake and Small Bitter Lake on the way to Suez Canal, are such types of lakes.

Braunite: This word has been taken from Arles in France and it is used for cultivation of aluminum oxide deposit. It is a synonym for aluminum. In tropical regions, aluminum hydroxide earth deposits are called bauxite. It contains iron oxide and phosphate. Its colour is white yellow to brown red.

Capillary Flow: Rising of water through spaces found in the soil above water level, is known as capillary flow. They are the attraction of voids in soil level.

Capillary Fring: Layer of soil immediately above water level, from which water rises up through capillary action.

Capillary Moisture: Water found in pores around soil granules is known as capillary water. This water comes up by surface tension and becomes available for plants and trees.

Cloud Burst: Sudden heavy quantity of rainfall in a short duration of time. It causes floods in rivers and causes soil as well as gully erosion. Low lying lands are submerged under water. When air full of water vapour rises up suddenly by the side of ranges of hills, condensation occurs to a great extent, causing heavy rainfall.

Condensation Nuclei: They are micro granules which are required for the process of condensation. Droplets are formed around them. They are concentration center or humidity centers. These micro granules exist in air but their quantity increases with pollens, salts, dust, air pollution, etc.

Congelifraction: This is the process when water drops are frozen as snow in surface voids and fissures and subsequently convert into water after getting temperature. Size of surface joints widens, which leads to weathering of rocks by freeze-thaw process. This is called weathering by frost.

Crop requirement (WR): This is consumption of water for metabolic/physiological functions of a crop, transpiration (T) through its leaves (stomata) and evaporation (E) (WR)/consumptive use (CU) of that particular crop. As water utilized in metabolic/physiological functions is very small (less than 1%), the evapotranspiration (ET) is considered to be almost equal to CU or the WR in the practical sense.

Crop rotation: Productivity of field decreases by cultivating the same crop continuously. Hence, two or more crops are grown one after the other in series according to climatic conditions. This is called crop rotation. At least one crop should be leguminous in this rotation. Such crops are like gram, methi, peas, gwar etc. which have rizeberium bacteria in their roots. By compounding with nitrogen of the atmosphere, they convert it into nitrate, which is made available to plants.

Cropping Intensity: This represents the ratio of gross cropped area to net sown area and multiplied by hundred.

Cropping Pattern: It is the share of individual crop area expressed in the percentage of total cropped area.

Disappeared Stream: This is the water of a river which goes underground by entering into voids and joints of porous rocks after flowing on the surface of earth for some distance. It happens mostly in case of limestone rocks.

Doab (Interfluves): In Indian context, alluvial plain between meeting place of two rivers is called doab (Two Waters), for example doab of Ganga and Yamuna rivers.

Drainage System: Such natural river system, in which water of the area is concentrated and exits at one point. It is considered parallel to watershed and it is delineated by water divide.

Ecotone: It is such an area or place where there is transition on the ground between two important plant communities. It may be a broad zone, which reflects a gradual blending of two communities divided by a sharp boundary line as per physical environmental conditions. Savanna ecotone area is located in the middle of Sahara region and equatorial region.

Edaphology: It is the science which studies role of soil in development of plants.

Ephemeral Plant: These plants are found in arid and semi-arid regions and they have a very short life cycle. Many times, their life cycle is for a few weeks.

Epiphytes: Epiphytes are plants which grow upon the surface of other plants but do not obtain food from the host plant. They use other plants for physical support. They have no roots on the ground.

Erimophytes: Plants developing on xerophyte and steep sloppy parts on the surface of land are called erimophytes.

Eutrophication: It is a natural process through which water or soil becomes more favourable for development of biota because of addition of mineral nutrients (nitrogen and phosphorus etc.) to the ecosystem. It very often relates to fresh water of rivers, lakes and tanks. Eutrophied water charged with nutrients is a better fertilizer. Oligotrophic lakes are converted into eutrophied lakes due to regular deposition of sediments according to geological time scale. In the end, sediment deposited in lakes, collects in the base of lakes and converts into bog mosses covered with waterlogged grasses. This natural process takes place very slowly and takes hundreds of years to complete. This complete process is known as eutrophication. This process sometimes becomes faster due to human intervention, when effluents as sewage and other domestic wastes are disposed off in it. Process of complete eutophication by human induced addition of elements is called 'cultural eutrophication'. The term 'Oligotrophic lakes' used above, means such lakes which are deep and have narrow steep banks. Such lakes get lesser quantity of oxygen due to deep water. Consequently, biotic diversity reduces due to lesser planktons. It is clear that algal blooms have the effect of deoxygenating water through consequent bacterial activity. Hence in due course of time, they convert into eutophied lakes due to process of eutrophication. Water in the lakes also reduces on account of it.

Evaporation: Evaporation is the process of conversion of water on the earth into water vapour due to solar radiation. Oceanic water reaches the atmosphere through this process. Evaporation is more from oceans as compared to continents. The main reason for evaporation is sufficient availability of water. According to an estimate, 4,55,000 cubic km water from seas reaches into the atmosphere every year through evaporation. As against it, 98,000 cubic km water is absorbed in atmosphere during the process of evaporation.

Faggare: It is an inland system by which water of an aquifer existing near a piedmont area, is brought to a low lying mountain area for irrigation. It is specially done in the Mauritania country in Africa.

Fathometer: It is an instrument for measuring depth of oceans. Speed of sound waves passing through water is used in it.

Fen: Fen is low lying marshy land partially inundated by water. Its soil is spongy and soft. It possesses minerals in ample quantity. Ancient fens were called 'Kaar', which later became dry, only shrubs were found there. Most of such areas have been declared as natural reserves.

Field capacity (*FC*): It is the maximum amount of water which can be retained by the soil.

Flood Tide: High tides causing flood on sea coasts every day, are called flood tides.

Flush: Sudden rise in water of river is known as flush. It causes a situation of flood in the river when the river water does not cross its banks.

Formal Region: Uniformity of one symptom that is found everywhere in the region. For example, Thar desert, Himalayan mountainous region, etc. are sufficiently expanded. Even in uniformity, sufficient diversity is found in human, economic and cultural activities, due to which full unity is visible but in changed forms.

Frost: When the activity of condensation is below freezing point, water vapour converts into ice crystals prior to water droplets. This is called frost. All those circumstances that are suitable for frost are essential for formation of dew drops.

Gale: Very fast flowing wind is called gale. Fast stormy winds blow at a speed of 70 km per hour. Normally, air blowing the figures at 11 or 12 on the Buefort scale is called storm or squall. It is also called storm or hurricane.

Geyser: In the language of Iceland, 'Goshur' means 'Sponter'. Geysers are sources of hot water from which after some intervals hot water and vapour come out with great speed. Temperature of such water remains up to 75°C to 90°C. There are 100 geysers in the world, including 'Old Faithful' in the world famous Yellowstone National Park situated in the Wayoming State of United States of America. Many geysers are found in Iceland, Caucasus and New Zealand. Geysers are also found at Sohna (Haryana), Rajgir (Rajgir hills), Purga valley (Ladakh) and Anantnag (Kashmir) in India.

Glacier: Moving snow is called glacier. Snow is solid form of water, which is obtained when temperature is below freezing point (0°C or 32°F). 2.7 per cent water of the world is found in the form of snow. Glaciers are of four types. Ice cap is situated at the top of the mountains, other types are continental glaciers, mountain glaciers and piedmont glaciers. Mobility of glaciers was first proved through examples by Swiss expert Hugo.

Gravitational Water: Water is found in the soil in different forms. When percolation of water takes place into the soil due to gravitational force, it is called gravitational water or independent water. Gravitational water moves in spaces within soil granules. This water is absorbed to a limited extent, after which its infiltration cannot take place.

Groundwater Basin: Groundwater basin is a geomorphic unit in which one vast aquifer or many aquifers are mutually interlinked. Thus, it is a sort of group of aquifers. Surface water basins in valleys located in the midst of mountain ranges are similar to groundwater basins. Hydrological drainage in hilly areas of limestone and sand, and arrangement of groundwater basins, can be completely different. Artesian basin is also considered parallel to it. During last decades, basin plan works have been undertaken by considering groundwater basin as a unit.

Groundwater: Surface water is collected in aquifers after going underground through percolation and infiltration during flow of water. 0.61 per cent of total available water found in nature is found at a depth of 400 metres.

Halophytes: Halophytes are plants developed in saline or marshy lands. Such plants are found in Kutch and Sunderbans in India.

Helophytes: Helophytes are plants developed at places having high temperature or sun's heat.

Herbicide: Herbicide is such a chemical group whose use can destroy plants. These plants prevent growth of plants favourable for agriculture or other special environment, hence they are destroyed by using herbicides, though harm to human and animal species is caused by them use.

Huerta: It is land situated in Spain where two or more crops are grown in high temperature and arrangement for irrigation is done every year since long time. It mainly extends in the southern parts of Valencia, Marcia and Grenada. Crops of pomegranate, orange and almond are grown there.

Hydrolysis: Rain water takes solutions of oxygen, carbon dioxide, many acids and carbonic elements from the rocks and creates new chemical compounds. Such a chemical reaction of water is called hydrolysis.

Hydrosere: A type of ecological sequence in which organisms are produced in water in various stages, is called hydrosere.

Hyetograph: It is an automatic type of rain recorder which is connected with pen filed with plawak ink. It records quantity of rain deposit in a rain guage.

Hygrophyte: Hygrophyte plants are found in areas having sufficient moisture or favourable plants. Mostly they are found in rain forests of tropical regions, where leaves are wide.

Hygroscope: It is an instrument which indicates humidity of air with the help of changes in extension of any material or form.

Hygroscopic Water: This water is obtained by dry soil through absorption of water vapour from the atmosphere. This water remains in the form of a thin film around soil granules or organic matter with relative force. This water has no movement nor is it destroyed by evaporation. It is also called waste water because organic substances of soil cannot use it. Its quantity is determined by structure of soil, quantity of colloidal material, humidity and temperature. Clay contains more hygroscopic water as compared to sand. Its quantity increases with increase in humidity and its quantity decreases with rise of temperature.

Illuviation: Collection of materials from upper A-horizon of soil profile towards lower side B-horizon, is called illuviation.

Infiltration: As soon as rainwater reaches the surface of the earth, a part of it is soaked by the earth. This activity of soaking of water by the earth is called infiltration. Rate of infiltration of water depends on structure and composition of soil.

Irrigation Intensity: The percentage of the area which receives irrigation.

Irrigation Requirement (IR): IR = WR/Irrigation efficiency in percentage

Isikeraunic: Lines indicating equal concentration or recurrence of thunder storms.

Isohyonene: Line for equal humid months on the map.

Isokrymene: Lines of continuous series indicating centers of equal minimum temperature of seas.

Isoneromene: Lines connecting points indicating equal atmospheric dryness.

Isonif: Lines indicating places of distribution of equal snow.

Isophene: Lines joining phenomenon of equal meteorological conditions eg. time for maturity of flowers.

Isophyte: Lines indicating equal height of vegetation.

Isopotential: Line indicating that level of artesian well up to which level improvement has been found.

Jokula Haup: This word has been taken from Iceland. When snow melts due to rise in temperature on account of volcanic eruption and water is stored in form of lakes and starts flowing as a river, it is called Jokul river.

La-Nina: As against the effective stream El-nino on the sea coast of Peru, La-nina is a new event, which is effective in the north middle portion of Pacific ocean. The literal meaning of La-nina is female child. Water near China sea becomes extraordinarily cold due to it and comes towards the equatorial line. The effect is that south eastern commercial winds enter into Asian Pacific ocean causing serious damage by heavy rains. The incident of La-nina mostly follows El-nin and duration of one year elapses between the two. It has been found that effect of El-nin is drought, whereas effect of La-nina is heavy rain.

Lotic Water: This includes rivers, springs, streams and soon, different types of fish and protozoa, sponges, celenterat, briyozoa are found in this water.

Mango Showers: In India, normally before monsoon the season, there is heavy rain along with thunder storms in the months of April-May. It is also called pre-monsoon rain. Since this is time for ripening of mango crop in India, and this rain is helpful in maturing mango crop, hence it is called Mango Showers.

Mulching: This method reduces wind erosion and water erosion. Root system, leaves and stalks of plants like pulses and mustard prevent soil moisture and evaporation. It supplies carbonic material to the soil, which increases, the proportion of moisture and soil productivity.

Nautical Mile: It is a sea mile of 1.852 km or 72,960 inches, which is larger than ground mile of 5,280 feet or 63,360 inches. One tenth part of a nautical mile is called *Cable*, which is equal to 608 feet.

Net Cropped Area: It is the areas sown with crops. The area sown more than once during the year is counted once only.

Nivation: Soil erosion by neve (ice crystals) in local mountain regions, which is caused by disintegration by frost action is called nivation.

Nivation Hollow: These are formation of small pits by ice crystals in snow area. They decrease or increase their size according to melting and freezing action.

Norwester: Stormy winds or cyclone mostly accompanied by damaging thunder storm causing heavy rainfall and sometimes showers of hail in plain areas of north India are called north western . In eastern India and Myanmar there is widespread rain during April-May from Baisakhi time. This rain is very useful for tea crop.

Oasis: This is a green area around water reservoirs in desert area due to availability of water. Oases are remnants of rivers. They can be big or small in size. Coconut and palm trees are found in oases. Wheat, barley and vegetables etc. are grown there.

Olimentation: Freezing of ice in fern area due to heavy quantity of snowfall is called olimentation. Formation of glaciers becomes easier due to such freezing.

Permeability: Capacity of rock by which liquid material and gases pass through it, which depends on size and extension of pores in rocks. It is the capacity of the rock to transmit fluids. The rate of permeability is high in sandstone and conglomerate because of the relatively large interconnected pore space between the grains. On the other hand, basalt is impermeable because of the presence of fractures and vesicles.

Piezometic Surface: A piezometric surface may be defined as the imaginary surface that coincides with the static level of the water in the aquifer.

Pinnate Drainage: It is a drainage in plant form in which companion rivers are very near to each other and they meet the main river at an acute angle.

Pluvial: This word is used for study related to rainfall or change in rainfall pertaining to the ancient period.

Polder: Lower portion of some lake or sea, from where water is shifted to the sea by pumping it out after constructing an artificial dam. Polder is a method of redeeming marshy land of coastal area from the sea. Such land is important from the point of view of agriculture. Such method is prevalent in the Netherlands.

Porosity: Porosity is the ratio of the aggregate volume of voids which are found in rocks or soil to the total volume of the rock or soil, and is expressed as a percentage. Porosity simply means how many pores a unit of rock or soil has. Thus, it is defined as the percentage of voids in the total volume of rocks in the soil. Porosity determines how much water a rock body can hold. The various factors affecting porosity are the shape, arrangement and degree of sorting of the constituent particles and the extent to which changes have occurred as a result of solution, compaction, cementation and faulting. The irregularity of the particle shape has a more important effect on the porosity than the particle size.

Quagmire: It is a part of humid and marshy or boggy land. If any person places his foot over it, the foot would sink. When the marshy portion of land shivers on keeping the foot, it is called shivering quagmire.

Quesec: Flow or release of water at the rate of one cubic feet per second. One quesecs flow delivers 3600 cubic feet of water in an hour.

Rabi: Hindustani word used for winter season or dry cold. Crops sown in winter and harvested in spring season are called rabi crops.

Rime: It is a form of frost which is formed by freezing of water drops into ice crystals on account of super cooling. It exists in form of kho on grass or other places on the Mediterranean Sea.

River Capture: Process of getting top water of one river by extending the drainage of another river is known as river capture.

Riviera: Ground strip near Geneva Gulf in Italy, and Italian word used for coastal parts of France. This is now used as coastal tourist spots for pleasure purposes.

Runoff: It is the surface flow of water from rainfall or melting of ice. It mainly flows from higher plane towards lower plane through rivers. Under ground flow of water is also included in it besides surface water.

Sahel Region: Arid region of North Africa situated in the southern corner of Sahara Desert. Desert and savanna type of vegetation is found here. Due to irregular rain for long durations, drought situation prevails there for most of the time during the year. People have a very hard life here due to repetitive droughts.

Seepage: Groundwater passing through some void or pore and flowing with slow speed on surface of the earth, is called seepage, whereas permeability of water through pores of rocks is called infiltration. Similarly, water flowing through pores, voids and fissures of soil and rocks downwards, is called percolation.

Senile River: It is surface flow of water received through rainfall or melting of ice. It is mainly in the form of rivers which flow horizontally from upper level to lower level.

Sleet: Such rain which is accompanied by ice and water droplets. Sometimes, melting ice and water drops rain together.

Snow Gauge: Instrument which measures depth of snow as rain gauge measures rainfall. Snow is measured after its melting.

Sub Soil: Level of soil immediately below the surface of soil is called sub-soil. It is the layer covered by materials of the soil. Analyzed deposits are mainly found in its structure.

Surface Flow: The part of rain water which cannot reach underground is called surface flow. Speed of flow depends on the slope of land. If slope is steep, flow is faster. If slope is less, flow would be slow. Infiltration is lesser in steep slope. Sub surface flow starts after infiltration of water. The drainage occurs through big voids. It is a very slow process.

Terrace Cultivation: Cultivation done in hilly areas by forming steps, is known as terrace cultivation. This also prevents soil erosion.

Terraroxa: Portuguese word used for soil in Parana Plateau in Brazil. These soils are deep and permeable and have organisms in sufficient quantity. The colour is dark red and violet and becomes sticky after wetting.

Territorial Water: It includes sea area adjacent to the coast of any country. It extends to 12 nautical miles (19 kilometres) from international point of view, where the country can exercise fishing rights.

Total (Gross) Cropped Area: It represents the total area covered with crops, i.e., the net sown area plus the area sown once only.

Transpiration: Transpiration is the process through which trees and plants release moisture into the atmosphere after taking it from roots and again drawing through leaves. This process is completed by stomata found in leaves. Groundwater is wasted to a great extent in this process mainly during the day hours.

Tropophyte: Tropophytes are those plants which become hygrophilous during rains and xerophytes during arid season.

Varve: Seasonal thin layer of sediment found in the base of any lake or sea in glacier region is called varve.

Vodose Water: Water found in between surface of land and groundwater level, is called vadose water. It reaches here after infiltration from the aeration sphere.

Wash Land: Such low land which is bounded to a river or estuary and is filled with water at the time of floods.

Wash: Wavy movement of sea water or back wash of some big water reservoir is called wash.

Water Gap: The narrow river valley which has developed by cutting of the middle part of hard rock where a stream is flowing at present.

Water Spout: It is a funnel type cyclonic depression developed in the maritime boundary in the form of clouds and water deposit. It is a tornado like cyclone found in warm seas.

Water Table: It is the upper part of the saturated zone which remains unequal and variable.

Watershed: Watershed is a geographical unit of area from which water flows through one outlet, or the water flows from the middle of the hill in the form of different streams and forms a drainage system and goes out from one normal flow point. Its size is different according to different geographical contexts. At present they are identified with the help of data and imageries obtained through remote sensing. In India, this information is given by remote sensing office situated at Hyderabad.

Wilting point (WP): The moisture level in the soil at which plants begin to wilt.

Xeric: Extent of moisture available for development of plants on ground level is called xeric.

Xerophyte: This word has been coined by combination of two Greek words, 'Xeros' which means 'dry' and 'Phyte' meaning 'plants'. Thus, xerophyte are plants able to adjust in arid conditions.

Xerosere: A situation of organic series when vegetation is established in sandy soil of arid regions is called xerosere.

Yazoo River: A companion river which flows parallel to the main river and cannot join the main river due to its raised banks is called Yazoo river. After flowing for a long distance, it is able to join the main river in some lower area.

Zonal Soil: This soil class is affected by local climate and vegetation because whenever earth is formed anywhere, remnants of local vegetation and climate both affect it. They are divided in two categories. First is Pedalfers (ped = soil, al = aluminium, fers = iron). This soil contains mainly aluminum and iron and soil is found in humid and forest areas where quantity of rainfall is more than evaporation. Second is Pedocal (ped = soil and cal = calcium). It is soil of semi-arid regions where evaporation is more than rainfall. Consequently, quantity of calcium carbonate is more.

Bibliography

Anon, (1983b), 'Contour Bunds Save Water in Semi-arid Ethiopian Highlands', United Nations University Newsletter, 7(2), 7.

Bale, J., (1977), *Hydrology and Water Resources in Tropical Africa*, Elsevier, Amsterdam.

Barrew, C.J., (1983), 'The Environmental Conservancies of Water Resource Development in the Tropics', in Ooijin Bee (ed.), *Natural Resources in Tropical Countries*, Singapore University Press, pp. 439-66.

Barrow, C., (1987), *Water Resources and Agricultural Development in the Tropics*, Longman, London.

Bedient, P.B. and Huber, W.C., (1988), *Hydrology and Flood Plain Analysis*, Addison-Wesley Pub. Co.

Berntein, L. and Francois, L.E., (1973), 'Comparisons of Drip, Burrow, and Sprinkler Irrigation', *Soil Science*, 115(1), 73-86.

Biswas, A.K., (1979a), 'Management of Traditional Resource Systems in Marginal Areas', *Environmental Conservation*, 6(4), 257-63.

Boffrall, A.F., (1981c), 'Water, Land and Conflict Management', *ODI Review*, No. 1, 73-83.

Bouwer, H., (1978), *Ground Water Hydrology*, McGraw-Hill, New York.

Boyko, H., (ed.) (1968), *Saline Irrigation Bar Agriculture and Forestry*, Drw. Junk, The Hague.

Bras, R.L., (1990), *Hydrology: An Introduction to Hydrologic Science*, Addison Wesely Pub. Co.

Canter, L.W., (1986), *Environmental Impacts of Water Resources Projects*, Lewis Publishers, Inc., Chelsea, Michigan 48118.

Capanera, D.A., (1983), 'International River Law', in Zaman, M. et al., (eds.), *River Basin Development*, Tycooly International Publishing, Dublin, pp. 173-84.

Carruthers, I.D., and Clark, C., (1981), *The Economics of Irrigation*, Liverpool University Press, Liverpool.

Chow, U.T. Maidment, D.R. and Mays, L.W., (1988), *Applied Hydrology*, McGraw-Hill Book Co.

Conklin, H.C., (1961), 'The Study of Shifting Cultivation', *Current Anthropology*, 2 (1), 27-30.

Dennel, R.W., (1982b), 'Dryland Agriculture and Soil Conservation: An Archaeological Study of Check-dam Forming and Wadi Siltation', in Spooner, B. and Mann, H.S. (eds.), *Desertification and Development: Dryland Ecology in Social Perspective*, Academic Press, London, pp. 171-200.

Deom, J., (1976), *Water Resources Development and Health: A Selected Bibliography*, World Health Organization, Geneva.

Doorenbos, J. and Pruitt, W.O., (1977), *Guidelines for Predicting Crop Water Requirements*, FAO, Rome.

Falkenmark, M. and Lindh, G., (1976), *Water for a Starving World*, Westview Press, Boulder, Colo.

Fauchon. J., (1980), 'Oasis Agriculture: How Pumps Divide the Peasantry', *Ceres*, 13(4), 41-5.

Ferderick, K.D., (1975), *Water Management and Agricultural Development: A Case Study of the Cuyo Region of Argentina*, Published for Resources of the Future, Johns Hopkins University Press, Baltimore.

Finkel, H.J. (ed.), (1977), *Handbook of Irrigation Technology*, Vol. 1 CRC Press, Boca Ratan, F1.

Finkel, H.J. (ed.), (1983), *Handbook of Irrigation Technology*, Vol. 2 CRC Press, Boca Ratan, F1.

Furon, R., (1963), *The Problem of Water: A World Study*, Faber and Faber, London.

Garzon, C.E., (1984), 'Water Quality in Hydroelectric Projects: Considerations for Planning in Tropical Forest Regions', *World Bank Technical Paper* No. 20, The World Bank, Washington, DC.

Gujar, R.K. and Jat, B.C., (2005), *Jal Sansadhan Bhugol*, Rawat Publications, Jaipur.

, (2001), *Jal Prabandh Vigyan*, Pointer Publishers, Jaipur.

Howe, C.W. and Easter, K.W., (1971), *Interbasin Transfers of Water: Economic Issues and Impacts*, Johns Hopkins Press, Baltimore and London.

Huntington, E., (1915), *Civilization and Climate,* Yale University Press, New Haven, Conn.

Jackson, I.J., (1977), *Climate, Water and Agriculture in the Tropics,* Longman, London and New York.

Jat, B.C., (2006), *Dictionary of Geography,* Malik & Co., Jaipur.

—, (2006), *Economic Geography,* Panchsheel Prakashan, Jaipur.

—, (2004), *Geomorphology,* Rawat Publications, Jaipur.

Karanth, K.R., (1989), *Hydrogeology*, Tata McGraw Hill Pub. Co., New Delhi.

Kates, R.W., (1981), 'Drought in the Sahel: Competing Views as to What Really Happened in 1910-14 and 1968-74', *Mazingira*, 5(2), 72-83.

Kayastha, S.L., (1981), 'An Appraisal of Water Resources of India and Need for National Water Policy', *Geojournal*, 5.6, 563-72.

Kremer, P.J., (1969), *Plant and Soil Water Relations: A Modern Synthesis,* Mc Graw-Hill, New York.

Linsley, R.K., et. al., (1988), *Hydrology for Engineers,* (S.I. Metric Edition), McGraw Hill, Singapore.

Lox, G.W., and Atkins, M.D., (1979), *Agricultural Ecology: An Analysis of World Food Production*, W.H. Freeman, San Francisco.

Maltby, E. and Turner, R.E., (1983), 'Wetlands of the World', *Geographical Magazine*, 55(1), 12-17

Mandal, R.C., (1974), 'Pitcher Farming', *Appropriate Technology*, 1(3), 7.

Mutreja, K.N., (1986), *Applied Hydrology,* Tata McGraw-Hill Pub. Co., New Delhi.

Oliver, H., (1972), *Irrigation and Water Resources Engineering,* Edward Arnold, London.

Pacey, A. and Cvllis, A., (1986), *Rainwater Harvesting: The Collection of Rainfall and Run-off in Rural Areas*, Intermediate Technology Publications, I King Street, London, WCZE 8HW.

Park, C.C., (1981), 'Man, Systems and Environmental Impacts', *Progress in Physical Geography*, 5(1), 1-31

Schultz, T.W., (1964), *Transforming Traditional Agriculture*, Yale University Press, Newhaven, Con.

Schulze, F.E. and Van Staveren, J.M., (1980), 'Land and Water Development in the Third World', in *International Institute for Land Reclamation and Improvement, Land Reclamation and Water Management: Developments, Problems and Challenges*, IILRI CPO Box 45, 6700 AA). Wageningen, The Netherlands, pp. 13-28.

Shainberg, I. and Oster, J.D., (1978), *Quality of Irrigation Water,* Pergamon Press, Oxford and New York, (International Irrigation Information Centre Publication No. 2).

Shaw, E.M. (1988), *Hydrology in Practice*, Van Nostrand (International).

Sidharth, K., (2000), *Oceanography: A Brief Introduction*, Kisalaya Publications, New Delhi.

Spooner, B. and Mann, H.S. (eds.), (1975), *Desertification and Development: Dryland Ecology in Social Perspective*, Academic Press, London and New York.

Subramanyaa, K., (1994), *Engineering Hydrology*, Tata McGraw-Hill, Delhi.

Tolman, C.F., (1937), *Ground Water*, McGraw-Hill, New York.

Utton, A.E. and Teclaff, L. (eds.), (1978), *Water in a Developing World: the Management of a Crucial Resource*, Westview Press, Boulder, Colo.

Varallyay, G., (1977), 'Soil Water Problems Related to Salinity and All Latinity in Irrigated Lands', in Washington, E.B. (ed.), *Arid Land Irrigation in Developing Countries: Environmental Problems and Effects*, Pergamn Press, Oxford and New York, pp. 251-64.

Viessman, W., Lewis, G.L., and Knapp, J.W. (1989), *Introduction to Hydrology*, 3rd ed., Harper and Row, New York.

Wagner, E.G. and Lanoix, J.N., (1959), *Water Supply for Rutal Areas and Small Communities*, World Health Organization, Geneva.

Ward, R.C., (1967), *Principles of Hydrology*, McGraw-Hill, London and New York.

Withers, B. and Vipond, S., (1980), *Irrigation: Design and Practice*, (Indian), Cornell University Press, Ithaca, New York.

Zaman, M., (1982), 'The Ganges Basin and the Water Dispute', *Water supply and Management*, 6(4), 321-8.

Zonn, I.S., (1979), 'Ecological Aspects of Irrigated Agriculture', *Water Supply and Management*, 3(4), 285-95.

Index